Infrared and Raman Spectroscopy of Polymers

PRACTICAL SPECTROSCOPY

A SERIES

Edited by Edward G. Brame, Jr.
Elastomer Chemicals Department
Experimental Station
E. I. du Pont de Nemours and Co., Inc.
Wilmington, Delaware

ADDITIONAL VOLUMES IN PREPARATION

Infrared and Raman Spectroscopy of Polymers

H. W. SIESLER   K. HOLLAND-MORITZ

MARCEL DEKKER, INC.   New York and Basel

Library of Congress Cataloging in Publication Data

Siesler, H. W. [date]
   Infrared and Raman spectroscopy of polymers.

   (Practical spectroscopy; v. 4)
   Includes bibliographical references and index.
   1. Polymers and polymerization--Analysis.
2. Infra-red spectrometry. 3. Raman spectroscopy.
I. Holland-Moritz, K., [date] joint author.
II. Title. III. Series.
QD139.P6S48   547.8'4046   80-14739
ISBN 0-8247-6935-X

MARCEL DEKKER, INC.
270 Madison Avenue, New York, New York 10016

Current printing (last digit):
10 9 8 7 6 5 4 3 2 1

PRINTED IN THE UNITED STATES OF AMERICA

The aim of this book is to help scientists in universities and in industry to make effective use of vibrational spectroscopy in solving problems in polymer physics and polymer analysis. Although some excellent early monographs on the infrared (IR) spectroscopy of polymers are available, the publication of this book would appear justified, and indeed necessary, in view of both the introduction of new techniques, such as Fourier transform IR (FTIR) spectroscopy, and the resulting widening of the field of application for vibrational spectroscopy and also to deal at the same time with both IR and Raman spectroscopy of polymers.

Given the amount of material involved, the presentation of theoretical principles, experimental techniques, and application examples in a single volume must inevitably result in the individual reader's regretting the absence of certain items while feeling others to be superfluous. A detailed discussion of the results of vibrational spectroscopy for individual classes of polymers has deliberately been omitted.

The authors would like to thank the Board of Management of Bayer AG for the permission to publish this book. Work on the monograph was made possible by the generous assistance of the Central Analytical Department in the Corporate Research and Development Division of Bayer AG (Dr. H. Walz) and the Institute of Physical Chemistry at the University of Cologne (Prof. D. O. Hummel), who permitted their equipment to be used to carry out the necessary investigations.

Our special thanks are due to our co-workers B. Bonneck, W. Kremer, H. P. Schlemmer, and W. Stach, who - directly or indirectly - helped to make it possible for this book to be written. We are also indebted to those colleagues who so kindly gave permission for their experimental data to be reproduced. We would also like to thank Dr. G. Bayer, H. Devrient, E. Knoll, Dr. T. Werner, and Dr. W. Hoffmann (Bayer AG, Leverkusen) and Dr. G. Spilgies (Bayer AG, Dormagen) for their assistance with problems of FTIR spectroscopy, mechanical measurements, and x-ray diffraction.

We owe our greatest debt of gratitude to our wives and children for their patience and constant encouragement.

H. W. Siesler

K. Holland-Moritz

CONTENTS

Chapter 1

INTRODUCTION

Infrared (IR) and Raman spectroscopy have become most important tools
for the characterization of the chemical and physical nature of polymers. Until the late 1960s, almost all investigations were based on
information derived from IR spectra alone, but since the introduction
of laser sources, Raman spectroscopy has increasingly contributed to
the elucidation of polymeric structure.

In principle, IR and Raman spectra provide qualitative and quantitative information about the following structural details of the
polymer under examination:

1. Chemical nature: structural units, type and degree of branching,
    end groups, additives, impurities
2. Steric order: *cis-trans* isomerism, stereoregularity
3. Conformational order: physical arrangement of the polymer chain,
    e.g., planar zigzag or helix conformation
4. State of order: crystalline, mesomorphous, and amorphous phases;
    number of chains per unit cell; intermolecular forces; lamellar
    thickness
5. Orientation: type and degree of preferential polymer chain and
    side group alignment in anisotropic materials

As a consequence of the sensitivity of IR and Raman spectroscopy to changes in the dipole moment and polarizability, respectively, of the vibrating group under consideration, IR spectroscopy generally yields more useful information for the identification of po-

1

lar groups, whereas Raman spectroscopy is especially helpful in the characterization of the homonuclear polymer backbone. Furthermore, the complementary nature of IR and Raman analysis is of particular importance in connection with symmetry considerations of the investigated structure. Thus, generally speaking, asymmetric vibrations give rise to strong IR absorptions, while symmetric modes yield prominent Raman bands.

There exist two basic approaches to the study of polymers by vibrational spectroscopy. The empirical interpretation of IR and Raman spectra is based on the concept of nearly independently vibrating atomic groups in the macromolecule and collects information mainly on the single structural features of the polymer, e.g., chemical composition, configuration, conformation, crystallinity. The theoretical treatment, frequently supported by spectral data obtained from isotope-substituted polymer analogues and polarization measurements on specimens showing directional properties, focuses on the complete assignment of IR and Raman spectra in terms of the vibrational behavior of the polymeric system. Both treatments have their drawbacks and limitations which have to be kept in mind; while the empirical method is less comprehensive (e.g., the phase relations between the motions of individual groups are neglected), the idealized model of polymer structure and the associated inter- and intramolecular forces assumed in the theoretical treatment cannot be materialized in real polymers.

Despite the uncontested importance of IR and Raman spectroscopy for the characterization of macromolecular structure, it should be emphasized that only a limited number of problems may be solved by the exclusive application of these spectroscopic techniques. Thus, in the majority of analytical investigations of polymer constitution and any additives, chemical separation of the components is inevitable; a more complete picture of the sequence distribution and stereoregularity of structural units in polymers is obtained by a combination of vibrational and nuclear magnetic resonance (NMR) spectroscopy; the results of IR and Raman spectroscopic investigations at elevated temperature are advantageously correlated with differential thermal analysis (DTA) or differential scanning calorimetry (DSC) measurements; and last but not least, a thorough knowledge of the structure of crystalline polymers cannot be attained without application of x-ray diffraction. These few, far from comprehensive, examples demonstrate that maximum information on the structural details in question can be obtained only by an appropriate choice and combination of chemical and physical methods.

Chapter 2

THEORETICAL AND EMPIRICAL ASPECTS OF
INFRARED AND RAMAN SPECTROSCOPY

In a first approximation the energy E of a molecule can be separated
into four additive terms belonging to various motions of this molecule:

1. Translation of the molecule (This motion, however, does not lead
    to any interaction with electromagnetic radiation and will be
    neglected in this discussion.)
2. Motions of the electrons in the molecule (E_{el})
3. Vibrations of the atoms or atomic groups in the molecule (E_{vib})
4. Rotations of the entire molecule (E_{rot})

$$E = E_{el} + E_{vib} + E_{rot} \tag{2.1}$$

This simplification is justified because energies of the electronic,
vibrational, and rotational motions differ considerably ($E_{el} \gg E_{vib}$
$\gg E_{rot}$). Only in the case of gases do the absorption bands due to
vibrations of the atoms show a fine structure caused by rotational
transitions. In solids (crystals, polymers) free rotation of the molecule is restricted, and the rotational energy term in Eq. (2.1) can
be neglected. Thus, the energy of such a molecule is determined by
electronic and vibrational contributions only.

A molecule can interact with electromagnetic radiation when
Bohr's frequency relation

$$\Delta E = h\nu \qquad \text{with } \Delta E = E_k - E_m \tag{2.2}$$

is fulfilled. In this equation, ΔE represents the energy difference
between two allowed energy levels k and m, h Planck's constant, and
ν the frequency of the absorbed or emitted electromagnetic radiation.
Although IR and Raman spectroscopies are based on the same physical
origin - the vibrations of the atoms of a molecule which in quantum
mechanics correspond to allowed transitions between different vibrational energy levels - the interaction between electromagnetic radiation and the sample differs considerably in both spectroscopic
methods. In IR spectroscopy specific frequencies of polychromatic
radiation are absorbed by the sample, whereas in Raman spectroscopy
the monochromatic, generally visible radiation can be scattered elastically with the same frequency (Rayleigh scattering) or inelasti-
cally with higher or lower frequencies (Raman scattering). Energies,
wavenumbers, and wavelengths of the radiation used in IR and Raman
spectroscopy are listed in Table 2-1.

TABLE 2-1   Energies, Wavenumbers, and Wavelengths of the Radiation
            Used in IR and Raman Spectroscopy

	Excitation source (nm)	Absolute wavenumber range (cm^{-1})	Relative wavenumber range (Δcm^{-1})	Wavelength range (nm)	Vibrational energy (kJ/mol)
IR	Hg, Globar		10-10000		$1.3 \cdot 10^{-1}$ - 120
Raman	Ar^{+}, 488	20486-10486	3-10000	488 - 953	
	Ar^{+}, 514.5	19430- 9430	3-10000	514.5-1060	
	Ar^{+}, 568.2	17595- 7595	3-10000	568.2-1316	$4 \cdot 10^{-2}$-120
	He-Ne,632.8	15798- 5798	3-10000	632.8-1724	
	Kr^{+}, 530.8	18839- 8839	3-10000	530.8-1131	
	Kr^{+}, 647.1	15449- 5449	3-10000	647.1-1834	

2.1 INTERACTION OF MOLECULES WITH ELECTROMAGNETIC RADIATION[†]

Before discussing the theory of molecular interaction with electromagnetic radiation, we give an example from electronics to illustrate
the effect of absorption. The electromagnetic radiation of a defined
frequency irradiated by a broadcasting station can be received in a
radio by means of a simple oscillatory circuit with antenna, capac-
itor, and coil. This circuit selectively "absorbs" only that frequency which corresponds to its *eigen-* or *resonance* frequency. From
this example a very simplified model for the interaction of IR radiation with matter can be derived. Let us assume in a first approximation that the sample consists of vibrating dipoles which interact
with the incident radiation. Each dipole can interact with that frequency which corresponds to its eigenfrequency. Thus, the molecule
absorbs only those frequencies from the incident radiation which coincide with the frequencies of the atomic oscillators. The residual
nonabsorbed radiation is reflected or transmitted.

In a more theoretical consideration we have to examine how the
electromagnetic radiation can perturb the potential energy of a molecule and induce a transition from an initial stationary state. In
quantum mechanics this problem is solved by introduction of an additive term, the *interaction operator* H_{int}, to the Hamiltonian H_o of
the unperturbed system. Thus, the Schrödinger equation of the stationary system can be written as follows:

$$(H_o + H_{int})\Psi = -\frac{\hbar}{i}\frac{\partial\Psi}{\partial t} \tag{2.3}$$

Since the perturbation alters the state of the system, a superposition of the solutions $\Psi_k(r,t)$ of the unperturbed system can be used
to provide a solution of Eq. (2.3):

$$\Psi(r,t) = \sum_k a_k(t)\Psi_k(r,t) \tag{2.4}$$

The coefficients $a_k(t)$ are time-dependent weighting factors, where
$a_k^*(t)a_k(t)$ gives the importance of the state k. From Eqs. (2.3) and
(2.4) it follows that

[†]For further comprehensive studies see Refs. 1 to 10.

$$(H_o + H_{int}) \sum_k a_k(t) \Psi_k(r,t) = -\frac{\hbar}{i} \sum_k \dot{a}_k(t) \Psi_k(r,t) \qquad (2.5)$$

$$-\frac{\hbar}{i} \sum_k a_k(t) \dot{\Psi}_k(r,t)$$

Since $\Psi_k(r,t)$ are solutions of the Schrödinger equation for the conservative system, Eq. (2.5) simplifies to

$$H_{int} \sum_k a_k(t) \Psi_k(r,t) = -\frac{\hbar}{i} \sum_k \dot{a}_k(t) \Psi_k(r,t) \qquad (2.6)$$

Because of the orthogonality of the wavefunctions Ψ_k, left multiplication with Ψ_m^* and integration over all space gives

$$\dot{a}_m(t) = -\frac{i}{\hbar} \sum_k a_k(t) \int \Psi_m^*(r,t) H_{int} \Psi_k(r,t) \, d\tau \qquad (2.7)$$

The residual terms in the sum $\sum_k \dot{a}_k(t) \Psi_k(r,t)$ cancel because the initial state of the system is characterized by $a_m(0) = 1$ and $a_k(0) = 0$ for $k \neq m$. Thus, upon separation of $\Psi_k(r,t)$ into $\Psi_k(r) \exp(\frac{i}{\hbar} E_k t)$, it follows with Eq. (2.3) for the rate at which a system can change from one stationary state to another under the influence of a perturbing electromagnetic field:

$$\dot{a}_m(t) = -\frac{i}{\hbar} \sum_k a_k(t) \exp\{\frac{i}{\hbar}(E_m - E_k)t\} \int \Psi_m^*(r) H_{int} \Psi_k(r) \, d\tau \qquad (2.8a)$$

The expression

$$\int \Psi_m^*(r) H_{int} \Psi_k(r) \, d\tau = \langle m | H_{int} | k \rangle \qquad (2.9)$$

is called the *transition moment*. With this notation we can rewrite Eq. (2.8a) as follows:

$$\dot{a}_m(t) = -\frac{i}{\hbar} \sum_k a_k(t) \exp\{\frac{i}{\hbar}(E_m - E_k)t\} \langle m | H_{int} | k \rangle \qquad (2.8b)$$

Putting

$$a_k(t) = b_k(t) \exp(\frac{i}{\hbar} E_k t) \qquad (2.10)$$

we obtain a set of homogeneous differential equations

$$\dot{b}_m(t) = -\frac{i}{\hbar}\{E_m b_m(t) + \sum_k b_k(t)<m|H_{int}|k>\} \quad (m=1, 2,..,n) \quad (2.11a)$$

Assuming that the time-dependent factor of H_{int} is either constant within the interval $0 \leq t \leq \theta$ or proportional to $\exp(i\omega t) + \exp(-i\omega t)$, this system can be resolved by any standard method [11, 12], since $\exp[\frac{i}{\hbar}(E_m - E_k)t]$ in Eq. (2.8b) can be replaced by the following expression $\exp[\frac{i}{\hbar}(E_m - E_k - \hbar\omega)t] + \exp[\frac{i}{\hbar}(E_m + E_k - \hbar\omega)t]$. Introduction of $b_k(t) = c_k\exp(\alpha t)$ leads to:

$$-(\frac{\hbar}{i}\alpha + E_m)c_m = \sum_k c_k<m|H_{int}|k> \quad (m = 1, 2, .., n) \quad (2.11b)$$

This system can be solved only if the determinant of the coefficients vanishes:

$$\begin{vmatrix} -E_1 - \frac{\hbar}{i}\alpha & & \cdots & & <n|H_{int}|1> \\ <1|H_{int}|2> & \ddots & & & \\ \cdots\cdots & & -E_k - \frac{\hbar}{i}\alpha & \ddots & \cdots\cdots \\ <1|H_{int}|n> & & \cdots & & -E_n - \frac{\hbar}{i}\alpha \end{vmatrix} = 0 \quad (2.12)$$

Since the perturbation caused by the interaction of the molecule with electromagnetic radiation can be adopted to be small, we have to expand Eq. (2.12) only up to second-order terms. Thus, we obtain

$$\frac{\hbar}{i}\alpha_t = -E_1 + \sum_{k\neq1}^{n} \frac{<m|H_{int}|1> <1|H_{int}|k>}{E_k - E_1} \quad (1 = 1, 2,..., n) \quad (2.13)$$

With these eigenvalues of Eq. (2.12) we can finally derive, after determination of c_k and $b_k(t)$, the expression for $a_k(t)$:

$$a_k(t) = \frac{<m|H_{int}|k>}{E_m - E_k} \{1 - \exp[\frac{i}{\hbar}(E_k - E_m)t]\} \quad (2.14)$$

$$+ \frac{1}{E_m - E_k} \sum_{1>m}^{k} \frac{<m|H_{int}|1> <1|H_{int}|k>}{E_m - E_1} \{1 - \exp[\frac{i}{\hbar}(E_k - E_m)t]\}$$

$$+ \sum_{1>m}^{k} \frac{<m|H_{int}|1> <1|H_{int}|k>}{(E_m - E_1)(E_1 - E_k)} \{1 - \exp[\frac{i}{\hbar}(E_k - E_m)t]\}$$

The first term of Eq. (2.14) describes a transition from an initial
state m to the state k and is due to emission or absorption of radiation, respectively. The second and third terms are connected with
transitions through intermediate states as Raman scattering and resonance fluorescence.

2.1.1 Infrared Absorption

In IR spectroscopy the interaction energy H_{int} between the electric
field $E = E_o[\exp(-i\omega t) + \exp(i\omega t)]$ and a system of charged particles
$\sum_i e_i r_i = \vec{\mu}$

$$H_{int}(t) = E \sum_i e_i r_i = E \vec{\mu} = E_o[\exp(-i\omega t) + \exp(i\omega t)]\vec{\mu} \qquad (2.15)$$

has to be introduced. Maximum interaction takes place when the electric vector E is parallel to the dipole moment $\vec{\mu}$. No light will be
absorbed when E is perpendicular to $\vec{\mu}$. The later discussion of orientation effects (Sec. 4.3.4) is based on this property.

After the introduction of H_{int} in Eq. (2.8b) and separation of
the time-dependent factor E, the first term of Eq. (2.14) can be re-
written as

$$a_k(t) = \frac{<m|\vec{\mu}|k>}{E_m - E_k + \hbar\omega} \left\{1 - \exp[\frac{i}{\hbar} (E_k - E_m + \hbar\omega)t]\right\} E_o$$

$$+ \frac{<m|\vec{\mu}|k>}{E_m - E_k - \hbar\omega} \left\{1 - \exp[\frac{i}{\hbar} (E_k - E_m - \hbar\omega)t]\right\} E_o \qquad (2.16)$$

For the absorption of radiation, e.g., a transition from k to m, the
expression of Eq. (2.16) simplifies to

$$a_k(t) = <m|\vec{\mu}|k> E_o \frac{1 - \exp[\frac{i}{\hbar} (E_k - E_m - \hbar\omega)t]}{E_k - E_m - \hbar\omega} \qquad (2.17)$$

With $2i \cdot \sin x = \exp(ix) - \exp(-ix)$, the probability for the transition at a given frequency can be expressed as

$$a_k^*(t)a_k(t) = 4<m|\vec{\mu}|k>^2 E_o^2 \frac{[\sin\frac{1}{2\hbar}(E_k - E_m - \hbar\omega)t]^2}{(E_k - E_m - \hbar\omega)^2} \qquad (2.18)$$

Integration over the whole frequency range yields

$$a_k^*(t)a_k(t) = \frac{1}{\hbar^2} <m|\vec{\mu}|k>^2 E_o^2 t \qquad (2.19)$$

Expanding $\vec{\mu}$ in a power series in r about the equilibrium value r_o we find from Eq. (2.9)

$$<m|\vec{\mu}|k> = \int \Psi_m^*[\vec{\mu}_o + \left(\frac{\partial\vec{\mu}}{\partial r}\right)_{r=r_o}(r - r_o)]\Psi_k d\tau$$
$$= \frac{\partial\mu}{\partial r} \int \Psi_m^* r\Psi \, d\tau \qquad (2.20)$$

Owing to the orthogonality of the wavefunctions, the contribution of the integral differs from zero when k - m = 1, which corresponds to a transition from a lower energy level m to the next higher level k = m + 1. In this case we get a nonzero transition moment if a change of the dipole moment occurs during the vibrations:

$$<m|\vec{\mu}|k> \neq 0 \qquad \text{if } \frac{\partial\vec{\mu}}{\partial r} \neq 0 \qquad \text{for k - m = 1} \qquad (2.21)$$

Thus, introducing the result into Eq. (2.19), we find that the probability for the described transition within time t depends on $(d\vec{\mu}/dr)^2$ and the square of the maximum amplitude of the incident radiation. To relate this result with the experimental quantities, we have to substitute the energy density (the energy per unit volume)

$$\rho = \frac{3}{2\pi} E_o^2 \qquad (2.22)$$

in Eq. (2.19):

$$a_k^*(t)a_k(t) = \frac{2\pi}{3\hbar^2} <m|\vec{\mu}|k> \rho t \qquad (2.23)$$

If we now change over from a single molecule to the number N_o of molecules per unit volume (cm^3), we obtain the following relation for the change in intensity dI of the radiation passing through a sample of thickness dr:

$$-dI = \frac{2\pi}{3\hbar^2} <m|\vec{\mu}|k>^2 \rho \hbar\omega_{km} N_o \, dr \qquad (2.24)$$

where $\hbar\omega_{km}$ is the energy absorbed in a single transition. Since I is the energy passing through the unit volume per time unit, the following relation between the propagation velocity v of the radiation and the energy density ρ is valid:

$$I = v\rho \qquad (2.25)$$

From Eqs. (2.24) and (2.25) we obtain

$$-dI = \frac{2\pi}{3\hbar^2} <m|\vec{\mu}|k>^2 \frac{I}{v} \hbar\omega_{km} N_o \, dr \qquad (2.26)$$

With Avogadro's constant and the molecular concentration c from Eq. (2.26) it follows that

$$-dI = \left(\frac{3\times10^{-3}}{2\pi h v} N\omega_{km} <m|\vec{\mu}|k>^2\right) Ic \, dr \qquad (2.27)$$

Using a'(ω) as the abbreviation for the expression in parentheses we recognize Beer's law

$$-dI = a'(\omega)Ic \, dr \qquad (2.28)$$

which relates the intensity decrease of the incident radiation to the effective path length dr in an absorbing sample. At a given frequency, integration of Eq. (2.28) for a homogeneous medium results in

$$\ln \frac{I_o}{I} = \ln \frac{1}{T} = a'(\omega)cb \qquad (2.29)$$

where $T = I/I_o$ is the *transmittance* and b the sample path length. With the logarithm to the base 10 Eq. (2.29) becomes

$$A = \log \frac{I_o}{I} = \log \frac{1}{T} = a(\omega)cb \qquad (2.30)$$

where A is called *absorbance* and a(ω), *absorptivity* [13].

2.1.2 Raman Scattering

According to Smekal, the Raman effect can be understood as a collision process between photons and molecules [14]. Prior to the collision, the photon possesses the energy E_o, the momentum p_o, and the frequency ν_o. The corresponding notations after the collision are E_o', p_o', and ν_o':

$$E_o = h\nu_o \qquad p_o = \frac{h\nu_o}{c} \qquad E_o' = h\nu_o' \qquad p_o' = \frac{h\nu_o'}{c} \qquad (2.31)$$

With E_k, v_o and E_m, v_o' for the energy levels and the velocities of the molecule with the mass M before and after the collision, respectively, we get, for the conversion-of-energy principle,

$$h\nu_o + \frac{Mv_o^2}{2} + E_k = h\nu_o' + \frac{Mv_o'^2}{2} + E_m \qquad (2.32)$$

Together with the conversion-of-momentum principle

$$\frac{h\nu_o}{c} + Mv_o = \frac{h\nu_o'}{c} + Mv_o' \qquad (2.33)$$

it follows that the change in the kinetic energy during the collision is negligibly small, since the term $h(\nu_o - \nu_o')$ is much greater than $M/2(v_o^2 - v_o'^2)$ because of the values for M and ν_o. Thus, Eq. (2.32) can be simplified to

$$\Delta\nu = \nu_o - \nu_o' = \frac{E_k - E_m}{h} \qquad (2.34)$$

In the Raman experiment, we observe scattered light with the frequency ν_o'. In comparision to the frequency of the exciting radiation, the frequency of the scattered light may be larger, equal, or smaller:

$$\Delta\nu = \nu_o - \nu_o' < 0$$

This implies that the initial energy level was higher than the final level: $E_k > E_m$. Thus, the scattered photon possesses higher energy $h\nu_o + (E_k - E_m)$ than the incident photon. The additional energy is

due to a transition from k to m. In the Raman spectrum, we observe
the *anti-Stokes* lines.

$$\Delta v = 0$$

The energy levels of the molecule before and after the collision are
the same (*Rayleigh scattering*).

$$\Delta v = v_o - v_o' > 0$$

The incident radiation induces a transition from a lower level E_k to
a higher level E_m which is accompanied by a loss of energy of the
scattered radiation. Therefore, the scattered light must possess a
lower frequency, and we observe the *Stokes* lines.

Figure 2-1 illustrates the three possible transitions. From this
simplified model of elastic and inelastic collisions between photon
and molecule, we realize the basic physical difference between the
effect of infrared absorption of polychromatic light and the Raman
scattering of monochromatic (generally visible) light. However, the

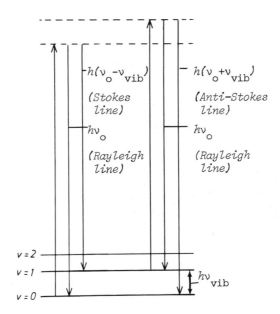

FIGURE 2-1   Energy diagram for Rayleigh, Stokes, and anti-Stokes
scattering.

phenomena common to both effects are the vibrations of molecules
which correspond in quantum mechanics to changes in the vibrational
levels of the molecules.

In Sec. 2.1.1 we showed that a necessary condition for the occurrence of IR absorption is a change in the dipole moment during the
vibration under consideration. To derive the equivalent condition for
the Raman effect, let us examine the influence of monochromatic electromagnetic radiation incident on a molecule. The incident radiation
can be represented by the electric field vector E,

$$E = E_o \cos 2\pi \nu_o t \tag{2.35}$$

which excites the molecular electron cloud to vibrations with frequency $2\pi \nu_o$ and induces a dipole moment $\vec{\mu}_i$ which can be related to
the electric field vector E via

$$\vec{\mu}_i = \alpha E = \alpha E_o \cos 2\pi \nu_o t = \alpha E_o \cos \omega_o t \tag{2.36}$$

With the exception of isotropic molecules (for example, CCl_4) the
molecular polarizability α shows directional properties, and the induced dipole moment $\vec{\mu}_i$ is not parallel to the electric field vector E.

The components of the induced dipole moment $\vec{\mu}_i$ are connected via
linear vector functions with the components of the electric field
vector E

$$\begin{pmatrix} \mu_x \\ \mu_y \\ \mu_z \end{pmatrix} = \begin{pmatrix} \alpha_{xx} & \alpha_{xy} & \alpha_{xz} \\ \alpha_{yx} & \alpha_{yy} & \alpha_{yz} \\ \alpha_{zx} & \alpha_{zy} & \alpha_{zz} \end{pmatrix} \begin{pmatrix} E_x \\ E_y \\ E_z \end{pmatrix} = \vec{\mu}_i = \alpha E \tag{2.37}$$

where the α_{ij} are the components of the tensor α. The component α_{ij}
determines the i-th component of the induced dipole moment which is
caused by the j-th component of the electric field vector. After expanding α in a power series of the coordinates $r = r_o \cos 2\pi \nu t$,

$$\alpha = \alpha_o + \left(\frac{\partial \alpha}{\partial r}\right)_{r=r_o} (r - r_o) \tag{2.38}$$

we find from Eqs. (2.36) and (2.38) for the induced dipole moment
that

$$\vec{\mu}_i = \alpha_o E_o \cos 2\pi \nu_o t + \frac{1}{2}\left(\frac{\partial \alpha}{\partial r}\right)_{r=r_o} E_o (r - r_o) \cos[2\pi(\nu_o - \nu_{vib})t]$$

$$+ \frac{1}{2}\left(\frac{\partial \alpha}{\partial r}\right)_{r=r_o} E_o (r - r_o) \cos[2\pi(\nu_o + \nu_{vib})t] \quad (2.39)$$

From Eq. (2.39) we see that the vibrating atoms modulate the frequency of the induced dipole moment. Besides the original incident frequency, the sum and difference frequencies $\nu_o + \nu_{vib}$ and $\nu_o - \nu_{vib}$ can be observed. The latter ones occur only if the polarizability changes during a vibration:

$$\frac{\partial \alpha}{\partial r} \neq 0 \quad (2.40)$$

Because of the tensor properties of α this means that at least one tensor component must change under the vibration.

A more theoretical treatment of the Raman effect is based on the results derived in Sec. 2.1. The second term of Eq. (2.14) describes a transition from state m through an intermediate state l to a final state k:

$$a_k^{Raman}(t) = \frac{1}{E_m - E_k} \sum_{l>m}^{k} \frac{\langle m|H_{int}|l\rangle \langle l|H_{int}|k\rangle}{E_m - E_l}$$

$$\cdot \left\{1 - \exp[\frac{i}{\hbar}(E_k - E_m)t]\right\} \quad (2.41)$$

With Eq. (2.15) $a_k^{Raman}(t)$ can be expressed as

$$a_k^{Raman}(t) = \frac{E_o^2\left\{1 - \exp[\frac{i}{\hbar}(E_k - E_m + \hbar\omega)t]\right\}}{E_m - E_k + \hbar\omega} \sum_{l>m}^{k} \frac{\langle m|\vec{\mu}|l\rangle \langle l|\vec{\mu}|k\rangle}{E_m - E_l + \hbar\omega}$$

$$\quad (2.42)$$

$$+ \frac{E_o^2\left\{1 - \exp[\frac{i}{\hbar}(E_k - E_m - \hbar\omega)t]\right\}}{E_m - E_k - \hbar\omega} \sum_{l>m}^{k} \frac{\langle m|\vec{\mu}|l\rangle \langle l|\vec{\mu}|k\rangle}{E_m - E_l - \hbar\omega}$$

Thus, the probability for a transition from a lower state m to a higher state k is given by

$$a_k^*(t)a_k(t) = \frac{E_o^4}{(E_m - E_k - \hbar\omega)^2} \alpha_{mk}^2$$

$$\cdot \left\{ 2 - \exp[\frac{i}{\hbar}(E_k - E_m - \hbar\omega)t] - \exp[\frac{-i}{\hbar}(E_k - E_m - \hbar\omega)t] \right\}$$

$$= \frac{4E_o^4 \sin^2[\frac{1}{2\hbar}(E_k - E_m - \hbar\omega)t]}{(E_m - E_k - \hbar\omega)^2} \alpha_{mk}^2 \qquad (2.43)$$

where

$$\alpha_{mk} = \sum_1 \frac{<m|\vec{\mu}|1> \; <1|\vec{\mu}|k>}{E_m - E_1 + \hbar\omega} \qquad (2.44)$$

is called the *polarizability* tensor for the discussed transition from m via l to k. The components of this tensor can be expressed as

$$(\alpha_{ij})_{mk} = \sum_1 \frac{<m|\vec{\mu}_i|1> \; <1|\vec{\mu}_j|k>}{E_m - E_1 + \hbar\omega} \qquad (2.45)$$

The tensor α_{mk} can be separated into a symmetric and an antisymmetric part. Born [15] pointed out that the antisymmetric tensor describes the optical activity. For nonoptical active molecules the tensor is always symmetric.

2.2 NORMAL VIBRATIONS WITH RESPECT TO MACROMOLECULES[†]

2.2.1 Basic Theory

An N-atomic molecule possesses 3N degrees of freedom. Subtraction of the pure translation and rotations of the whole molecule leaves 3N - 6 vibrational degrees of freedom (3N - 5 for linear molecules). Thus, an ethane molecule can vibrate in 18 different modes, but a polyethylene chain with 10^5 methylene groups could vibrate in about 9×10^5 modes. Fortunately, Raman and IR spectra of polymers never show such a large number of corresponding bands.

In what follows, the reasons for this enormous reduction of observable absorption bands in the vibrational spectra of polymers is

---

[†]For further comprehensive studies, see Refs. 16 to 30.

discussed. However, let us first regard an N-atomic isolated molecule. In a cartesian coordinate system $x = (x_1, x_2, x_3)$, the kinetic and potential energy for an atom of mass m is given by

$$2E_{kin} = \dot{x}'M\,\dot{x} \qquad M = \begin{pmatrix} m & 0 & 0 \\ 0 & m & 0 \\ 0 & 0 & m \end{pmatrix} \qquad (2.46)$$

$$\text{with } M = \begin{pmatrix} m & 0 & 0 \\ 0 & m & 0 \\ 0 & 0 & m \end{pmatrix}$$

$$2E_{pot} = x'\hat{F}\,x \qquad (2.47)$$

where $\dot{x}'$ and $\dot{x}$ are the time derivatives of the row and column vector x, respectively. Eq. (2.46) also holds for an N-atomic molecule, where the position of the atoms can be represented by $x = (x_1, x_2, \ldots, x_{3N})$. Then M becomes a 3N × 3N diagonal matrix of the atomic masses. Generally, it is much easier and more useful to introduce *internal coordinates* r instead of cartesian coordinates. These internal coordinates describe changes in bond lengths and bond angles. Upon transformation of the cartesian coordinates into internal coordinates

$$r = B\,x \qquad (2.48)$$

the kinetic energy E_{kin} and the potential energy E_{pot} can be expressed by

$$E_{kin} = \dot{r}'K\,\dot{r} = \sum_{i,j} k_{ij}\dot{r}_i\dot{r}_j \qquad (2.46b)$$

and

$$E_{pot} = r'F\,r = \sum_{i,j} f_{ij}r_i r_j \qquad (2.47b)$$

The elements $k_{ij} = \partial^2 E_{kin}/\partial\dot{r}_i\partial\dot{r}_j$ of the matrix K are called *reduced mass factors*, and the coefficients $f_{ij} = \partial^2 E_{pot}/\partial r_i\partial r_j$ are the *force constants*. The application of Lagrange's equation on Eqs. (2.46b) and (2.47b) results in the differential equation

$$F\,r + K\,\ddot{r} = 0 \qquad (2.49)$$

Substituting $r = l_j\cos(\omega_j t + \phi)$ in Eq. (2.49), we obtain

$$(F - \lambda_j K)l_j = 0 \qquad (j = 1, 2, \ldots, N) \qquad (2.50a)$$

With $K = G^{-1}$ Eq. (2.50a) can now be transformed to

$$(F G - \lambda_j E) l_j = 0 \tag{2.50b}$$

where λ_j are the eigenvalues and l_j the corresponding eigenvectors. Eq. (2.50b) can be solved only if the *secular determinant* vanishes

$$\det(F G - \lambda_j E) = 0 \tag{2.51a}$$

Using the solutions λ_j of Eq. (2.51a), the frequencies of the vibrations can be determined by

$$\nu_j = \frac{\lambda_j^2}{2\pi} \tag{2.52}$$

After evaluation of the eigenvalues λ_j, we can derive from Eq. (2.50a) the eigenvectors l_j and form the diagonal matrix Λ of the eigenvalues and the matrix L of the corresponding eigenvectors l_j. With this notation we obtain from Eq. (2.50b)

$$G F L = L \Lambda \tag{2.53}$$

Thus, the determination of normal vibrations requires calculation of the eigenvalues of GF, which generally proves to be the most difficult procedure of the mathematical treatment.

To derive Λ and L from GF we have to take into account the fact that GF is not symmetric, although G and F are symmetric. Because G is a symmetric matrix, it can be diagonalized by an orthogonal transformation

$$A'G A = D D \tag{2.54}$$

Before discussing the introduction of the normal coordinates by means of the matrix L, let us consider two possibilities for solving the secular equation.

D is a diagonal matrix whose elements are the square roots of the eigenvalues of G. Since G has, in practice, real eigenvalues from Eq. (2.54) it follows

$$D^{-1}A'G A D^{-1} = E \tag{2.55}$$

With the matrices A and D we can derive from Eq. (2.53)

$$G \; A \; D^{-1}D \; A'F = L \; \Lambda \; L^{-1} \tag{2.56}$$

Left multiplication by $D^{-1}A'$ and right multiplication by AD results in

$$\underbrace{D^{-1}A'G \; A \; D^{-1}}_{E}\underbrace{D \; A'F \; A \; D}_{H} = D^{-1}A'L \; \Lambda \; L^{-1}A \; D \tag{2.57}$$

According to Eq. (2.55), the first term is equal to E. The second term, called H, is

$$H = D \; A'F \; A \; D \tag{2.58}$$

Because $D' = D$, the matrix H is symmetric and can be easily diagonalized by means of an orthogonal transformation

$$C'H \; C = \Lambda' \tag{2.59}$$

By left and right multiplication of Eq. (2.58) by $L^{-1}AD$ and $D^{-1}A'L$, respectively, we obtain

$$L^{-1}A \; D \; H \; D^{-1}A'L = \Lambda \tag{2.60}$$

Comparison of Eqs. (2.59) and (2.60) results in

$$C' = L^{-1}A \; D \qquad \text{and} \qquad C = D^{-1}A'L \qquad \text{with } \Lambda' = \Lambda \tag{2.61}$$

Thus, the matrix L of the eigenvectors l_j and its inverse L^{-1} can be evaluated from Eq. (2.61)

$$L = A \; D \; C \qquad \text{and} \qquad L^{-1} = C'D^{-1}A' \tag{2.62}$$

The eigenvalues of the nonsymmetric GF matrix can be calculated in three steps. First G is diagonalized by means of Eq. (2.54), then F is transformed according to Eq. (2.58) to H, and finally H is diagonalized via Eq. (2.60). The preceding treatment does not take into account any symmetry of the molecule under consideration.

Often, for simplification, other types of coordinates are suitable constructed by linear combinations of the cartesian, the mass-adjusted cartesian, or the internal coordinates x, x_m, or r, respectively. Thus, commonly *symmetry coordinates* s are introduced in order to take full advantage of molecular symmetry. Let us adopt for our consideration the following relation between symmetry coordinates s and the mass-adjusted cartesian coordinates $x_m = m^{-1/2}x$

$$s = U\, x_m \qquad (U\ \text{orthogonal}) \tag{2.63}$$

which is preferred by computer methods to determine the eigenvalues λ_j. The linear combinations are determined in such a way that each symmetry coordinate is symmetric or antisymmetric with respect to each symmetry operation. With these coordinates the secular determinant (2.51a) can be factored into blocks of smaller size.

The secular equation of the mass-adjusted cartesian coordinates can be derived from Eq. (2.51a) by the introduction of $r = BM^{1/2}U^{-1}s$. Then the secular determinant reads

$$\det(F\, B\, M^{1/2}U^{-1} - \lambda_j(G^{-1}B\, M^{1/2}U^{-1})E) = 0$$

$$\det(U\, M^{-1/2}B^{-1}M^{1/2}B\, M^{-1}B'M^{1/2}F\, B\, M^{1/2}U^{-1} - \lambda_j E) = 0 \tag{2.51b}$$

$$\det(U\, B'_m F\, B_m U^{-1} - \lambda_j E) = \det(A - \lambda_j E) = 0$$

where

$$B\, M^{1/2} = B_m \qquad \text{and} \qquad U\, B'_m F\, B_m U^{-1} = A$$

Since the result of the similarity transformation $B'_m F B_m$ of the symmetric matrix F of the force constants is again a symmetric matrix, the following orthogonal transformation results in a matrix A which is also symmetric, and in addition factored into smaller blocks, each of which concerns a single species to which the corresponding symmetry coordinates belong. Then each of the blocks of the corresponding secular determinant can be separately set equal to zero. Thus, by introducing symmetry coordinates instead of mass-adjusted cartesian coordinates x_m, the secular determinant can be factored into smaller determinants, each of which belongs to a symmetry species of the molecular point group. A similar treatment can be performed by use of the cartesian or internal coordinates.

By means of the eigenvector matrix L [Eq. (2.62)] we can introduce the *normal coordinates* q and relate them to the internal coordinates r

$$r = L \, q \tag{2.64}$$

In the case of N different eigenvalues, the kinetic and potential
energies are expressed in terms of normal coordinates by the following equations, which do not contain mixed products of coordinates as
in Eqs. (2.46b) and (2.47b)

$$2E_{kin} = \dot{q}'\dot{q} = \sum_i \dot{q}_i^2 \tag{2.46c}$$

$$2E_{pot} = q'\Lambda q = \sum_i \lambda_i q_i^2 \tag{2.47c}$$

For degenerated eigenvalues the potential energy reads

$$2E_{pot} = \sum_i \lambda_i q_i^2 + \frac{1}{2} \sum_k \lambda_k \sum_{j=1}^{n_k} q_{jk}^2 \tag{2.47d}$$

In this equation the first term is valid for nondegenerate eigenvalues and the second for degenerate values. By the introduction of
normal coordinates the potential and kinetic energies can be represented in a quadratic form containing no mixed products of the coordinates for different eigenvalues. Applying Lagrange's equation to
Eqs. (2.46c) and (2.47c), we can resolve the coupled differential
equation system (2.49) into 3N - 6 (or 3N - 5) independent differential equations

$$q_i + \lambda_i q_i = 0 \qquad i = 1, 2, \ldots, 3N - 6 \text{ (or } 3N - 5) \tag{2.65}$$

which can be solved separately, yielding the *normal frequencies* of
the normal vibrations.

For practical use it is often not necessary to calculate the
normal frequencies for the determination of the normal coordinates,
since the properties of the normal vibrations can be deduced by
group-theoretical considerations. The application of group theory to
the classification of normal vibrations is based on the fact that a

symmetry operation S (a mathematical operation) does not change the
potential and kinetic energies of the molecule. So from

$$2E_{pot} = 2SE_{pot} = \sum_i \lambda_i Sq_i^2 + \frac{1}{2} \sum_k \lambda_k \sum_{j=1}^{n_k} Sq_{jk}^2 \qquad (2.47e)$$

it can be deduced that because $Sq_i^2 = q_i'^2$ for nondegenerate eigenvalues, yields

$$q_i^2 = q_i'^2 \qquad or \qquad q_i = \pm q_i' \qquad (2.66)$$

Thus, for nondegeneracy only symmetric or antisymmetric normal coordinates are possible. The corresponding symmetric or antisymmetric
normal vibrations are designated with A and B, respectively. For
degeneracy, it follows that

$$\sum q_i^2 = \sum q_i'^2 \qquad (2.67)$$

With respect to the degree of degeneracy for these vibrations
the symbols E, F, G, etc., are used. The indices in Table 2-2 indicate symmetry or antisymmetry relative to the symmetry element.
Group-theoretic considerations reveal that all realizable symmetry
operations of a molecule which leave at least one point of the molecule unchanged form a mathematical group, the so-called *point group.*

TABLE 2-2   Designation of the Irreducible Representations

Races	A	symmetric	with respect to the main rotation axis
	B	antisymmetric	
	E	degenerated vibrations	
	F		
Indices	g	symmetric	with respect to a center of inversion
	u	antisymmetric	
	'	symmetric	with respect to a mirror plane; for more
	"	antisymmetric	than one mirror plane, ' or " refer to σ_n
	1, 2,.. counting indices		

For every possible symmetry operation of an N-atomic molecule, a reducible matrix representation R can be found which describes the
transformational behavior of the whole molecule. In the matrix R, a
matrix of order 3 occurs at positions which depend on the changes in
the positions of the atoms under the corresponding symmetry operation.
This matrix A describes the transformation of the cartesian coordinates for this symmetry operation.

The characters or traces of the matrix representations for all
symmetry operations are given by

$$\chi(R_S) = N_S(\pm 1 + 2\cos\alpha) \tag{2.68}$$

Here N_S represents the number of atoms which remain unchanged under
the symmetry operation and α the rotation angle. The + sign stands
for *proper* operations as the identity E and the rotation C_n; the
– sign stands for *improper* operations such as reflection σ, inversion
i , and rotation-reflection S_n. The expression in parentheses is the
trace of the matrix A. The list of the characters $\chi(R_S)$ for all symmetry operations of the point group of a given molecule is called
reducible representation Γ.

By means of the transformation in Eq. (2.64) between internal
coordinates r and normal coordinates q, we can relate the matrix R_S
to a matrix R_S^{irr} by the orthogonal transformation

$$R_S^{irr} = L\, R_S L' \tag{2.69}$$

Since L is an orthogonal matrix, the reducible representation R_S is
diagonalized by the coordinate transformation in the case of nondegenerate eigenvalues. For degenerate eigenvalues, R_S is transformed
into a block diagonal matrix which possesses only in the main diagonal one- or higher-dimensional submatrices, whose order depends on
the degree of degeneracy of the eigenvalues of R_S. For the characterization of these submatrices by their characters, the reducible representation Γ can be separated into irreducible representations Γ_i:

$$\Gamma = \sum_{i=1}^{k} n_i \Gamma_i \tag{2.70}$$

where n_i gives the multiplicity of the irreducible representation Γ_i. The i-th irreducible representation Γ_i contains the characters of the submatrices, which describe the transformational behavior of the i-th coordinate for all symmetry operations of a point group. In the character tables - generally to be found in the appendix of most books on group theory and basic vibrational spectroscopy - the characters of the irreducible matrix representations are listed, which characterize the symmetry behavior of a class of symmetry operations. In the i-th row one finds the characters of the symmetry operations belonging to the i-th irreducible representation of the point group under consideration, which reflects the behavior of the i-th normal coordinate. The characters of the submatrices of the irreducible matrix representation of one symmetry operation are arranged in columns.

With the normal coordinates (representing the relative distortions of the atoms during a normal vibration) as a basis, from Eq. (2.69) the number n_i of normal coordinates (or vibrations) of an irreducible representation or vibrational race can be deduced:

$$n_i = \frac{1}{h} \sum_S \chi(R_S)\chi_i(S) \tag{2.71}$$

Here, $\chi(R_S)$ can be calculated from Eq. (2.68), and $\chi_i(S)$ represents the character of the i-th irreducible matrix representation of the h symmetry operations S. To obtain the number of proper vibrations the number of pure translations n_i^{trans} and rotations n_i^{rot},

$$n_i^{trans} = \frac{1}{h} \sum_S (\pm 1 + 2\cos\alpha_S)\chi_i(S) \tag{2.72}$$

$$n_i^{rot} = \frac{1}{h} \sum_S (1 \pm 2\cos\alpha_S)\chi_i(S) \tag{2.73}$$

have to be subtracted from n_i.

Because the dipole moment vector can be transformed like a translation vector, the occurrence of infrared activity requires n_i^{trans} to differ from zero. The transformational behavior of the polarizability tensor α can be described by

$$
\begin{pmatrix} \alpha'_{xx} & \alpha'_{xy} & \alpha'_{xz} \\ \alpha'_{yx} & \alpha'_{yy} & \alpha'_{yz} \\ \alpha'_{zx} & \alpha'_{zy} & \alpha'_{zz} \end{pmatrix} = \begin{pmatrix} \cos\phi & \sin\phi & 0 \\ -\sin\phi & \cos\phi & 0 \\ 0 & 0 & \pm 1 \end{pmatrix} \begin{pmatrix} \alpha_{xx} & \alpha_{xy} & \alpha_{xz} \\ \alpha_{yx} & \alpha_{yy} & \alpha_{yz} \\ \alpha_{zx} & \alpha_{zy} & \alpha_{zz} \end{pmatrix} \begin{pmatrix} \cos\phi & -\sin\phi & 0 \\ \sin\phi & \cos\phi & 0 \\ 0 & 0 & \pm 1 \end{pmatrix} \quad (2.74)
$$

Evaluation of Eq. (2.74) results in $(\alpha_{xy} = \alpha_{yx},\ \alpha_{xz} = \alpha_{zx},\ \alpha_{yz} = \alpha_{zy})$

$$
\begin{pmatrix} \alpha'_{xx} \\ \alpha'_{yy} \\ \alpha'_{zz} \\ \alpha'_{xy} \\ \alpha'_{xz} \\ \alpha'_{yz} \end{pmatrix} = \begin{pmatrix} \cos^2\phi & \sin^2\phi & 0 & 2\cos\phi\sin\phi & 0 & 0 \\ \sin^2\phi & \cos^2\phi & 0 & -2\cos\phi\sin\phi & 0 & 0 \\ 0 & 0 & 1 & 0 & 0 & 0 \\ -\sin\phi\cos\phi & \sin\phi\cos\phi & 0 & \cos^2\phi-\sin^2\phi & 0 & 0 \\ 0 & 0 & 0 & 0 & \pm\cos\phi & \pm\sin\phi \\ 0 & 0 & 0 & 0 & \mp\sin\phi & \pm\cos\phi \end{pmatrix} \begin{pmatrix} \alpha_{xx} \\ \alpha_{yy} \\ \alpha_{zz} \\ \alpha_{xy} \\ \alpha_{xz} \\ \alpha_{yz} \end{pmatrix} \quad (2.75)
$$

Thus, the number of different components α_{ij} or their combinations belonging to each symmetry species is given by

$$
n_i^{\text{Raman}} = \frac{1}{h} \sum_S 2\cos\alpha_S \cdot (\pm 1 + 2\cos\alpha_S) \cdot \chi_i(S) \quad (2.76)
$$

Let us now extend our consideration to crystallizable polymers [24-30]. The theoretical treatment of IR and Raman spectroscopy is based on an idealized model of a regular infinitely long polymer chain which is packed into a perfect crystal with well-defined unit cells. All the unit cells - regarded as isolated units - possess the same symmetry elements S_1, S_2,..., S_N. In addition to the unit cell symmetry operations, translations of a multiple or a combination of the lattice vectors are allowed. The set of all translations forms the *translation group* T of infinite order. All the possible combinations of pure translations with the unit cell operations S_n yield the elements of the *space group* of infinite order or (in the case of a one-dimensional infinitely long, isolated chain) a *line group*.

Introduction of Born's cyclic boundary condition permits this treatment to be applied to real finite crystals and highly ordered polymers. In this case we obtain a space group of infinite order. By multiplication of each element of the translation group with one ele-

ment S_n of the unit cell operations, we can form a set of elements
(coset) which represents an element F_n of the *factor group* F. The order of the factor group is given by the number N of the possible symmetry operation in a unit cell. The resulting factor group is isomorphous to one of the 32 point groups and therefore possesses the same
character table. Thus, a factor group analysis can be carried out that
is quite similar to a point group analysis of the normal vibrations of
simpler molecules discussed earlier. Once the factor group of the
space or line group and the isomorphous point group have been determined, the number of IR- and Raman-active fundamentals of each species
can be deduced by means of the corresponding character table via Eqs.
(2.71) to (2.73).

The vibrational behavior of polymer crystals or a hypothetical
isolated chain may be represented in terms of the normal vibrations
and normal coordinates of the repeating unit, since in a geometrically
regularly arranged polymer chain the fundamentals arise only if the
atoms of the repeating units vibrate with a distinct phase difference
$\delta = 0$, or $\delta = 2p\pi/1$ in the case of a helix conformation. Here, p
stands for the number of full turns of a helix and l for the number
of chemical repeating units.

A general approach to the calculation of the normal frequencies
of conformational regular infinite polymer chains, which includes the
treatment of helical polymers with a distinct phase difference δ
between neighboring chemical repeating units, is based on the assumption that all chemical repeating units describe the same motion in
space but may be out-of-phase by ϕ [24, 25]. By definition of complex
cartesian coordinates $x^C(\phi)$, complex B_m^C and F^C matrices,

$$x^C(\phi) = \sum_{k=-\infty}^{+\infty} x_k \exp(ik\phi) \tag{2.77}$$

$$B_m^C(\phi) = \sum_{k=-\infty}^{\infty} B_{m,k} \exp(ik\phi) \tag{2.78}$$

$$F^C(\phi) = \sum_{k=-\infty}^{+\infty} F_k \exp(ik\phi) \tag{2.79}$$

this assumption can be satisfied. The subscript k refers to the k-th chemical repeating unit. The secular determinant (2.51b) then reads

$$\det\{U\ B_m^{'C}(\phi)\ F^C(\phi)\ B_m^C(\phi)\ U^{-1} - \lambda_j(\phi)\ E\} = 0 \qquad (2.51c)$$

and is a function of the phase angle ϕ. However, according to the selection rules, optical activity occurs in an infinite polymer chain only for those modes with $\phi = 0$, $\phi = \delta$, and $\phi = 2\delta$. For all other phase angles ϕ, the vibrations are neither Raman- nor IR-active but contribute to the thermodynamic properties of the polymer. Generally, the secular determinant (2.51c) will be complex and the corresponding complex quantities $x^C(\phi)$, $B_m^C(\phi)$, and $F^C(\phi)$ have to be transformed to their real equivalents by a similarity transformation. For this purpose let us define a matrix $X'(\phi)$ of order $3N \times 3N$ (N = number of atoms in the chemical repeating unit),

$$X'(\phi) = \begin{pmatrix} x^C(\phi) & 0 \\ 0 & x^{C*}(\phi) \end{pmatrix} \qquad (2.80)$$

where $x^{C*}(\phi)$ is the conjugated complex to $x^C(\phi)$, and perform the following similarity transformation:

$$X(\phi) = \frac{1}{2}\begin{pmatrix} E & E \\ -iE & iE \end{pmatrix}\begin{pmatrix} x^C(\phi) & 0 \\ 0 & x^{C*}(\phi) \end{pmatrix}\begin{pmatrix} E & iE \\ E & -iE \end{pmatrix}$$

$$= \begin{pmatrix} \sum\limits_k x_k(\phi)\ \cos k\phi & -\sum\limits_k x_k(\phi)\ \sin k\phi \\ \sum\limits_k x_k(\phi)\ \sin k\phi & \sum\limits_k x_k(\phi)\ \cos k\phi \end{pmatrix} \qquad (2.81)$$

$X(\phi)$ is a real matrix of order $6N \times 6N$. After applying similar transformations to the $B_m^C(\phi)$ and $F^C(\phi)$ matrices, the final secular determinant can be written as follows:

$$\det(U\ B_m'(\phi)\ F(\phi)\ B_m(\phi)\ U^{-1} - \lambda_j(\phi)\ E) = 0 \qquad (2.51d)$$

For phase angles $\phi = 0$ and π the performance of the orthogonal transformation

$$U\{B_m'(\phi) \ F(\phi) \ B_m(\phi)\}U^{-1} = A(\phi) \tag{2.82}$$

results again in a factored matrix $A(\phi)$, since under these conditions
local symmetry coordinates can be defined within the chemical repeating unit. Because of the transformation of the complex vector $x^C(\phi)$
and the complex matrices $B_m^C(\phi)$ and $F^C(\phi)$ into their real equivalents
$X(\phi)$, $B_m(\phi)$, and $F(\phi)$, the eigenvalues of $A(\phi)$ occur in pairs and we
obtain only 3N distinct eigenvalues $\lambda_j(\phi)$ even if the order of $A(\phi)$
is 6N × 6N. The eigenvalues $\lambda_j(\phi)$ and the corresponding vibrational
frequencies $\nu_j(\phi)$

$$\nu_j(\phi) = \frac{1}{2\pi} \sqrt{\lambda_j(\phi)} \tag{2.52b}$$

are functions of the phase angle. Thus, by introduction of the complex quantities [Eqs. (2.77) to (2.79)] in addition to the vibrational frequencies for a distinct phase angle, the dispersion curves
$\nu_j = f(\phi)$ for an infinite polymer chain can be calculated by varying
the phase angle ϕ.

For real crystals we have to take into account the distortions
which result in the finite length of regular segments, but for most
spectroscopic studies these effects arising from the finite extension
of the crystals are negligible. A theoretical approach to the calculation of the vibrational spectrum of polymer chains containing defect structures has been given by Zerbi [31-33].

In polymers where sequences of chemical groups (for example, CH_2)
are regularly arranged between heavier groups (for example, CO, CONH,
O, SO_2), coupling effects within the sequence have to be taken into
account. The normal vibrations of such a system can be approximated
by a linear array of N coupled oscillators, where every isolated
oscillator would vibrate with the frequency ω_o. However, the system
of coupled oscillators produces a set of different frequencies according to the possible phase differences between the oscillators which
in practice results in the *band progression* series.

If there are N identical groups in the geometrical repeating
unit of the chain, the phase difference for optically active vibrations of adjacent groups is given by

$$\phi = \frac{2\pi k}{N} \qquad (k = 0, 1, 2, \ldots, N - 1) \qquad (2.83)$$

In polymers with more than one chain per unit cell intermolecular coupling may lead to a splitting of bands due to in-phase and
out-of-phase vibrations of adjacent chains in the unit cell. This
correlation or *Davidov splitting* finds its group-theoretic explanation by turning from a line group analysis, which is valid for isolated, conformationally regular chains with negligible intermolecular forces, to a space group analysis, which takes into account
the intermolecular interactions between adjacent chains in a unit
cell.

2.2.2   Application to Polymers with Methylene Sequences

2.2.2.1   Factor Group Analysis of Polyester-x,y

To illustrate the concept of factor group analysis, the series of
linear, aliphatic polyesters PE-x,y is discussed in more detail.
Furthermore, this discussion points out all the basic requirements
necessary for understanding the vibrational behavior of polymers
with methylene sequences.

The polyesters with an even number of methylene groups in the
alcoholic and acidic part (PE-e,e) crystallize quite readily, while
the odd-even or even-odd types (PE-o,e or PE-e,o) and the polyesters
with an odd number of methylene groups in both parts (PE-o,o) sometimes crystallize only with difficulties. Polyesters with an even
number of methylene groups in the alcoholic and acidic parts occur
preferentially in a monoclinic unit cell, assuming an essentially
planar zigzag conformation of the backbone atoms [34-39]. The unit
cell contains two molecular chains with orthogonal planes (Fig. 2-2)
of the skeletal atoms. But only in the spectra of polyesters recorded
at low temperature (153 K) do some bands split into doublets, proba-

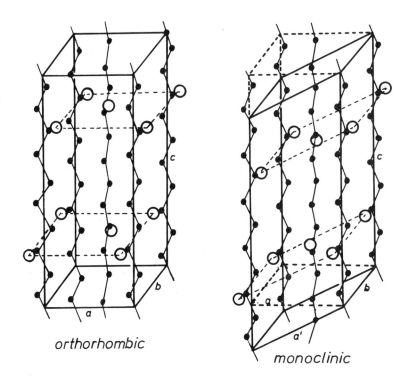

orthorhombic

monoclinic

FIGURE 2-2   Arrangement of the polyester skeleton in the crystal-
             lographic unit cell. ●: oxygen and carbon atoms of the
             skeleton, O: oxygen of the carbonyl group.

bly because of the coupling of the in-phase and out-of-phase vibrations of the two chains within the unit cell. Therefore, a factor
group analysis based on the line group of the isolated, one-dimensional chain can be employed instead of a factor group analysis
based on the space group which takes into consideration the three-
dimensional geometrical order of the polymer.

     According to the odd or even number of methylene groups in the
alcoholic or acidic part, the repeating unit of the polyesters possess different symmetry elements (Fig. 2-3). The corresponding symmetry operation, the order of the factor group, and the isomorphous
point group are listed in Table 2-3. Because of the mass differences,

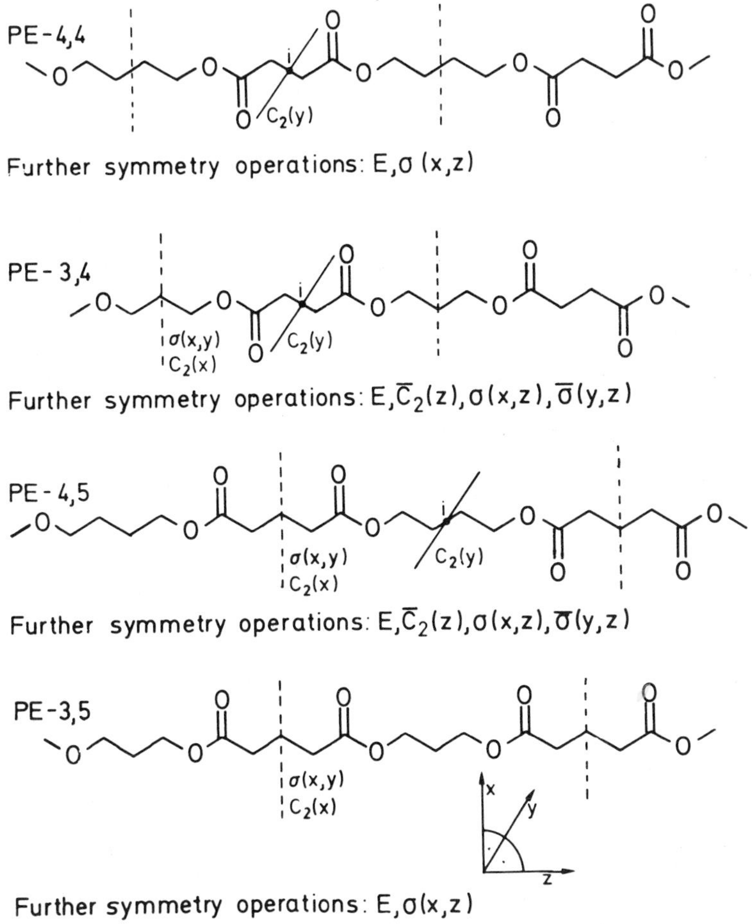

FIGURE 2-3 Symmetry operations of linear aliphatic polyesters. The
dashed vertical lines indicate the repeating unit necessary for normal coordinate analysis when phasing the
polymer according to Eq. (2.77) either by 0 (PE-e,e and
PE-o,o) or by π (PE-o,e and PE-e,o).

TABLE 2-3   Point Group and Symmetry Operations of the Polyester-x,y

Type	Order of factor group	Symmetry operations	Isomorphous point group
PE-e,e	4	E, $C_2(y)$, $\sigma(x,z)$, i	C_{2h}
PE-o,o	4	E, $C_2(x)$, $\sigma(x,z)$, $\sigma(x,y)$	C_{2v}
PE-e,o	8	$\begin{cases} E, C_2(y), \bar{C}_2(z), C_2(x), i, \\ \sigma_v(x,z), \bar{\sigma}_v(y,z), \sigma_h(x,y) \end{cases}$	D_{2h}
PE-o,e			

TABLE 2-4   Calculation of the Normal Vibrations of the Polyester Backbone

C_{2h}	E	$C_2(y)$	$\sigma(x,z)$	i
A_g	1	1	1	1
A_u	1	1	-1	-1
B_g	1	-1	-1	1
B_u	1	-1	1	-1
$\cos\alpha_S$	1	-1	1	-1
$N_R(C,O)$	10	0	10	0
$\chi_{trans} = \pm 1 + 2\cos\alpha_S$	3	-1	1	-3
$\chi_{rot} = 1 \pm 2\cos\alpha_S$	3	-1	-1	3
$\chi(R_S) = N_R(\pm 1 + 2\cos\alpha_S)$	30	0	10	0
$\chi(R_S) - \chi_{trans} - \chi_{rot}$	24	2	10	0
$n(A_g)$	$(1/4)[24\cdot1 + 2\cdot1 + 10\cdot1 + 0\cdot1] = 9$			
$n(A_u)$	$(1/4)[24\cdot1 + 2\cdot1 + 10\cdot(-1) + 0\cdot(-1)] = 4$			
$n(B_g)$	$(1/4)[24\cdot1 + 2\cdot(-1) + 10\cdot(-1) + 0\cdot1] = 3$			
$n(B_u)$	$(1/4)[24\cdot1 + 2\cdot(-1) + 10\cdot1 + 0\cdot(-1)] = 8$			

the vibrations of the backbone can be treated separately from the vibrations of the hydrogen atoms.

The discussion of the polyester PE-4,4, which is assumed to form a planar zigzag chain of the carbon-oxygen backbone, will enable us to generalize the results of the PE-e,e types. Here, the backbone of the repeating unit consists of eight carbon and two oxygen atoms.

The number n of normal modes of this backbone can be calculated on
the basis of the character table of the point group C_{2h} and Eqs.
(2.73), (2.76), (2.79), and (2.80). To provide a better survey, Table
2-4 is used for the evaluation.

From this table the number of normal modes of the backbone and
their distribution on the irreducible representations can be derived:

$$n(C,O) = 9A_g + 4A_u + 3B_g + 8B_u \tag{2.84}$$

However, in solid polymers rotations about the x- and y-axis are impossible because of Born's boundary condition. These rotations of
the repeat unit around an axis perpendicular to the chain axis now
correspond to deformation vibrations of the skeleton. Owing to the
restricted rotations the degrees of freedom increase to 3N − 4.
Since a rotation about the x-axis is symmetric with respect to all
symmetry operations and a rotation, the y-axis is symmetric with
respect to the identity E and the inversion i, but antisymmetric to
the rotation $C_2(y)$ and the reflection $\sigma(x,z)$ (indicated by the parentheses in the character table). These rotations have to be added
to the A_g and B_g representation, respectively, and then (2.84) reads

$$n(C,O) = 10A_g + 4A_u + 4B_g + 8B_u \tag{2.85}$$

It can be easily verified by use of (2.79) and (2.80) that the
A_g and B_g modes are Raman-active and the A_u and B_u modes, IR-active.
Taking into account the oxygen atom of the carbonyl group [the carbon atom has been already considered in Eq. (2.80)], the additional
modes can be represented by

$$n(C=O) = 2A_g + 1A_u + 1B_g + 2B_u \tag{2.86}$$

The two A_g vibrations are the symmetrical stretching and in-plane
vibrations of the carbonyl group. The out-of-plane vibrations of the
carbonyl, which is symmetric with respect to the $C_2(y)$ but antisymmetric to $\sigma(x,z)$ and i, belongs to the A_u representation. The
corresponding antisymmetric out-of-plane vibration can be assigned
to the B_g mode. The antisymmetric stretching (with respect to $C_2(y)$)
and in-plane carbonyl vibrations belong to the B_u representation.
The 12 hydrogen atoms give rise to further 36 normal modes, whose

distribution on the four representations can easily be deduced from
Table 2-4 by changing the values of $N_R(C,O)$ into those of $N_R(H)$.
Thus, it follows for the number of the corresponding normal vibra-
tions

$$n(H) = 9A_g + 9A_u + 9B_g + 9B_u \qquad (2.87)$$

In the case of an even number of methylene units, it is convenient
to treat two methylene groups as an ethylene unit. Besides the vibrations between the carbon atoms such an ethylene unit can carry out
12 vibrations, which are shown in Fig. 2-4.

The vibrations of Fig. 2-4 are typical for all polymers containing an even number of methylene units in an all-trans conformation. In the case of intermolecular coupling of chains within one
crystallographic unit cell, additional in phase and out-of-phase

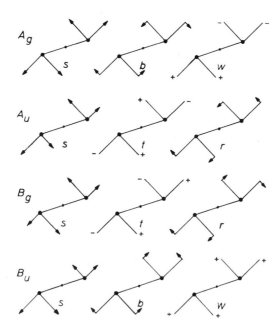

FIGURE 2-4   The 12 possible methylene vibrations of a $-CH_2-CH_2-$unit.
(The comparatively small motions of the carbon atoms are
not indicated.)

vibrations which reflect the symmetry behavior of the unit cell have
to be considered. The six typical vibrations of a single methylene
unit can easily be derived, since we no longer distinguish between
symmetrical and antisymmetrical vibrations with respect to the center
of symmetry. Thus, we find two C-H-stretching modes (symmetric and
antisymmetric with respect to the C_2-operation) and one bending,
wagging, twisting, and rocking mode.

In Table 2-5 these vibrations and some other frequently used
modes for describing the vibrational behavior of polymers are listed.

TABLE 2-5   Some Commonly Encountered Vibrations

Geometry	Vibration	Nomenclature
	Symmetric CH_2-stretching	$\nu_s(CH_2)$
	Antisymmetric CH_2-stretching	$\nu_{as}(CH_2)$
	CH_2-bending	$\delta(CH_2)$
	CH_2-twisting	$\gamma_t(CH_2)$
	CH_2-wagging	$\gamma_w(CH_2)$
	CH_2-rocking	$\gamma_r(CH_2)$
	OH in-plane bending	$\delta(OH)$
	OH out-of-plane bending	$\gamma_w(OH)$
	CCC-bending	$\Delta(CCC)$

The stretching, bending, and wagging vibrations which are symmetric with respect to all symmetry operations belong to the A_g-representation, whereas those which are antisymmetric to $C_2(y)$ and i belong to B_u. The twisting, rocking, and stretching vibrations symmetric with respect to $C_2(y)$ but antisymmetric to i are assigned to A_u. The B_g representation contains the vibrations antisymmetric to $C_2(y)$ and symmetric to i.

The possible normal vibrations for the other PE-(e,e) can be determined by stepwise addition of further ethylene groups. Every ethylene group contributes 18 additional vibrations (6 skeletal and 12 hydrogen vibrations). Generally, the number of the normal vibrations of the skeletal and the hydrogen atoms can be calculated from the following equations

$$
\begin{aligned}
n(C=O) &= 2A_g + 1A_u + 1B_g + 2B_u \\
n(C,O) &= (2m+4)A_g + (m+1)A_u + (m+1)B_g + (2m+2)B_u \\
n(H) &= 3mA_g + 3mA_u + 3mB_g + 3mB_u \\
\hline
n(PE\text{-}e,e) &= (5m+6)A_g + (4m+2)A_u + (4m+2)B_g + (5m+4)B_u
\end{aligned} \tag{2.88}
$$

where m stands for the number of $(CH_2)_2$ units.

In the case of the polyesters of the type PE-e,o or PE-o,e, two chemical repeating units have to be taken into consideration. The number of normal vibrations can be analogously derived with the aid of the character table for the point group D_{2h}

$$
\begin{aligned}
n(PE\text{-}o,e;\ PE\text{-}e,o) =\ & \frac{5m+13}{2} A_g + (2m+2)B_{1g} + \frac{5m+9}{2} B_{2g} \\
& +(2m+2)B_{3g} + (2m+2)A_u + \frac{5m+9}{2} B_{1u} \\
& +(2m+4)B_{2u} + \frac{5m+11}{2} B_{3u}
\end{aligned} \tag{2.89}
$$

where m represents the number of methylene groups within a chemical repeating unit. For the polyesters PE-o,o it follows

$$
n(PE\text{-}o,o) = (5m+6)A_1 + (4m+1)A_2 + (5m+4)B_1 + (4m+3)B_2 \tag{2.90}
$$

where m stands for the number of methylene groups divided by 2. All
representations are Raman-active; the A_1, B_1, and B_2 representations
are IR-active.

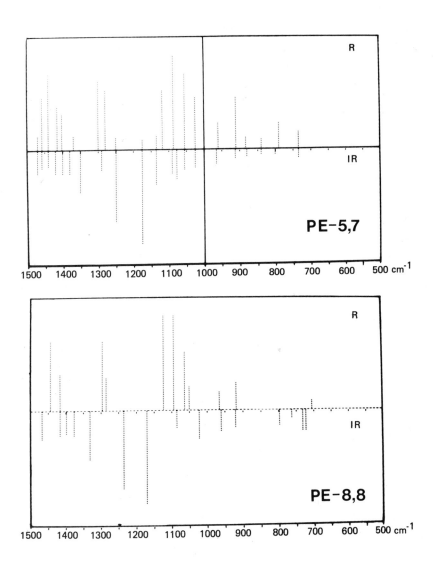

FIGURE 2-5 Spectral changes of the polyester-8,8 and polyester-5,7.

Figure 2-5 shows as an example the spectral changes of the IR
and Raman spectra of polyester-8,8 and polyester-5,7. According to
Table 2-3 and Fig. 2-3, the polyester-8,8 possesses a center of symmetry and belongs to a factor group isomorphous to the point group
C_{2h}. However, polyester-5,7 whose factor group is isomorphous to the
point group C_{2v}, does not possess a center of symmetry. Thus,
according to the previous results, for polyester-8,8 the vibrations
cause either IR or Raman bands, whereas in polyester-5,7 besides A_2,
which is forbidden in infrared, all vibrations are IR- and Raman-
active. This behavior can be qualitatively verified by the inspection
of Fig. 2-5 [39].

2.2.2.2 Normal Vibrations of Single-Chain Polyethylene

To provide a basis for the discussion of the normal coordinate analysis and the assignment of the observed frequencies of some more complicated polymers this section gives a detailed description of the
methods to calculate the vibrational frequencies of an infinitely
long, isolated, planar zigzag model of a polyethylene chain. This
model does not take into account intermolecular forces between adjacent chains.

The fundamentals of the theory on which the formalism of the calculation is based were derived in Sec. 2.1.1. In principle, there are
two different approaches to the calculation of the frequencies and
normal coordinates:

1. The symmetry properties of the polyethylene chain are not con-
    sidered and therefore the secular determinant will not be fac-
    tored into smaller blocks representing the respective symmetry
    species.

2. Symmetry properties are taken into account in the solution of the
    problem and the secular determinant is factored into smaller sub-
    determinants which can be separately diagonalized. As a basic
    unit either an ethylene unit or a methylene unit is used. In the
    latter case successive methylene units differ by a phase angle
    of 180°.

For illustration and for a better understanding of the advantages of
symmetry considerations we have chosen the second method which starts
from a methylene unit. The vibrations of the geometrical repeating
unit (ethylene unit) will be obtained by "phasing" the polymer. The
data necessary for the calculation can be seen from Eq. (2.51b). In
this equation the matrices U and B_m reflect the transformation from
cartesian coordinates into symmetry coordinates and from cartesian
coordinates into mass adjusted internal coordinates, respectively.
F is the matrix of the intramolecular force constants. The diagonal
elements can be associated with the forces which are responsible for
changes in bond lengths, bond angles, and torsions whilst the off-
diagonal elements belong to certain interactions between different
internal coordinates.

To solve Eq. (2.51b) we first have to consider the molecular geometry. Fig. 2-6 shows the planar polyethylene chain with those
atoms which have to be defined because of force constant consistency
(see below).

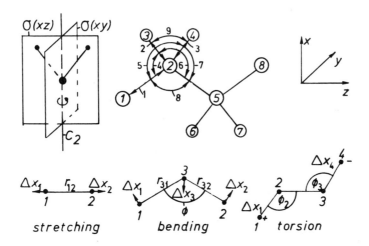

FIGURE 2-6   Designation of the atoms and symbolism necessary for the
calculation of the B matrix of polyethylene.

TABLE 2-6   Atomic Weights and Cartesian Coordinates of the Eight
            Atoms Necessary for the Normal Coordinate Analysis of
            Polyethylene

Atom	Atomic weight	x (nm)	y (nm)	z (nm)
1	12.011149	-0.889165	0.0	-1.257364
2	12.011149	0.0	0.0	0.0
3	1.007970	0.631076	0.892402	0.0
4	1.007970	0.631076	-0.892402	0.0
5	12.011149	-0.889165	0.0	1.257364
6	1.007970	-1.520240	0.892402	1.257364
7	1.007970	-1.520240	-0.892402	1.257364
8	12.011149	0.0	0.0	2.514728

The cartesian coordinates and the atomic weights of the atoms are
listed in Table 2-6. The ten defined internal coordinates are indicated in Fig. 2-6 with the exception of the torsions around the C-C
bond between atom 2 and 5. The transformation matrix B between the
internal coordinates r and the cartesian coordinates x can be derived with the help of the formulae in Ref. 22, which are reproduced
here without any further deduction. Using the designation of Fig.
2-6 the displacements expressed in cartesian coordinates of the respective atoms during the change in bond length, bond angle and rotational angle (torsion) can be written as follows:

1. Change in bond length

$$\Delta x_1 = e_{21} = -e_{12} \quad \text{and} \quad \Delta x_2 = e_{12} = -e_{21} \quad (2.91)$$

where

$$\Delta x_i = (\Delta x_i, \Delta y_i, \Delta z_i) \quad \text{and} \quad e_{ij} = \frac{(r_{ijx}, r_{ijy}, r_{ijz})}{\sqrt{r_{ijx}^2 + r_{ijy}^2 + r_{ijz}^2}}$$

2. Changes in the bond angle

$$\Delta x_1 = \frac{e_{31}\cos\phi - e_{32}}{r_{31}\sin\phi}, \quad \Delta x_2 = \frac{e_{32}\cos\phi - e_{31}}{r_{32}\sin\phi} \quad \text{and}$$

$$\Delta x_3 = \frac{e_{31}(r_{31} - r_{32}\cos\phi) + e_{32}(r_{32} - r_{31}\cos\phi)}{r_{31}r_{32}\sin\phi} \quad (2.92)$$

3. Changes in the rotational angle (torsion)

$$\Delta x_1 = - \frac{e_{12} \times e_{23}}{r_{12}\sin^2\phi_2}$$

$$\Delta x_2 = \frac{r_{23} - r_{12}\cos\phi_2}{r_{23}r_{12}\sin^2\phi_2} (e_{12} \times e_{23}) - \frac{\cos\phi_3}{r_{23}\sin^2\phi_3} (e_{43} \times e_{32})$$

$$\Delta x_3 = \frac{r_{32} - r_{43}\cos\phi_3}{r_{32}r_{43}\sin^2\phi_3} (e_{43} \times e_{32}) - \frac{\cos\phi_2}{r_{32}\sin^2\phi_2} (e_{12} \times e_{23})$$

$$\Delta x_4 = - \frac{e_{43} \times e_{32}}{r_{43}\sin^2\phi_3} \qquad\qquad (2.93)$$

Several torsional coordinates are possible. The most reasonable coordinate involves an average value of all n possible *trans* torsions
(three for polyethylene). Thus, the Δx_i have to be calculated for
three combinations of atoms, where atom 2 and 3 (Fig. 2-6) are common to all *trans* torsions and atom 1 and 4 have to be always in the
trans conformation. The desired torsional coordinate can be obtained
by averaging:

$$\Delta x_i^s = \frac{1}{\sqrt{3}} \sum \Delta x_i (p,2,3,q) \qquad\qquad (2.94)$$

where $\Delta x_i (p,2,3,q)$ represents the respective *trans* conformation of
atom p and q. For polyethylene the following *trans* torsions are
allowed:

$$\Delta x_i (1,2,5,8), \qquad \Delta x_i (3,2,5,7), \qquad \Delta x_i (4,2,5,6)$$

Application of the above Eqs. (2.91) to (2.94) to the cartesian coordinates of Table 2-6 yield the B matrix which can be transformed
into the mass-adjusted B_m matrix by division of the square roots of
the respective atomic masses. The elements of the B matrix are listed
in Table 2-7.

The next step is to build up the U matrix which transforms - according to the formalism chosen in Eq. (2.51b) - the cartesian coordinates into external local symmetry coordinates. These coordinates are defined as components of the vector s^i and can be calcu-

TABLE 2-7 Elements of the B' Matrix of Polyethylene

Internal coordinates →

0	0	0	-.265	-.265	0	0	.530	0	0
0	0	0	-.562	.562	0	0	.0	0	.397
.577	-.577	-.577	.187	.187	0	0	-.374	0	0
0	-.816	.816	-.108	-.108	-.108	-.108	-1.060	1.494	0
-.816	0	0	.826	-.826	.826	-.826	0	0	.162
0	0	0	-.979	-.979	.979	.979	0	0	0
0	.577	.577	.373	0	.373	0	0	-.747	.396
0	.816	-.816	-.264	0	-.264	0	0	.528	-.280
0	0	0	.792	0	-.792	0	0	0	-.280
0	0	0	0	.373	0	.373	0	-.747	-.396
-.577	0	0	0	.264	0	.264	0	-.528	-.280
0	0	0	0	.792	0	-.792	0	0	-.280
.816	0	0	0	0	-.265	-.265	.530	0	0
0	0	0	0	0	-.562	.562	0	0	.162
0	0	0	0	0	-.187	-.187	.374	0	0
0	0	0	0	0	0	0	0	0	-.396
0	0	0	0	0	0	0	0	0	-.280
0	0	0	0	0	0	0	0	0	-.280
0	0	0	0	0	0	0	0	0	-.396
0	0	0	0	0	0	0	0	0	-.280
0	0	0	0	0	0	0	0	0	.280
0	0	0	0	0	0	0	0	0	0
0	0	0	0	0	0	0	0	0	.397
0	0	0	0	0	0	0	0	0	0

Cartesian coordinates →

lated via the equation

$$s^i = N \sum_S \chi_i(S) R_S x \tag{2.95}$$

Here, N is a normalization factor, $\chi_i(S)$ the character of the symmetry operation S in the i-th irreducible representation, R_S the matrix representation of the symmetry operation S, and $x = (x,y,z)$. In the practical application of the above formula the transformational behavior of the cartesian coordinates of each atom has to be analyzed for every symmetry operation. Since we intend to start with only one CH_2 group, we can focus our considerations on the symmetry properties of the methylene unit. Fig. 2-6 shows the basic symmetry operations E, $C_2(x)$, $\sigma(xy)$, and $\sigma(xz)$. In Table 2-8 the transformational behavior of the cartesian coordinates for these symmetry operations and the character table of the point group C_{2v} is listed.

TABLE 2-8   Transformation Behavior of the Cartesian Coordinates for the Individual Symmetry Operations of the Methylene Group and Character Table of the Point Group C_{2v}

Coordinate	E	$C_2(x)$	$\sigma_v(xy)$	$\sigma_v(xz)$
$x(C_2)$	$x(C_2)$	$x(C_2)$	$x(C_2)$	$x(C_2)$
$y(C_2)$	$y(C_2)$	$-y(C_2)$	$y(C_2)$	$-y(C_2)$
$z(C_2)$	$z(C_2)$	$-z(C_2)$	$-z(C_2)$	$z(C_2)$
$x(H_3)$	$x(H_3)$	$x(H_4)$	$x(H_3)$	$x(H_4)$
$y(H_3)$	$y(H_3)$	$-y(H_4)$	$y(H_3)$	$-y(H_4)$
$z(H_3)$	$z(H_3)$	$-z(H_4)$	$-z(H_3)$	$z(H_4)$
$x(H_4)$	$x(H_4)$	$x(H_3)$	$x(H_4)$	$x(H_3)$
$y(H_4)$	$y(H_4)$	$-y(H_3)$	$y(H_4)$	$-y(H_3)$
$z(H_4)$	$z(H_4)$	$-z(H_3)$	$-z(H_4)$	$z(H_3)$
A_1	1	1	1	1
A_2	1	1	-1	-1
B_1	1	-1	1	-1
B_2	1	-1	-1	1

Multiplication and summation of the corresponding values of Table 2-8 according to Eq. (2.95) and subsequent normalization results in the following external local symmetry coordinate vector components:

$$s_1^{A_1}: \quad x_{C_2} + x_{C_2} + x_{C_2} + x_{C_2} = 4x_{C_2} \quad \rightarrow \quad s_1^{A_1} = x_{C_2}$$

$$s_2^{A_1}: \quad 2x_{H_3} + 2x_{H_4} \quad \rightarrow \quad s_2^{A_1} = 1/\sqrt{2} \, (x_{H_3} + x_{H_4})$$

$$s_3^{A_1} = 1/\sqrt{2} \, (y_{H_3} - y_{H_4})$$

$$\tag{2.96}$$

$$s_4^{A_2} = 1/\sqrt{2} \, (z_{H_3} - z_{H_4})$$

$$s_5^{B_1} = y_{C_2} \qquad s_6^{B_1} = 1/\sqrt{2} \, (x_{H_3} - x_{H_4}) \qquad s_7^{B_1} = 1/\sqrt{2} \, (y_{H_3} + y_{H_4})$$

$$s_8^{B_2} = z_{C_2} \qquad s_9^{B_2} = 1/\sqrt{2} \, (z_{H_3} + z_{H_4})$$

The normalization factor N is calculated via $N = 1/\sqrt{\sum_i a_i^2}$, where a_i represents the respective factor of the cartesian coordinate x_i (e.g. $1/\sqrt{16}$ for $s_1^{A_1}$ and $1/\sqrt{8}$ for $s_2^{A_1}$).

The factors of the cartesian coordinates in the above Eqs. (2.96) form the U matrix of the CH_2 group. The number of the external symmetry coordinates in each irreducible representation is equal to the number of normal vibrations in this representation including pure translations and rotations. To these external local symmetry coordinates of the methylene group we have to apply Eqs. (2.77) and (2.81). The conformational regularity of the polymer chain can be taken into account by introduction of a phase difference 0, ψ, or 2ψ (ψ = helix angle) between adjacent chemical repeat units. Mathematically, this is accomplished for the cartesian or the symmetry coordinates via the equations

$$x(\phi) = \sum_k x_k \exp(ik\phi) \qquad \text{and} \qquad s(\phi) = \sum_k s_k \exp(ik\phi) \tag{2.97}$$

where ϕ is the phase angle. Only those vibrations with $\phi = 0$, ψ, or

2ψ can be Raman or IR-active. As we will see below, for most polymers
the summation is only necessary for $k = -2, -1, 0, 1$, and 2. For
polyethylene we obtain ($\phi = 0$ or π)

$$s^i(\phi) = \sum_{k=-1}^{2} s_k^i \exp(ik\phi) = \sum_{k=-1}^{2} s_k^i \cos(ik\phi) \qquad (2.98)$$

Explicitly, we obtain the external symmetry coordinates $s_1^{A_{1g}}(0)$
and $s_1^{B_{3u}}(\pi)$ from the corresponding external local symmetry coordinate $s_1^{A_1}$ of the CH_2 group

$$s_1^{A_{1g}}(0) = \sum_{k=-1}^{2} s_1^{A_1} \cos(k \cdot 0)$$

$$= -x_{C_1} \cos(-1 \cdot 0) + x_{C_2} \cos(0 \cdot 0) - x_{C_5} \cos(1 \cdot 0) + x_{C_8} \cos(2 \cdot 0)$$

$$= -x_{C_1} + x_{C_2} - x_{C_5} + x_{C_8} \qquad (2.99)$$

and

$$s_1^{B_{3u}}(\pi) = \sum_{k=-1}^{2} s_1^{A_1} \cos(k \cdot \pi) = x_{C_1} + x_{C_2} + x_{C_5} + x_{C_8} \qquad (2.100)$$

The splitting of the irreducible representation A_1 into the species
A_{1g} and B_{3u} finds its explanation in the fact that on phasing we
change from the point group C_{2v} of the CH_2 unit to a factor group
which is isomorphous to the point group D_{2h}. Since on phasing only
the character of the symmetry operation $C_2(x)$ of the CH_2 group is
influenced and those of the operations E, $\sigma(xy)$, and $\sigma(xz)$ remain
unchanged, it can easily be seen from Table 2-8 and the character
table of the point group D_{2h} that the following correlations hold between the species of the point groups C_{2v} and D_{2h}

Analogous calculations for the other external local symmetry coordinates result in the following equations:

$$s_2^{A_{1g}}(O) = \frac{1}{2}(x_{H_3} + x_{H_4} - x_{H_6} - x_{H_7})$$

$$s_2^{B_{3u}}(\pi) = \frac{1}{2}(x_{H_3} + x_{H_4} + x_{H_6} + x_{H_7})$$

$$s_3^{A_{1g}}(O) = \frac{1}{2}(y_{H_3} - y_{H_4} + y_{H_6} - y_{H_7})$$

$$s_3^{B_{3u}}(\pi) = \frac{1}{2}(y_{H_3} - y_{H_4} - y_{H_6} + y_{H_7})$$

$$s_4^{A_{1u}}(O) = \frac{1}{2}(z_{H_3} - z_{H_4} - z_{H_6} + z_{H_7})$$

$$s_4^{B_{3g}}(\pi) = \frac{1}{2}(z_{H_3} - z_{H_4} + z_{H_6} - z_{H_7})$$

$$s_5^{B_{1g}}(O) = -y_{C_1} + y_{C_2} - y_{C_5} + y_{C_8}$$

$$s_5^{B_{2u}}(\pi) = \quad y_{C_1} + y_{C_2} + y_{C_5} + y_{C_8}$$

$$s_6^{B_{1g}}(O) = \frac{1}{2}(x_{H_3} - x_{H_4} + x_{H_6} - x_{H_7})$$

$$s_6^{B_{2u}}(\pi) = \frac{1}{2}(x_{H_3} - x_{H_4} - x_{H_6} + x_{H_7})$$

$$s_7^{B_{1g}}(O) = \frac{1}{2}(y_{H_3} + y_{H_4} - y_{H_6} - y_{H_7})$$

$$s_7^{B_{2u}}(\pi) = \frac{1}{2}(y_{H_3} + y_{H_4} + y_{H_6} + y_{H_7})$$

$$s_8^{B_{1u}}(O) = \quad z_{C_1} + z_{C_2} + z_{C_5} + z_{C_8}$$

$$s_8^{B_{2g}}(\pi) = -z_{C_1} + z_{C_2} - z_{C_5} + z_{C_8}$$

(2.102)

$$s_9^{B_{1u}} = \frac{1}{2} (z_{H_3} + z_{H_4} + z_{H_6} + z_{H_7})$$

$$s_9^{B_{2g}} = \frac{1}{2} (z_{H_3} + z_{H_4} - z_{H_6} - z_{H_7})$$

(2.102)

From the Eqs. (2.102) we can deduce the U matrix for each symmetry species. Each U matrix of the respective irreducible representation possesses as many columns as symmetry coordinates. The column vectors of the matrix are the symmetry coordinates with the factors of the cartesian coordinates [Eqs. (2.102)] as components. Table 2-9 lists the complete U matrix for polyethylene.

The next step is to build up the force constant matrix F. For a nonpolymeric compound this is the easiest part of the normal coordinate analysis provided the values of the force constants are well-known. In polymers, however, interactions to the neighboring units have to be considered. As discussed in Sec. 2.1.1 the potential energy of an infinitely long polymer chain can be written in terms of force constants determining the interactions between one chemical unit (which is labeled 0) and the other chemical units. Since the force constants between remote units are negligible we have to consider only the units -1, 0, and 1. Thus, Eq. (2.79) simplifies to:

$$F(\phi) = \sum_{k=-1}^{1} F_{0k} \cos(k\phi) \qquad \text{with } \phi = 0 \text{ or } \pi \qquad (2.103)$$

In Table 2-10 the force constants were defined in terms of the internal coordinates (Fig. 2-6). Interactions to the neighboring CH_2 units are marked by a dash. With these definitions of the force constants and the corresponding values listed in the last column of Table 2-10 we can work out the F matrices for the phase angles $\phi = 0$ and $\phi = \pi$ (Table 2-11). Now the matrices B, U, and F of Eq. (2.51b) are prepared (this procedure is normally done by computer) and we have to form U^{-1}, B_m, and B_m' and to carry out the multiplication as indicated in Eq. (2.51b). This can either be done by using step by step one of the eight U matrices of the respective irreducible

TABLE 2-9 Elements of the U' Matrix of Polyethylene

Column group labels: A_{1g} B_{3u} A_{1u} B_{3g} B_{1g} B_{2u} B_{1u} B_{2g}

Cartesian coordinates

TABLE 2-10   Definitions and Values of the Force Constants

Force constant	Involved internal coordinates				Value[†]
1	1				4.532
2	2	3			4.538
3	4	5	6	7	0.663
4	8				1.032
5	9				0.533
6	10				0.024
7	2 - 3				0.019
8	1 - 1'				0.083
9	1 - 6	1 - 7	1 - 4'	1 - 5'	0.174
10	1 - 4	1 - 5	1 - 6'	1 - 7'	-0.097
11	1 - 8	1 - 8'			0.303
12	4 - 5	6 - 7			-0.019
13	4 - 6	5 - 7			0.021
14	8 - 4	8 - 5	8 - 6	8 - 7	-0.022
15	6 - 4'	7 - 5'			0.073
16	6 - 5'	7 - 4'			-0.058
17	6 - 6'	7 - 7'	4 - 4'	5 - 5'	-0.009
18	6 - 7'	7 - 6'	4 - 5'	5 - 4'	-0.004
19	4 - 6'	5 - 7'			0.010
20	5 - 6'	4 - 7'			0.012
21	6 - 8'	7 - 8'	8 - 4'	8 - 5'	-0.064
22	8 - 8'				0.097

[†]Units of force constants are stretch (mdyn/Å), angle deformation and torsion (mdyn Å/rad), and angle-stretch (mdyn Å)

representations and the corresponding F matrices or by using the complete U and F matrices. In the first case we obtain successively eight matrices A which have to be diagonalized separately. In the other case the multiplication results in a matrix A' which is composed of eight submatrices. Now, the eigenvalues can be calculated by diagonalization of A' or, in other words, by determination of the eigenvalues of Eq. (2.51b). These are related to the eigenfrequencies by $\lambda = \sqrt{2\pi\nu}$. The results for all symmetry species are listed in Table 2-12. The potential energy distribution in terms of the force constants is tabulated in the last column. The potential energy distribution (PED) of each vibration of the molecule can be calculated as shown by Mikawa [30]. We can define the force constant matrix as:

$$F = \sum Z_j F_j \tag{2.104}$$

TABLE 2-11 Elements of the F Matrix for Polyethylene

$\phi = 0$

					Internal coordinates					
Internal coordinates	4.698	0	0	0.077	0.077	0.077	0.077	0.606	0	0
	0	4.538	0.019	0	0	0	0	0	0	0
	0	0.019	4.538	0	0	0	0	0	0	0
	0.077	0	0	0.645	-0.027	0.104	-0.046	-0.086	0	0
	0.077	0	0	-0.027	0.645	-0.046	0.104	-0.086	0	0
	0.077	0	0	0.104	-0.046	0.645	-0.027	-0.086	0	0
	0.077	0	0	-0.046	0.104	-0.027	0.645	-0.086	0	0
	0.606	0	0	-0.086	-0.086	-0.086	-0.086	1.226	0	0
	0	0	0	0	0	0	0	0	0.533	0
	0	0	0	0	0	0	0	0	0	0.024

$\phi = \pi$

					Internal coordinates					
Internal coordinates	4.366	0	0	-0.271	-0.271	0.271	0.271	0	0	0
	0	4.538	0.019	0	0	0	0	0	0	0
	0	0.019	4.538	0	0	0	0	0	0	0
	-0.271	0	0	0.681	-0.011	-0.062	0.046	0.042	0	0
	-0.271	0	0	-0.011	0.681	0.046	-0.062	0.042	0	0
	0.271	0	0	-0.062	0.046	0.681	-0.011	0.042	0	0
	0.271	0	0	0.046	-0.062	-0.011	0.681	0.042	0	0
	0	0	0	0.042	0.042	0.042	0.042	0.838	0	0
	0	0	0	0	0	0	0	0	0.533	0
	0	0	0	0	0	0	0	0	0	0.024

TABLE 2-12   Calculated Frequencies of an Isolated Polyethylene Chain
             and Their Assignment

Species	Frequency (cm^{-1})	Assignment	Potential energy distribution[†]
A_{1g}	1133	ν(C-C)	54% (1), 48% (4)
	1439	δ(CH$_2$)	80% (5), 18% (3)
	2863	ν_s(CH$_2$)	98% (2)
B_{3u}	0	T(x)	
	1472	δ(CH$_2$)	76% (5), 24% (3)
	2850	ν_s(CH$_2$)	99% (2)
A_{1u}	1063	t(CH$_2$)	99% (3)
B_{3g}	1301	t(CH$_2$)	83% (3)
B_{1g}	0	R(z)	
	1170	r(CH$_2$)	80% (3)
	2928	ν_{as}(CH$_2$)	99% (2)
B_{2u}	0	T(y)	
	716	r(CH$_2$)	96% (3), 16% (6)
	2910	ν_{as}(CH$_2$)	99% (2)
B_{1u}	0	T(z)	
	1176	w(CH$_2$)	100% (3)
B_{2g}	1060	ν(C-C)	63% (1), 18% (3), 14% (9)
	1385	w(CH$_2$)	90% (3), 52% (1)

[†]The potential energy distribution is given in terms of force constants (Table 2-10). Due to the contribution of off-diagonal elements,
the percentage may be greater than or less than 100%. Only the major
contributions to the potential energy are listed.

Z_j is a matrix with the same dimension as F and possesses the element 1 at those positions where the force constant F_j occurs in F.
Using the above expression, the matrix A of Eq. (2.51b) can be expressed as

$$A = \sum U\, B'_m Z_j B_m U^{-1} F_j = \sum Z^* F_j = U\, B'_m F\, B_m U^{-1} \qquad (2.105)$$

The elements of the *Jacobian* matrix may then be formed by

$$J_{ij} = \frac{\delta \lambda_i}{\delta F_j} = (S' Z_j^* S)_{ii} \qquad (2.106)$$

where S is the eigenvector matrix. Then the PED reads

$$(\text{PED})_{ij} = \frac{J_{ij} F_j}{\lambda_i} \qquad (2.107)$$

2.3 GROUP FREQUENCIES

The interpretation of IR and Raman spectra can be based on theoretical or empirical considerations. For the theoretical approach the
polymer system under consideration has to be idealized, in order to
take full advantage of the simplification introduced by the molecular
symmetry and the periodicity of potential functions and force constants. Crystallizable polymers, in actual fact, never possess the
ideal regular chemical and conformational composition which was
assumed in the theoretical discussion. The selection rules are no
longer strictly applicable. Often additional absorption bands occur.
Depending on the nature of the irregularities (e.g., nonuniformity
in chain length, steric configuration), the observed spectrum may be
interpreted as a superposition of the spectral characteristics of
the different chemical, sterical, and conformational species within
the sample.

   In polymers with irregular, randomly coiled chains a normal coordinate analysis is not possible, since there is no regularity in
the force fields of the molecules. Nevertheless, the number of bands
remains restricted because of the similarity of the force fields of
equivalent chemical groups. Whereas vibrations in geometrically well-
defined polymer chains give rise to sharp bands because of nearly
identical force fields within the chains, vibrations of chemical
groups in a comparable neighborhood of random chains generally cause
broader bands because of the differences in the force fields caused
by the random conformation of the polymer chain.

   A semiempirical method uses the results of theoretical considerations and computations on small molecules or simple polymers for
the interpretation and assignment of the spectra of polymers with
more complicated structures. This method is connected with the concept of group frequencies and is necessarily more qualitative and
less rigorous then the mathematical treatment outlined in Sec. 2.2.
The concept of group frequencies [40-42] and their application to
IR spectroscopic analysis of polymers are based on a minimum mechanical and electrical coupling of the vibrations of atomic groups with
those of the rest of the molecule. Hence their bands always occur at

about the same characteristic frequencies. The lower the degree of
coupling of a vibrating group with the remainder of the molecule,
the more characteristic is the corresponding absorption band. The
vibrational coupling between different atomic groups is considerably
reduced when atoms with markedly differing masses or bonds with large
force constants are involved. Thus, the approximate invariance and
transferability of force constants is more pronounced for stretching
than for deformation vibrations. Deformation force constants are
generally about an order of magnitude smaller than stretching constants. The bands of weakly coupling groups (for example, C=O, C-O-C,
C≡N) or larger structural units (e.g., phenyl, cyclopropyl, methyl,
amide) which always occur in the same wavelength regions are called
characteristic bands or band combinations, respectively. Thus, detailed information on the structure of the investigated material can
be gained by correlating characteristic bands and band combinations
with chemical groups present in the molecule. Apart from a large
number of characteristic band tables [21, 43-46], a simplified procedure has been proposed by considering the IR spectrum of a polymer
alongside the data from qualitative elementary analysis. With the
aid of a collection of characteristic band tables, divided according
to the chemical elements present in the sample [47], a preliminary
analysis will reveal the general nature of the polymer.

A comparison of the IR spectra of polyamide-6, polyamide-7, and
polyamide-8 (α-forms) (Fig. 2-7) shows that the characteristic absorption bands of the amide group at 3300 cm^{-1} [ν(NH)], 1635 cm^{-1}
[ν(C=O), amide I], 1540 cm^{-1} [δ(NH) + ν(C-N), amide II], 690 cm^{-1}
(amide V), and 580 cm^{-1} (amide VI) occur at about the same wavenumbers in the spectra of the different polyamides irrespective of
the length of the hydrocarbon segments. On the other hand, the structural change of one CH_2 group in going from polyamide-6 to poly-
amide-7 to polyamide-8 affects the wavenumber region between 1400
and 800 cm^{-1} characteristically, thus ensuring the possibility of
analytical differentiation.

The concept of group frequencies and characteristic band combinations is also frequently utilized in Raman spectroscopy. For prac-

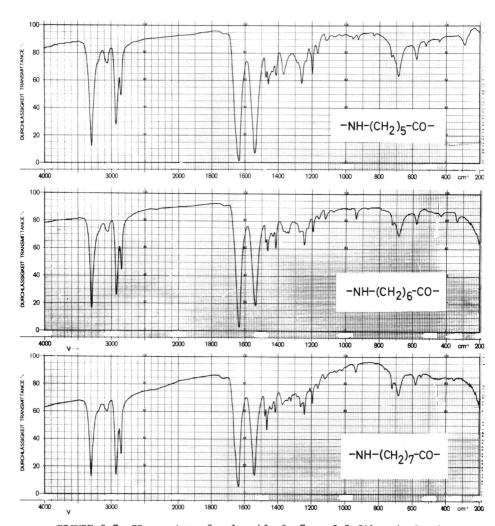

FIGURE 2-7 IR spectra of polyamide 6, 7, and 8 films (α form).

tical applications the intensity considerations play the major role
in determining whether an absorption band is useful in one method or
the other. Owing to the dependence of Raman band intensity on the
change in polarizability, correlations for readily polarizable bonds
such as C=C , -C≡C-, -N=N-, -S-S-, or -C-S- can be used to advantage in Raman spectroscopy, especially in cases where IR is relatively insensitive. Generally, although some overlap exist, the two techniques are highly complementary in this regard.

TABLE 2-13 Fundamental Vibrations of the Olefinic Double Bond[†]

Wavenumber region cm^{-1}	Intensity[††] IR	Raman	Assignment Nomenclature	Geometry	Trans	Cis	Vinyl
3095–3075	m	w	$\nu_a(CH_2)$		–	–	+
3040–3010	w	w	$\nu(CH)$		+	+	+
3010–3000	–	m	$\nu_s(CH_2)$		–	–	+
1675–1670	–	s	$\nu(C=C)$ *trans*		+	–	–
1662–1650	w	s	$\nu(C=C)$ *cis*		–	+	–
1648–1638	s	s	$\nu(C=C)$ vinyl		–	–	+
1420–1410	w	w	$\delta(CH_2)$		–	–	+
1425–1400	w	–	$\delta(CH)$ *cis*		–	+	+
1310–1290	vw	s	$\delta(CH)$ *trans*		+	–	+
995–985	s	–	$\gamma_w(CH)$ *trans*		–	–	+
980–965	vs	–	$\gamma_w(CH)$ *trans*		+	–	–
910–905	vs	–	$\gamma_w(CH_2)$		–	–	+
730–650	s	–	$\gamma_w(CH)$ *cis*		–	+	–
630–620	s	–	$\gamma_w(CH)$ *cis*		–	–	+

[†]Courtesy of G. Peitscher, Chemische Werke Hüls, Marl, West Germany.
[††]vw: very weak, w: weak, m: medium, s: strong, vs: very strong.

The compilation of the wavenumber positions and relative intensities of the stretching vibrations involving the carbon atoms of
the *cis*, *trans*, and vinyl olefinic C=C double bond and the deformation vibrations of hydrogen atoms attached to these carbon atoms
may serve as an illustration (Table 2-13).

As an additional feature of Raman lines, it has been observed
that their intensity is less sensitive to changes in the molecular
environment of a chemical bond than the intensity of IR absorption
bands. The intensity of the $\nu(C\equiv N)$ absorption band, for example, is
drastically reduced in the infrared upon α-halogenation but retained
in the Raman [42, 48]. Similarly to IR, in Raman spectroscopy there
are a number of spectral regions useful for the determination of
aromaticity and the type of ring substitution [21]. In the Raman
spectrum of an acrylonitrile-butadiene-styrene (ABS) terpolymer
(Fig. 2-8) several absorptions can be correlated with single structural features of the composing units. The most intense Raman lines
can be assigned to the aromatic C-H- and ring-stretching (3060 and

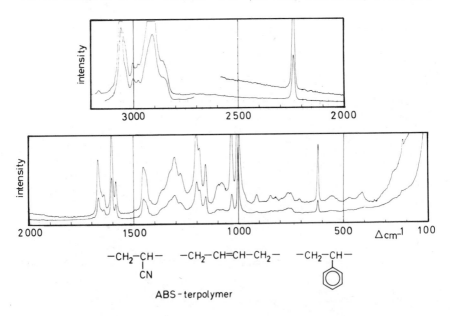

FIGURE 2-8   Raman spectrum of an ABS terpolymer.

1000 cm^{-1}), ring deformation (1600 and 1580 cm^{-1}), and in-plane aromatic C-H-bending and ring-bending modes (1020 and 615 cm^{-1} respectively), where the bands at 1020, 1000, and 615 cm^{-1} are particularly characteristic of monosubstitution. The absorptions at 1665,
1650, and 1640 cm^{-1} belong to the ν(C=C) vibrations of the *trans*-1,4-,
cis-1,4-butadiene and terminal vinyl units, respectively, in the
polymer chain. Finally, the band at 2240 cm^{-1} can be assigned to the
ν(C≡N) vibration of the acrylonitrile units. Thus, combination of
the results derived from the corresponding Raman and IR spectra
(Fig. 4-15) will provide an almost complete qualitative and quantitative picture of the polymer composition.

In a previous investigation of various polyamides (6, 6,10, and
11) Gall and Hendra [49] have demonstrated that the Raman spectra
are complementary to the infrared in that the Raman bands arise predominantly from the vibrations of the backbone CH$_2$ sequences, and
considerable differences in the 1300-1000 cm^{-1} region enable them
to be easily distinguished. The bands of the IR spectra, however,
arise mainly from motions of the amide groupings.

Apart from the characteristic group frequencies a large number
of additional bands occur in the spectrum arising from coupling
between adjacent groups. Because of the unique arrangement of atoms
in a particular molecule, the spectrum may be regarded as a molecular fingerprint and a more accurate identification of the structure
of the polymer can be established by reference to one of the published collections of polymer spectra [43, 5o]. Let us consider as
an example the methylene bending vibration of linear aliphatic polyesters. These polyesters contain three types of methylene groups
with different intramolecular environment [Fig. 2-9(a)]: the CH$_2$
groups next to the oxygen atom of the ester linkage (type 1), the
CH$_2$ groups adjacent to the carbonyl groups (type 3), and the methylene groups as next neighbors (type 2). Because of the different
degrees of coupling, the vibrational behavior of these CH$_2$ groups
is different. Comparing the spectra of the monomeric diols, dicar-
boxylic acids, and polyethers with the spectra of the polyester and

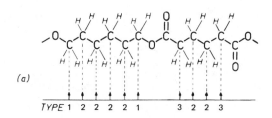

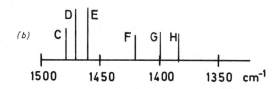

FIGURE 2-9   (a) Different types of methylene groups in aliphatic
polyesters, (b) IR bands of the CH_2-bending (C: type 1,
D and E: type 2, F: type 3) and CH_2-wagging vibrations
(G: type 1, H: type 2 of the alcoholic part).

finding the characteristic changes of bands when altering the number
of methylene groups in only one component, it was possible to make
the following assignment [Fig. 2-9(b)]. The CH_2-wagging vibrations
also exhibit a characteristic dependence of the molecular surround-
ings [Fig. 2-9(a)]. According to Tadokoro [51, 52], the bands G and
H of Fig. 2-9(b) can be assigned to the wagging vibrations of the
methylene groups of the alcoholic part. Because of the ester linkage,
the adjacent CH_2 groups (type 1) vibrate with a higher frequency
(band G) than the other CH_2 groups (band H) of the alcoholic part.

2.4   STATE OF ORDER

Any interpretation of IR spectra of polymers must take into account
the physical and chemical state of order of the macromolecular system
investigated. For the discussion of the various phenomena of order
in polymers, the basic terms will be defined shortly (see also Refs.
53 to 56).

2.4.1 Definitions

Constitution

The constitution of a molecule defines the nature of the atoms and their type of bonding irrespective of their spatial arrangement.

Configuration

The configuration of chemical groups characterizes a chemical state of a molecule. Different configurations constitute different chemica individuals and cannot be converted into others without rupture of chemical bonds.

Conformation

The conformation of chemical groups characterizes a physically cause geometrical state of a molecule. Different conformations of a molecule can be produced by rotation about single bonds without rupture of chemical bonds.

Polymer chains are formed by a sequence of chemical repeating units which may be arranged regularly or irregularly. To describe the order of a polymer chain the following definitions are introduced.

Regularity

Regularity characterizes a state of a (theoretically infinitely long and isolated) polymer chain where chemical repeating units are arranged in a regular manner. This definition may be specified according to the chemical configuration and the physical state of the polymer under consideration.

Stereoregularity

Stereoregularity characterizes a chemically caused state of a polymer chain in which every chemical repeating unit or every sequence of chemical repeating units with a well-defined configuration is followed by the same chemical repeating unit or the same sequence of chemical repeating units with the same configuration.

The occurrence of polymers with different stereoregularity is based on the existence of stereoisomeric centers such as

Double bonds, resulting in cis and trans configuration.
Tetrahedral stereoisomeric centers giving rise to

a. Isotactic polymers with a sequence of chemical repeating units of identical configurations along the chain

b. Syndiotactic polymers with a sequence of regularly alternating configurations

c. Atactic polymers with no regularity in the sequence of configurations

Conformational Regularity

Conformational regularity characterizes a physically caused geometrical state of the polymer chain where every chemical repeating unit or every sequence of chemical repeating units of a well-defined conformation can be transformed by means of a srew axis symmetry operation into a chemical repeating unit or a sequence of chemical repeating units of the same conformation. Some polymers can occur in more than one regular conformation.

Crystallinity

The crystalline state of a polymer is characterized by a regular arrangement of conformationally regular polymer chains. Thus, conformational regularity (or one-dimensional crystallinity) is a necessary condition for crystallinity. For every crystalline polymer there must exist a crystallographic unit cell whose translations in the three directions of its edges produce the macroscopic solid.

The amorphous state of a polymer can be defined experimentally by the absence of sharp x-ray reflections. However, the crystalline and amorphous states in real polymers are difficult to characterize experimentally and often the results depend on the test methods, which frequently use different sample preparation techniques. The deviation from regularity caused by polymerization or thermal or mechanical pretreatment leads to defects in the three-dimensional

order of the polymer chains. The link between a crystal of a low-molecular-weight compound and a crystalline polymer can be more readily
understood if it is taken into account that many materials occur in
polycrystalline form, that is, in aggregates of single crystals, separated by grain boundaries grown from a large number of nuclei. The
structure of semicrystalline polymers can be derived from this picture by increasing the amount of grain boundaries, thus bringing the
amorphous content up to 50% and more. According to this concept, a
semicrystalline polymer involves the presence of complex crystalline
structures embedded in an amorphous matrix with single molecular
chains running through a series of crystalline and amorphous regions.
For a detailed discussion of crystallinity in polymers, the reader
is referred to the literature [57-71].

Vibrational spectroscopy can help us to derive from spectral
changes (e.g., wavenumber shift, band feature, intensity, bandwidth)
information about the structure and the state of order in the polymer
under examination in relation to the parameters of composition, constitution, stereoregularity, pretreatment, temperature, etc.

An illustrative example showing how constitutional isomers may
be differentiated by IR spectroscopy is given in Fig. 2-10. Apart
from the common spectral features of the aromatic $\nu(CH)$ (3100 to
3000 cm^{-1}), aliphatic $\nu(CH)$ (3000 to 2800 cm^{-1}), and aromatic $\nu(CC-$
ring) and aliphatic $\delta(CH_2)$ (1520 to 1430 cm^{-1}) vibrations, the spectra can be readily assigned with the aid of the intense bands of
the aromatic $\gamma_w(CH)$ at 820 cm^{-1} and 700 and 750 cm^{-1} and their corresponding overtone and combination bands in the 1700 to 2000 cm^{-1}
wavenumber region to poly-p-xylylene and polystyrene, respectively.

2.4.2 Classification of Bands

For the characterization of configuration, conformation, and the
state of order, basically three types of bands may be distinguished.
They are caused by

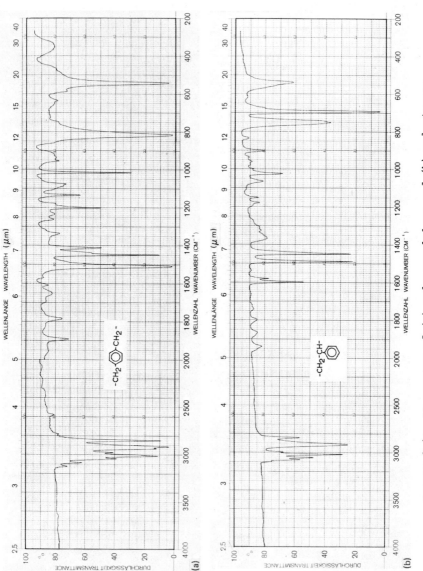

FIGURE 2-10 IR spectra of (a) poly-*p*-xylylene and (b) polystyrene.

1. The stereoregularity of the polymer chain

2. The conformational regularity of the polymer chain, which depends mainly on the intramolecular interactions between the chemical groups of the same chain

3. The crystallinity of the polymer, which depends on the intermolecular forces between adjacent chains

Pure stereoregularity bands have to occur in the same number in all possible phases, since the stereoregularity of a polymer chain depends only on the nature of the stereospecific polymerization and the applied catalyst system. Thus, the C=C-stretching vibration causes in the Raman spectra of polybutadienes (Fig. 2-11) characteristic stereoregularity bands: 1664 cm^{-1} in *trans*-1,4-polybutadiene, 1650 cm^{-1} in *cis*-1,4-polybutadiene, and 1639 cm^{-1} in syndiotactic 1,2-polybutadiene [72]. Analogous absorption bands which are characteristic of the individual stereoregular isomers can be observed in the IR spectra. The intense γ_w(CH) out-of-plane vibrations at 910, 967, and 740 cm^{-1} are representative of the syndiotactic 1,2-, *trans*-1,4-, and *cis*-1,4-structures, respectively [73-75]. The corresponding ν(C=C) Raman bands in polyisoprenes (Fig. 2-12) are observed at 1662 cm^{-1} in the *cis*- and *trans*-1,4-polyisoprene, at 1641 cm^{-1} in 3,4-polyisoprene and at 1639 cm^{-1} in 1,2-polyisoprene [76].

In the literature absorption bands are currently characterized as stereoregularity bands, whose occurrence can be related to a certain stereoregular structure of the chains. However, these bands vanish or show very weak intensity in the spectra of the molten or dissolved sample. The assignment is based on the assumption that stereoregular polymer chains prefer certain conformations in the solid state and each of the corresponding structures has different selection rules for IR and Raman activity. Thus, isotactic polypropene forms a 3:1 helix upon crystallization [77] and therefore belongs to a factor group which is isomorphous to the point group D_3; syndiotactic polypropene crystallizes in dependence on pretreatment in a 4:1 helix or a planar zigzag conformation [78, 79], whose symmetry behavior can be characterized by D_2 or C_{2v}, respectively.

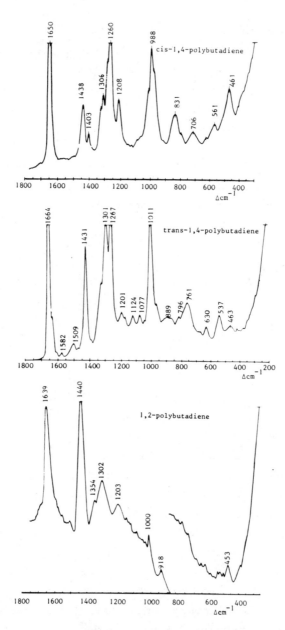

FIGURE 2-11   Raman spectra of *cis*-1,4-polybutadiene, *trans*-1,4-poly-
butadiene, and 1,2-polybutadiene. [Reproduced with permission from S. W. Cornell and J. L. Koenig, Macromolecules, 2:540 (1969). Copyright by the American Chemical Society.]

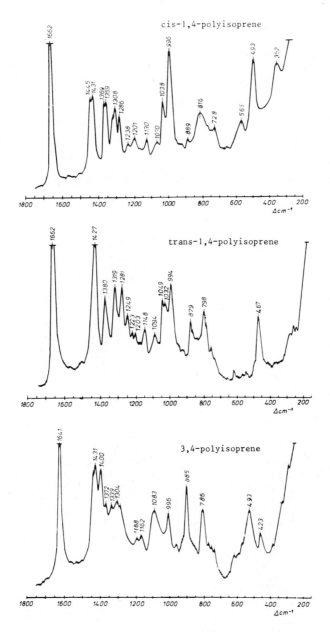

FIGURE 2-12   Raman spectra of *cis*-1,4-polyisoprene, *trans*-1,4-poly-
isoprene, and 3,4-polyisoprene. [Reproduced with permission from S. W. Cornell and J. L. Koenig, Macromolecules 2:546 (1969). Copyright by the American Chemical
Society.]

Figure 2-13 shows the Raman spectra of atactic, syndiotactic, and
isotactic polypropene. Remarkable spectral differences can be observed
in the spectra below 1000 cm^{-1}. This is also the case in the IR spectra (Fig. 2-14). However, the spectra of the molten or dissolved
stereoregular samples do not differ considerably. Most of the so-
called stereoregularity bands are caused by a regular geometrical arrangement of the chemical repeating units and should be characterized
as conformational regularity bands, although the conformational regularity is actually governed by the present stereoregularity.

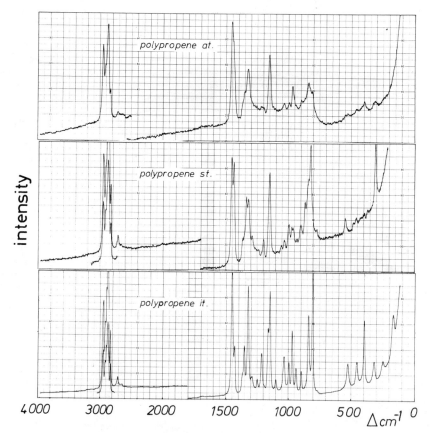

FIGURE 2-13   Raman spectra of atactic, syndiotactic, and isotactic
               polypropene.

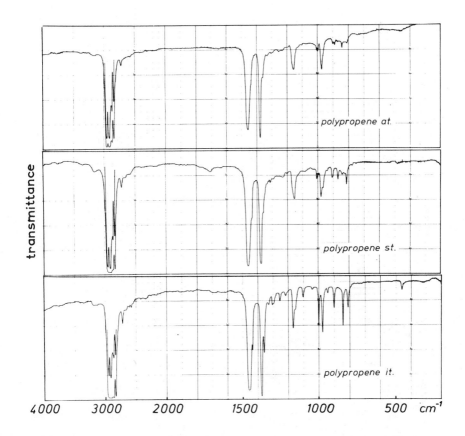

FIGURE 2-14   IR spectra of atactic, syndiotactic, and isotactic
             polypropene.

    While the stereoregularity of a polymer depends on the specific
polymerization conditions its configuration can be changed only by
chemical reactions. Conformational regularity and crystallinity are
strongly influenced by the pretreatment and the given experimental
conditions, since conformational regularity can be related to different conformations due to rotations about C-C single bonds in polymer
chains. The characteristic bands of the respective conformation can
be associated with a distinct intramolecular order in the chain.
When this order is changed, the corresponding bands must disappear,

or in the case of changes in conformational regularity, new bands
must appear. Thus, syndiotactic polypropene occurs in the more stable
modification as twinned 4:1 helix. A planar zigzag conformation can
be obtained on cold-drawing the quenched sample [78, 79]. Also, isotactic polybutene-1[†] is reported to crystallize in three polymorphous
modifications with different conformations of the main chains. The
tetragonal modification II with a 11:3 helix can be obtained on cooling the sample from the melt [80, 81]. At room temperature this less
stable modification converts to modification I [82]. Thereby the 11:3

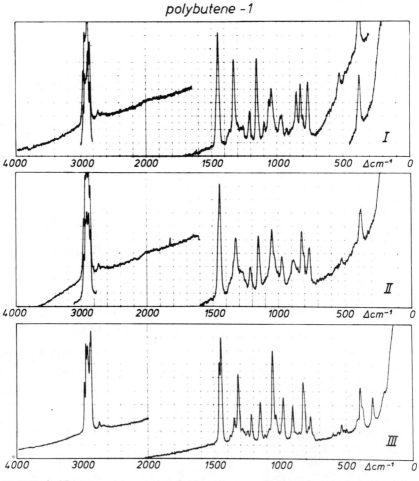

FIGURE 2-15   Raman spectra of the three modifications of isotactic
              polybutene-1.

[†]Systematic name: poly(1-ethylethylene) (see Table 4-1).

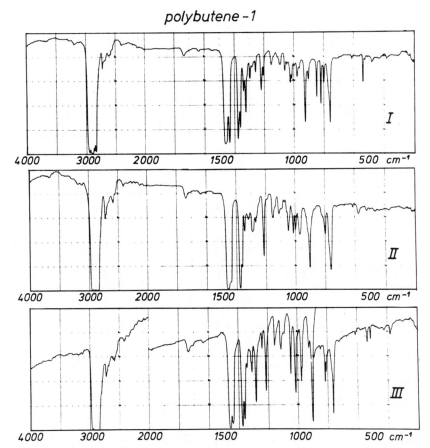

FIGURE 2-16   IR spectra of the three modifications of isotactic
              polybutene-1.

helix changes into a 3:1 helix [83, 84]. Modification III crystallizes
in a 4:1 helix [85] from a solution in benzene or decahydronaphthalin.
The Raman [86] and the IR spectra [87] of the different modifications
of isotactic polybutene-1 are excellent examples for conformational
regularity bands (Figs. 2-15 and 2-16). Similarly, isotactic poly-
(propylethylene) (or polypentene-1) can occur in three different
conformations depending on the thermal pretreatment [88-91]. Poly-
tetrafluoroethylene forms below 292 K a 13:6 helix and above 292 K
a 15:7 helix [92, 93]. Intensity and wavenumber of the conformational
regularity bands allow in the first instance assumptions about inter-

molecular interactions within a conformational regular chain. However, since conformational regularity is a necessary condition for
crystallinity conclusions on the three-dimensional order within the
polymers are often possible.

Some authors have assigned certain absorption bands in the IR
spectra of various crystalline polymers (polyethylene, polyethylene
terephthalate, polyamide-6, polyamide-6,6, amylose) to specific conformations in the fold regions [94-99]. On the basis of their IR data
Koenig et al. have proposed a mechanism for the α to γ transition in
polyamide 6 showing the role of the fold in maintaining a minimum of
molecular disordering during the transformation [99]. More recently
Painter et al. [100] have applied FTIR difference spectroscopy to
study the morphology of polyethylene. The band at 1346 cm^{-1} which has
been detected to be unique to polyethylene single crystals was assigned to a regular, tight fold structure with adjacent reentry.

For most of the so-called "crystallinity" bands empirical relationship between their intensity and the parameters of other techniques characterizing the state of order of the investigated polymer
(e.g., density, x-ray diffraction) have been established, irrespective of an unambiguous assignment of these absorption bands. In the
spectrum of polyvinyl alcohol, for example, the absorption band at
1141 cm^{-1}, whose origin has been the subject of much discussion [101-
103], shows a distinct intensity dependence on crystallization induced by thermal pretreatment (Fig. 2-17) and has been proposed for
the estimation of the degree of crystallinity in polyvinyl alcohol.

Similarly, in ethylene-propylene-diene (e.g., dicyclopentadiene)
terpolymers (EPDM) containing more than 65% ethylene, a correlation
has been found between the relative intensity of the ν(C-C) Raman
band at 1065 cm^{-1} and the amount of crystallinity determined from
x-ray diffraction and mechanical properties [104].

Real intermolecular interactions between adjacent chains can only
be observed by vibrational spectroscopy when two or more chains run
through a crystallographic unit cell. In this case in-phase and out-

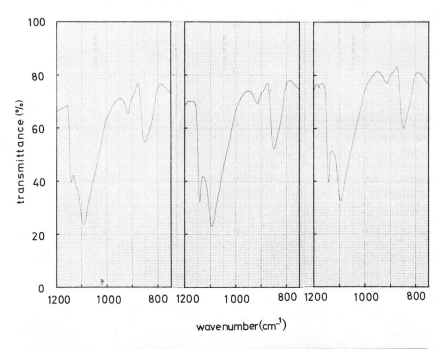

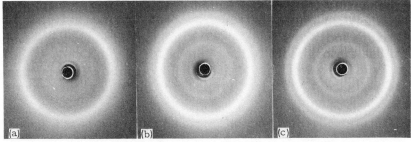

FIGURE 2-17   IR spectra and wide-angle x-ray diffraction patterns of
              polyvinyl alcohol (Elvanol 72-60) film cast from
              $C_2H_5OH/H_2O(30:70)$ solution: (a) no thermal treatment,
              density 1.2814 g/cm^3; (b) 30 min at 403 K, density
              1.2875 cm^3; (c) 30 min at 428 K, density 1.2997 cm^3.

of-phase vibrations of the adjacent chains are possible and may lead

to characteristic band splitting (*correlation splitting*, *Davidov*

splitting). However, since the intermolecular forces are much weaker

than the intramolecular forces, the spectra of polymers will be in-

fluenced more strongly by the intramolecular interactions within a

chain than by the coupling effects between neighboring chains. There-

fore, often crystallinity bands can be observed only at low temperatures. The most discussed splitting of this kind is encountered in
the IR spectra of polyethylene and polymers with long methylene sequences at 720 and 730 cm^{-1} and is assigned to the CH_2-rocking vibrations of methylene sequences with a planar zigzag conformation (B_{2u},
B_{3u}) [105, 106]. In the Raman spectra of such polymers the CH_2-bend-
ing vibrations show a considerable splitting into a B_{1g} (1443 cm^{-1})
and an A_g (1414 cm^{-1}) mode [107-110]. The splitting does not occur in
modifications with one chain per unit cell.

REFERENCES

1. G. M. Barrow, *Introduction to Molecular Spectroscopy*,
    McGraw-Hill, New York 1962.

2. L. Pauling and E. B. Wilson, *Introduction to Quantum Mechanics*,
    McGraw-Hill, New York 1935.

3. M. M. Sushchinski, *Raman Spectra of Molecules and Crystals*,
    Keter, New York 1972.

4. J. A. Konigstein, *Introduction to the Theory of the Raman Effect*,
    Reidel, Dordrecht 1972.

5. J. Tang and A. C. Albrecht, in *Raman Spectroscopy, vol. 2*,
    ed. H. A. Szymanski, 2:33 (1970), p. 33.

6. J. Brandmüller and H. Moser, *Einführung in die Ramanspektroskopie*,
    Steinkopff, Darmstadt 1962.

7. A. S. Dawidow, *Quantenmechanik*, VEB Deutscher Verlag der
    Wissenschaften, Leipzig 1967.

8. J. C. Evans, in *Infra-Red Spectroscopy and Molecular Structure*,
    ed. M. Davies, Elsevier, New York 1963, p. 199.

9. I. N. Levine, *Molecular Spectroscopy*, Wiley, New York 1975.

1o. G. Placzek, in *Handbuch der Radiologie, vol. VI/2*, ed. E. Marx,
    Leipzig 1934, p. 205.

11. E. L. Ince, *Ordinary Differential Equations*, Dover, New York
    1956.

12. A. Jeffrey, *Mathematics for Engineers and Scientists*, Nelson,
    London 1969.

13. L. May, Appl. Spectrosc. 27:419 (1973).

14. A. Smekal, Naturwissenschaften 11:873 (1923).

15. M. Born and E. Wolf, *Principles of Optics*, Pergamon Press, New
    York 1954.

16. P. Gans, *Vibrating Molecules*, Chapman & Hall, London 1971.

17. G. Turrell, *Infrared and Raman Spectra of Crystals*, Academic Press, London 1972.

18. F. A. Cotton, *Chemical Application of Group Theory*, Interscience, New York 1963.

19. D. M. Bishop, *Group Theory and Chemistry*, Clarendon, Oxford 1973.

2o. W. G. Fateley, F. R. Dollish, N. T. McDevitt, and F. F. Bentley, *Infrared and Raman Selection Rules for Molecular and Lattice Vibrations*, Wiley, Interscience, New York 1972.

21. N. B. Colthup, L. H. Daly, and S. E. Wiberley, *Introduction to Infrared and Raman Spectroscopy*, Academic Press, New York 1964.

22. E. B. Wilson, J. C. Decius, and P. C. Cross, *Molecular Vibrations*, McGraw-Hill, New York 1955.

23. G. Herzberg, *Molecular Spectra and Molecular Structure II*, Van Nostrand, Princeton, N.J., 1960.

24. P. W. Higgs, Proc. Roy. Soc. A220:472 (1953).

25. F. J. Boerio and J. L. Koenig, J. Macromol. Sci. Revs. Macromol. Chem. 7:209 (1972).

26. H. Tadokoro and M. Kobayashi, in *Polymer Spectroscopy*, ed. D. O. Hummel, Verlag Chemie, Weinheim 1974, p. 3.

27. R. Zbinden, *Infrared Spectra of High Polymers*, Academic Press, New York 1964.

28. G. Zerbi, Appl. Spectrosc. Rev. 2:193 (1969).

29. S. Krimm, Advan. Polym. Sci. 2:51 (1960).

30. Y. Mikawa, J. W. Brasch, and R. J. Jakobsen, J. Mol. Spectrosc. 24:1314 (1967).

31. G. Zerbi, in *Phonons*, ed. M. A. Nusimovici, Flammarion Sciences, Paris 1971, p. 248.

32. A. Rubic and G. Zerbi, Chem. Phys. Lett. 34:343 (1975).

33. G. Zerbi, in *Lattice Dynamics and Intermolecular Forces*, ed. S. Califano, Academic Press, New York 1975, p. 384.

34. C. S. Fuller, J. Amer. Chem. Soc. 64:154 (1942).

35. T. Schoon, Z. Phys. Chem. B39:385 (1938).

36. K. H. Storks, J. Amer. Chem. Soc. 60:1753 (1938).

37. A. Turner Jones and C. W. Bann, Acta Cryst. 15:105 (1962).

38. S. Y. Hobbs and F. Y. Billmeyer, J. Polym. Sci. A2, 7:1119 (1969).

39. K. Holland-Moritz and D. O. Hummel, J. Mol. Struct. 19:289 (1973).

4o. H. A. Szymanski, *Theory and Practice of Infrared Spectroscopy*, Plenum Press, New York 1964.

41. L. J. Bellamy, Appl. Spectrosc. 33:439 (1979).

42. L. J. Bellamy, *Advances in Infrared Group Frequencies*, Methuen, London 1968.

43 D. O. Hummel and F. Scholl, *Atlas der Kunststoffanalyse, vol. I/1, 2*, Verlag Chemie, Weinheim 1968.

44. W. Otting, *Spektrale Zuordnungstafeln der Infrarot-Absorptions-banden*, Springer, Berlin 1963.

45. H. J. Hediger, *Infrarotspektroskopie*, Akadem. Verlagsgesell-schaft, Frankfurt 1971.

46. K. W. F. Kohlrausch, *Ramanspektren*, Heyden, London 1972.

47. J. Haslam and H. A. Willis, *Identification and Analysis of Plastics*, Iliffe, London 1967.

48. H. L. Sloane, in *Polymer Characterization*, ed. C. D. Craver, Plenum Press, New York 1971, p. 15.

49. M. J. Gall and P. J. Hendra, The Spex Speaker, 16 (1971).

50. Sadtler Commercial Spectra, *Monomers and Polymers (Series D)*, Sadtler Research Laboratories, Philadelphia 1970.

51. H. Tadokoro, Y. Chatani, M. Kobayashi, T. Yoshihara, S. Mura-hashi, and K. Imada, Rep. Progr. Polym. Phys. Jap. 6:303 (1963).

52. S. Kobayashi, H. Tadokoro, and Y. Chatani, Makromol. Chem. 112: 225 (1968).

53. G. Zerbi, F. Ciampelli, and V. Zamboni, J. Polym. Sci. C7:141 (1964).

54. D. O. Hummel, Sixth Wayne State University Polymer Conference Series, 1970.

55. K. Holland-Moritz and H. Siesler, Appl. Spectrosc. Rev. 11:1 (1976).

56. K. Holland-Moritz, in *Proceedings of the 5th European Symposium on Polymer Spectroscopy*, ed. D. O. Hummel, Verlag Chemie, Wein-heim 1979, p. 93.

57. W. Pechold and S. Blasenbrey, Kolloid-Z. Z. Polym. 241:955 (1970).

58. P. C. Hägele and W. Pechold, Kolloid-Z. Z. Polym. 241:977 (1970).

59. E. W. Fischer, Kolloid-Z. Z. Polym. 231:458 (1967).

60. G. C. Oppenlander, Science 159:1311 (1968).

61. P. Geil, *Polymer Single Crystals*, Krieger, Huntington 1973.

62. B. Wunderlich, *Macromolecular Physics, vols. I, II*, Academic Press, New York 1973.

63. A. J. Hopfinger, *Conformational Properties of Macromolecules*, Academic Press, New York 1973.

64. A. Keller, J. Polym. Sci. Polym. Symp. 51:7 (1975).

65. A. Peterlin, J. Polym. Sci. C9:61 (1965).

66. A. Keller, Rep. Progr. Phys. 31:623 (1968).

67. R. Hosemann, Ber. Bunsenges. Phys. Chem. 74:755 (1970).

68. R. Hosemann, Makromol. Chem. 1:559 (1975).

69. A. Keller, Kolloid-Z. Z. Polym. 231:386 (1969).

70. E. W. Fischer, Progr. Colloid Polym. Sci. 57:149 (1975).

71. F. P. Price, *The Meaning of Crystallinity in Polymers*, Wiley, New York 1966.

72. S. W. Cornell and J. L. Köenig, Macromolecules 2:540 (1969).

73. G. Natta, P. Pino, and G. Mazzanti, Gazz. Chim. Ital. 87:528 (1957).

74. M. A. Golub and J. J. Shipman, Spectrochim. Acta, 16:1165 (1960).

75. J. L. Binder, J. Polym. Sci. A1:47 (1963).

76. S. W. Cornell and J. L. Koenig, Macromolecules 2:564 (1969).

77. G. Natta and P. Corradini, Atti. Accad. Naz. Lincei Cl. Sci. Fis. Mat. Nat. Rend. 4:73 (1955).

78. G. Natta, Makromol. Chem. 35:94 (1960).

79. G. Natta, M. Peraldo, and G. Allegra, Makromol. Chem. 75:215 (1964).

80. G. Natta, P. Corradini, and L. Porri, Atti. Accad. Naz. Lincei Cl. Sci. Fis. Mat. Nat. Rend. 26:728 (1956).

81. F. Danusso and G. Gianotti, Makromol. Chem. 80:1 (1964).

82. J. P. Longo and R. Salovey, J. Polym. Sci. A2, 4:997 (1966).

83. F. J. Boerio and J. L. Koenig, J. Macromol. Sci. Rev. Macromol. Chem. C7 (2):209 (1972).

84. G. Natta, P. Corradini, and I.W. Bassi, Makromol. Chem. 21:246 (1956).

85. G. Cojazzi, V. Malta, G. Celotti, R. Zanetti, Makromol. Chem. 177:915 (1976).

86. S. W. Cornell and J. L. Koenig, J. Polym. Sci. A2, 7:1965 (1969).

87. G. Goldbach and G. Peitscher, J. Polym. Sci. Polym. Lett. Ed. 6:783 (1968).

88. J. Y. Decroix, M. Moser, and M. Boudeulle, Eur. Polym. J. 11: 357 (1976).

89. M. Moser and M. Boudeulle, J. Polym. Sci. Phys. Ed. 14:1161 (1976).

90. I. D. Rubin, J. Polym. Sci. A2, 5:1323 (1967).

91. K. Holland-Moritz, E. Sausen, P. Djudovic, M. M. Coleman, and P. C. Painter, J. Polym. Sci. Polym. Phys. Ed. 17:25 (1979).

92. G. Zerbi and M. Sacchi, Macromolecules 6:6 (1973).

93. C. W. Bunn and E. R. Howells, Nature 174:549 (1954).

94. J. L. Koenig and M. C. Agboatwalla, J. Macromol. Sci. Phys. Ed. B2:191 (1968).

95. J. L. Koenig and P. D. Vasko, J. Macromol. Sci. Phys. Ed. B4: 347 (1970).

96. H. Schonhorn and J. P. Luongo, Macromolecules 2:266 (1969).

97. D. C. Prevorsek and J. P. Sibilia, J. Macromol. Sci. Phys. B5:616 (1971).

98. W. O. Statton, J. L. Koenig, and M. Hannon, J. Appl. Phys. 41: 4290 (1970).

99. P. D. Frayer, J. L. Koenig, and J. B. Lando, J. Macromol. Sci. Phys. Ed. B6:129 (1972).

100. P. C. Painter, J. Havens, W. W. Hart, and J. L. Koenig, J. Polym. Sci. Polym. Phys. Ed. 15:1223 (1977).

101. H. Tadokoro, H. Nagai, S. Seki, and I. Nitta, Bull. Chem. Soc. Jap. 34:1504 (1961).

102. S. Krimm, Advan. Polym. Sci. 2:51 (1960).

103. J. F. Kenney and G. W. Willcockson, J. Polym. Sci. A1, 4:679 (1966).

104. G. Schreier and G. Peitscher, Z. Anal. Chem. 258:199 (1972).

105. S. Krimm, J. Chem. Phys. 22:567 (1954).

106. R. S. Stein and G. B. B. M. Sutherland, J. Chem. Phys. 21:370 (1953).

107. F. Boerio and J. L. Koenig, J. Chem. Phys. 52:4826 (1970).

108. K. Holland-Moritz and E. Sausen, Colloid Polym. Sci. 254:342 (1976).

109. K. Holland-Moritz, J. Appl. Polym. Sci. (Symp.) 34:49 (1978).

110. K. Holland-Moritz and K. van Werden, J. Polym. Sci. Polym. Phys. Ed. (in press).

Chapter 3

EXPERIMENTAL TECHNIQUES

3.1 SPECTROMETERS

The instrumentation for IR and Raman spectroscopy is different because of the different nature of the physical processes involved. A
sample prepared for IR spectroscopy absorbs those frequencies from
the incident polychromatic IR radiation ($\sim$1 to 300 μm) which can ex-
cite allowed vibrational transitions. Absorptions occur only when the
vibrational transitions are accompanied by changes in the local dipole moments. In the transmitted or totally reflected light the intensity of these frequencies is attenuated dependent on the concentration of the absorbing species. Thus, by measuring the intensity of
the incident and transmitted or reflected light in relation to the
frequency, the absorption intensity at a given frequency can be determined.

For Raman spectroscopy a sample is illuminated with monochromatic
radiation of frequency ν_o (generally between 350 and 700 nm). The
scattered light of frequency ν_o (Rayleigh stray light) is modulated
by the Raman light of frequency $\nu_o \pm \nu_{vib}$ ($-\; \hat{=}$ Stokes, $+\; \hat{=}$ anti-
Stokes), where $+\nu_{vib}$ or $-\nu_{vib}$ correspond to the frequency of a transition accompanied by a change in polarizability from one vibrational
state to the next lower or higher state, respectively. For the intensity ratio between exciting radiation I_{source}, Rayleigh scatter
$I_{Rayleigh}$, and Raman scatter I_{Raman}, the following relation holds:

$$I_{Raman} \approx 10^{-4}\; I_{Rayleigh} \approx 10^{-8}\; I_{source} \qquad (3.1)$$

Because of the rather low intensity of Raman scatter in comparison
to the intensity of the light source and the short wavelength of the
analyzed light (Table 2-1), the optical concept of the Raman spectrometers differs considerably from IR spectrometers. Sections 3.1.1
and 3.1.2 give a basic survey of the building principles of IR and
Raman spectrometers. For more comprehensive studies the reader is
referred to the literature [1-22].

3.1.1 Infrared Spectrometers

3.1.1.1 Dispersive Spectrometers

In Raman spectroscopy the most interesting part of the spectrum (0 to
4000 Δcm^{-1}, corresponding to 20486 to 19486 cm^{-1} or 488 to 606 nm for
Ar^+, Table 2-1) lies in the visible and near-infrared (NIR) region
of electromagnetic radiation. Therefore, in Raman spectroscopy for
the whole wavenumber range only one spectrometer with high resolution
is required. However, in IR spectroscopy the wavenumber region considered corresponds to wavelengths of 3×10^5 to 2.5×10^3 nm, and
generally at least two spectrometers with about five interchangeable
gratings are used. These difficulties arise mainly from the rapid
falloff in energy with increasing wavelength, high-order reflections
from diffraction gratings, absorption by the window and prism material, and the rotational spectrum of water vapor, which extends
over the whole far-infrared (FIR) range.

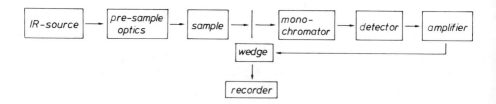

FIGURE 3-1 Block scheme of a dispersive IR spectrometer.

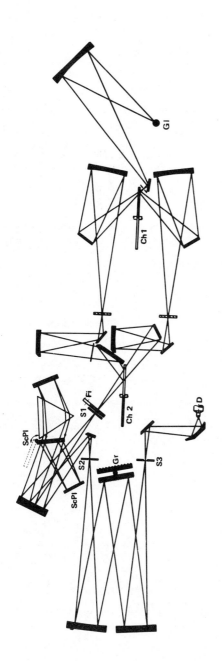

FIGURE 3-2 Optical layout of the IR spectrometer model 325 (with kind
permission of the Perkin Elmer GmbH, Überlingen, Germany).
Ch1, Ch2: choppers; S1, S2, S3: slits; Fi: filter; Gr: grating; D: detector; Gl: globar; Sc Pl : scattering plate.

Generally, IR spectrometers are purged with dry air and can be used down to a lower wavenumber limit of 400 or 200 cm^{-1}. Grating FIR spectrometers, commonly evacuable and equipped with a high-pressure mercury lamp, are used between 400 and 30 cm^{-1}.

The block scheme of Fig. 3-1 shows the building principle of a grating or prism double-beam NIR, IR, or FIR spectrometer, and in Fig. 3-2 the detailed optical layout of the Perkin-Elmer IR spectrophotometer model 325 is reproduced. The IR radiation transmitted or reflected by the sample and the radiation of the reference beam are chopped by a rotating sector mirror (10 to 20 s^{-1}) before entering the entrance slit of the monochromator.

For the measurement of IR spectra at low and high temperatures, it must be taken into account that radiation emitted by cells, sample, and detector produce additional signals which lead to shifts in the absorption background. To eliminate this source of errors a double-chopper system (with one chopper positioned immediately after the source and the other in front of the monochromator entrance slit) can be used. The synchronously driven choppers possess a phase difference of 90°. Thus, according to Fig. 3-3, the monochromator receives (dependent on the position of the two crossed choppers) the following

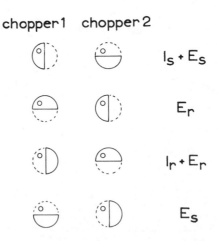

FIGURE 3-3   Observed intensities for a double-chopper system.
I: absorbed energy, E: emitted energy.

energies successively within one cycle:

1. The complete energy $E_S + I_S$ of the sample beam
2. The energy E_R resulting in the emission within the reference beam
3. The whole energy $E_R + I_R$ of the reference beam
4. The energy E_S resulting in the emission of the sample beam

By electronic means the portions E_R and E_S can be eliminated.

In the monochromator the incident, chopped radiation is dispersed by prisms or gratings into a spectrum. Each part of this spectrum can successively pass the exit slit by rotation of the dispersing devices. The dispersed light is then focused onto a detector. In most instruments thermocouples, golay cells, pyroelectric detectors (TGS), or bolometers are used to generate electrical signals whose intensities are proportional to the energy of the incident radiation. Since reference and sample beam are chopped, each difference in energy between reference and sample beam leads to two (one chopper) or four (double chopper) different signals at the detector. Thus, the detector output alters corresponding to the used chopper system. Proportional to these signals an alternating current (AC) amplifier produces an AC signal which drives a wedge system (optical comb attenuation) by means of a servo motor into the reference beam to minimize the energy differences between reference and sample beam. This "optical null" principle, where the reference beam is attenuated according to the absorption in the sample beam until the detector receives the same energy, is applied in most spectrometers. A suitable form of the wedge system produces a linear dependence between its position and the transmittance and therefore allows an easy mechanical or electrical coupling with a recorder.

Instead of an optical null balance system, a ratio recording system is used in some instruments which eliminates the limitations imposed by the accuracy of the wedge system and results in superior ordinate accuracy. Moreover, the ratio-recording electronics has the advantage that at 0% transmission no radiation is being attenuated by the wedge in the reference beam. Thus, the system remains "alive" and allows the zero to be accurately located for quantitative studies.

Dichroic measurements are commonly performed with the polarizer positioned in front of the monochromator slit in the recombined sample and reference beams, and the absorbances $A_{\parallel}$ and $A_{\perp}$ are determined from spectra recorded successively with the polarizer inclination to transmit radiation polarized parallel and perpendicular to the sample reference axis (e.g., direction of elongation), respectively. A detailed discussion of the various errors encountered in dichroic measurements and their correction procedures is given by Zbinden [22, 23]. However, since the introduction of the commercially available wire-grid polarizers [24] for the IR region, many sources of error such as polarizer inefficiency or beam convergence have been almost eliminated. Errors due to polarization effects of the monochromator can be minimized by orienting the sample with the stretch axis at 45° to the monochromator entrance slit and rotating the polarizer to bring the electric vector of the polarized light parallel and perpendicular to the draw direction, respectively. For low degrees of orientation, the magnitude of $A_{\parallel} - A_{\perp}$ is frequently small and the dichroic ratio R is close to unity. Hence, the accuracy in the determination of R from separate measurements lacks sensitivity, and a differential technique using two polarizers has been proposed for the direct measurement of $A_{\parallel} - A_{\perp}$ [25, 26]. In this method one polarizer is placed in the sample beam and one in the reference beam, each oriented at $+45^{\circ}$ or -45° to the monochromator slit with the two planes of polarization at right angles with respect to each other. The sample is positioned in the common beam with the reference direction parallel to the plane of polarization of one polarizer. The value of $A_{\parallel} - A_{\perp}$ is then directly obtainable from the recorded spectrum and additional measurements of either $A_{\parallel}$ or $A_{\perp}$ make possible the determination of R:

$$\frac{A_{\parallel} - A_{\perp}}{A_{\perp}} = R - 1 \tag{3.2}$$

The sensitivity of this method has been quantitatively discussed in relation to the sensitivity of conventional dichroism measurements by Read et al. [26].

3.1.1.2 Fourier Transform Infrared (FTIR) Spectroscopy

In recent years the range of applicability of IR spectroscopy to
chemical and physical problems has certainly been expanded by the re-
vival of Fourier transform spectroscopy. Although the most frequently
used basic optical component of FTIR instruments, the Michelson interferometer [27], has been known for almost a century, it was not until
the development of digital computers that the spectra of broad-band IR
sources could be accurately derived from the corresponding interferograms in a reasonable time, since the two are related by the complex
mathematical procedure of the Fourier transformation. The following
chapter is intended to give a brief introduction to the basic concepts of FTIR spectroscopy, and for more comprehensive treatments of
the theory and instrumentation the interested reader is referred to
the recent special literature [8-11, 13, 15, 28-31].

Basic Theory

In a conventional dispersive IR spectrometer the polychromatic radiation of the source is dispersed by a monochromator (prism, grating)
and detected in small wavenumber increments $\Delta \bar{\nu}$ as determined by the
slit width. Thus, the spectral information $I(\bar{\nu})$, i.e., the intensity
of the radiation in relation to wavenumber, is actually encoded in the
propagation direction of the dispersed radiation. In the FTIR technique the phase difference ϕ is utilized to encode $I(\bar{\nu})$. However, this
implies that the light waves coming from the source have to be split
into two partial waves. This procedure is accomplished in the interferometer, whose principal optical scheme is outlined in Fig. 3-4. It
consists of two mutually perpendicular plane mirrors, one of which is
stationary while the other moves at a constant velocity and a beam
splitter mounted at 45° to these mirrors. The incoming radiation is
partially transmitted to the movable mirror. After each beam has been
reflected back to the beam splitter, they are again partially transmitted and reflected and a certain fraction of the incoming beam (depending on the transmittance and reflectivity of the beam splitter)
will propagate in a direction perpendicular to the incoming beam toward the detector.

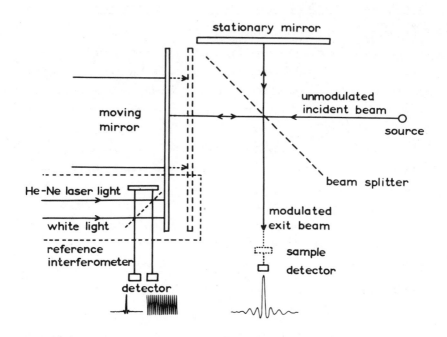

FIGURE 3-4   Optical scheme of a Michelson interferometer commonly
used in FTIR instruments.

For the derivation of the basic equations let us first consider
the simplest case where the radiation of a monochromatic source $(\bar{\nu}_o)$
has been successfully split into two partial waves of equal intensity
$I(\bar{\nu}_o)$ in a perfectly compensated interferometer. In the superposition
region the intensity $I(\phi)$ measured by the detector will then be [8,
9, 15, 28, 29]

$$I(\phi) = 2I(\bar{\nu}_o)(1 + \cos\phi) \qquad\qquad (3.3a)$$

The phase difference ϕ can be expressed by the optical path difference s of the two beams and the wavelength λ_o

$$\phi = \frac{2\pi s}{\lambda_o} \qquad\qquad (3.4)$$

and Eq. (3.3a) then reads

$$I(s) = 2I(\bar{\nu}_o)(1 + \cos2\pi\bar{\nu}_o s) \qquad (3.3b)$$

Constructive interference takes place when the optical path difference s (corresponding to twice the distance of mirror movement) is an
integral multiple of the wavelength λ_o,

$$s = n\lambda_o \qquad \text{with } n = 0, \pm1, \pm2,\ldots \qquad (3.5a)$$

and destructive interference takes place when

$$s = \left(n + \frac{1}{2}\right)\lambda_o \qquad (3.5b)$$

In a scanning Michelson interferometer the optical path difference s is varied by moving one mirror at a constant velocity v. The
effect of this sweeping is to modulate the output intensity sinusoidal-
ly in relation to time with a frequency f that depends on the wavenumber $\bar{\nu}$ and the mirror velocity v [8]:

$$f = 2\bar{\nu}v \qquad (3.6)$$

Each input frequency from a polychromatic or continuous source can be
treated independently and the output is the summation or integral, respectively, of all the oscillations due to each optical frequency. For
a continuous source Eq. (3.3b) can thus be written

$$I(s) = 2\int_0^\infty I(\bar{\nu})(1 + \cos2\pi\bar{\nu}s)d\bar{\nu} \qquad (3.7)$$

The function I(s) is separable in a constant component which is independent of s,

$$I(\infty) = 2\int_0^\infty I(\bar{\nu})d\bar{\nu} \qquad (3.7a)$$

and an oscillating component which can take positive or negative
values. This component is actually responsible for the characteristic
interferogram structure and is therefore called the interferogram
function F(s):

$$F(s) = 2\int_0^\infty I(\bar{\nu})\cos2\pi\bar{\nu}s \, d\bar{\nu} \qquad (3.7b)$$

or because of $I(\bar{\nu}) = I(-\bar{\nu})$

$$F(s) = \int_{-\infty}^{+\infty} I(\bar{\nu}) \cos 2\pi\bar{\nu}s \, d\bar{\nu} \qquad (3.7c)$$

In Fig. 3-5 the interferograms of a single discrete line and a broad-band source are shown. Owing to the inverse relation of the finite coherence length of light waves and the half-width of the corresponding spectral line, no interference will take place between the two beams outside a certain path difference s, and F(s) will rapidly disappear [that is, $I(s) = I(\infty)$] for a broad-band source [Fig. 3-5(b)]. If the optical properties of the two arms of the interferometer are equal, the interferogram is symmetrical about s = 0 and the main central peak is observable at this position where the light of all wavelengths is constructively interfering (white light position):

$$I(0) = 4 \int_{0}^{\infty} I(\bar{\nu}) d\bar{\nu} = 2I(\infty) \qquad (3.7d)$$

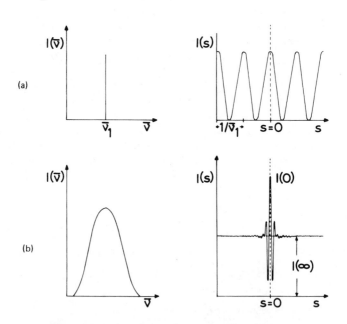

FIGURE 3-5   Interferogram and corresponding spectra of (a) monochromatic and (b) polychromatic radiation. [The units of I(s), $I(\bar{\nu})$, s, and $\bar{\nu}$ are arbitrary.]

While the corresponding wavenumbers can be readily derived from the interference patterns of one or two (beat patterns for example) discrete lines, spectral details become increasingly more difficult to obtain from the interferograms of polychromatic or continuous radiation, and the mathematical procedure of the Fourier transformation has to be applied. This procedure correlates the spectrum $I(\bar{\nu})$ with the interferogram function $F(s)$ and can be expressed for a perfectly compensated interferometer by [8, 9, 30]

$$I(\bar{\nu}) = \int_{-\infty}^{+\infty} F(s) \cos 2\pi \bar{\nu} s \, ds \tag{3.8a}$$

or because of $F(s) = F(-s)$,

$$I(\bar{\nu}) = 2 \int_{0}^{\infty} F(s) \cos 2\pi \bar{\nu} s \, ds \tag{3.8b}$$

For the calculation of the spectrum $I(\bar{\nu})$ with the aid of the cosine Fourier transform relation of Eq. (3.8), the knowledge of the interferogram for optical path differences up to infinity is required.

In practice, however, the mirror displacement is restricted to a finite value s_{max} of the optical path difference, and it is important to consider the effect of this geometrical restriction on the resulting spectrum (Fig. 3-6). Mathematically, this corresponds to the multiplication of $F(s)$ with a rectangular ("boxcar") truncation function $D_r(s)$, where

$$D_r(s) = \begin{cases} 1 & \text{for } -s_{max} < s < +s_{max} \\ 0 & \text{for } s < -s_{max} \text{ or } s > +s_{max} \end{cases} \tag{3.9}$$

According to the convolution theorem, the Fourier transform of the product $F(s) \cdot D_r(s)$ can be expressed by

$$I(\bar{\nu}) * D_r(\bar{\nu}) = \int_{-\infty}^{+\infty} F(s) \cdot D_r(s) \cos 2\pi \bar{\nu} s \, ds \tag{3.10a}$$

Here, $I(\bar{\nu})$ represents the true spectrum and $D_r(\bar{\nu})$ is the Fourier transformed of $D_r(s)$ generally referred to as the instrument line shape or

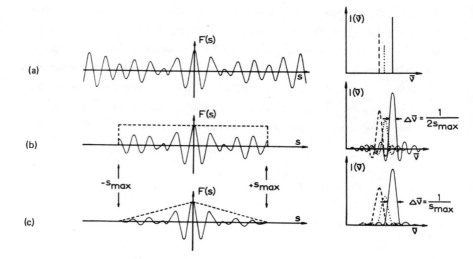

FIGURE 3-6   The effect of finite interferogram and apodization on the
resolution of the spectrum (demonstrated schematically on
three discrete lines): (a) infinite interferogram, (b)
finite interferogram with rectangular truncation and corresponding spectrum with line-shape function (sinx)/x,
and (c) finite interferogram with triangular apodization
and corresponding spectrum with line-shape function
$(sin^2x)/x^2$. [Reprinted with permission from R. Geick,
Topics in Current Chemistry 58:75 (1975).]

scanning function, which takes the form (sinx)/x. When the (sinx)/x
function is convolved with a single spectral line of wavenumber $\bar{\nu}_n$
and intensity $I(\bar{\nu}_n)$ the resultant function $I_i(\bar{\nu})$ can be expressed by
the relation [8-10, 28]:

$$I_i(\bar{\nu}) = I(\bar{\nu}_n)*D_r(\bar{\nu}) = 2I(\bar{\nu}_n)s_{max}\frac{\sin[2\pi(\bar{\nu} - \bar{\nu}_n)s_{max}]}{2\pi(\bar{\nu} - \bar{\nu}_n)s_{max}} \qquad (3.10b)$$

Such curves are shown for a spectrum consisting of three discrete
lines in Fig. 3-6(b).

The physical meaning of $I(\bar{\nu})*D_r(\bar{\nu})$ is that the true spectrum $I(\bar{\nu})$
is scanned with the line-shape function $D_r(\bar{\nu})$ and the true spectrum
will be modified by the scanning function in analogy to the slit func-

tion of a dispersive instrument. For the rectangular truncation function, the half width of the scanning function $D_r(\bar{\nu})$ is approximately

$$\Delta\bar{\nu} = \frac{1}{2s_{max}} \qquad\qquad (3.10c)$$

The negative side lobes inherent to the boxcar line-shape function can be eliminated when a triangular truncation function $D_t(s)$ is
applied where

$$D_t(s) = \begin{cases} 1 - \left|\dfrac{s}{s_{max}}\right| & \text{for } -s_{max} < s < +s_{max} \\ 0 & \text{for } s < -s_{max} \text{ or } s > +s_{max} \end{cases} \qquad (3.11)$$

The Fourier transform of this truncation function takes the form
$(\sin^2 x)/x^2$, and the operation is generally called apodization. The
corresponding relation to Eq. (3.10b) reads [see also Fig. 3-6(c)]:

$$I_i(\bar{\nu}) = I(\bar{\nu}_n)*D_t(\bar{\nu}) = I(\bar{\nu}_n)s_{max}\frac{\sin^2[\pi(\bar{\nu} - \bar{\nu}_n)s_{max}]}{[\pi(\bar{\nu} - \bar{\nu}_n)s_{max}]^2} \qquad (3.12a)$$

The elimination of the negative side lobes, however, goes at the cost
of an increase in the half width of the scanning function $D(\bar{\nu})$ by a
factor of about 2:

$$\Delta\bar{\nu} = \frac{1}{s_{max}} \qquad\qquad (3.12b)$$

The resolving power of a spectrometer is commonly defined as

$$R = \frac{\bar{\nu}}{\Delta\bar{\nu}} \qquad\qquad (3.13a)$$

Hence for a Fourier transform spectrometer we obtain

$$R = 2\bar{\nu}s_{max} \qquad\qquad (3.13b)$$

for the rectangular truncation function and

$$R = \bar{\nu}s_{max} \qquad\qquad (3.13c)$$

for the triangular apodization. Several other apodization functions
have been proposed in literature [8, 9].

Digitization of the Interferogram

The Fourier transformation of Eq. (3.10) can be performed by a computer according to the analogue or digital principle. For the most frequently applied digital analysis the interferogram is sampled in equally spaced intervals of the optical path difference Δs. Instead of a continuous function $F(s)$, a finite number of interferogram points are then obtained and the integral of Eq. (3.10) can be approximated by a sum (where $s_{max} = N\,\Delta s$):

$$I_{obs}(\bar{\nu}) = \sum_{n=-N}^{+N} F(n\cdot\Delta s)\cdot D(n\cdot\Delta s)\cdot\cos(2\pi\bar{\nu}n\cdot\Delta s)\cdot\Delta s \qquad (3.14a)$$

Here the line-shape function $\overline{D}(\bar{\nu})$ and the observed spectrum $I_{obs}(\bar{\nu})$ can be defined by infinite sums [30]:

$$\overline{D}(\bar{\nu}) = \sum_{m=-\infty}^{+\infty} D(\bar{\nu} + m\,\frac{1}{\Delta s}) \qquad (3.14b)$$

and

$$I_{obs}(\bar{\nu}) = \sum_{m=-\infty}^{+\infty} I(\bar{\nu} + m\,\frac{1}{\Delta s}) \qquad (3.14c)$$

with $m = 0, \pm1, \pm2,\ldots$, which means that the window function has an infinite number of main maxima each separated by $\Delta\bar{\nu} = 1/\Delta s$, and the true spectrum $I(\bar{\nu})$ is infinitely reduplicated equally spaced by this interval.

According to the sampling theorem of information theory harmonic functions can be digitized unambiguously only with a sampling frequency equal to twice the bandwidth of the system under consideration. Thus, in order to make $I_{obs}(\bar{\nu})$ unique, $I(\bar{\nu})$ must be nonzero only in the range $0 \leq \bar{\nu} \leq \bar{\nu}_{max}$, and the sampling interval is related to the upper wavenumber limit $\bar{\nu}_{max}$ of the spectrum (corresponding to the minimum wavelength λ_{min}) by the expression (Nyquist criteria):

$$\lambda_{min} = \frac{1}{\bar{\nu}_{max}} = 2\cdot\Delta s \qquad \text{or} \qquad \Delta s = \frac{1}{2\bar{\nu}_{max}} \qquad (3.15a)$$

Intensity contributions at $\bar{\nu} < \bar{\nu}_{max}$ from light waves with $\bar{\nu} \geq \bar{\nu}_{max}$ (so-
called "aliasing" or "folding") have to be eliminated by optical and/
or electrical filtering such that $I(\bar{\nu}) = 0$ for $\bar{\nu} \geq \bar{\nu}_{max}$.

When the spectrum from zero to $\bar{\nu}_{max}$ is required at a resolution
of $\Delta\bar{\nu}$, the number of points N to be sampled is given by

$$N = \frac{2\bar{\nu}_{max}}{\Delta\bar{\nu}} \qquad (3.15b)$$

Introducing Eq. (3.12c), N becomes

$$N = 2\bar{\nu}_{max} s_{max} \qquad (3.15c)$$

When the spectral range is limited to fall between a minimum and a
maximum wavenumber $\bar{\nu}_{min}$ and $\bar{\nu}_{max}$, respectively, $\bar{\nu}_{max}$ has to be replaced by $\bar{\nu}_{max} - \bar{\nu}_{min}$ in Eqs. (3.15b) and (3.15c).

For a reproducible collection of data points separated by the
specified sampling interval and started at the same mirror retardation for repeated scans, the interferogram produced by the white-light
and laser-light sources in the reference interferometer are used as
time basis. As a consequence of the simultaneous displacement of the
movable mirrors in the reference and main interferometers, the corresponding interferograms are locked together (Fig. 3-4). Since the
zero retardation position measured by the white-light detector always
occurs in the same place, this signal can be used as trigger to initi-
ate the data collection. The position of the fixed mirror of the reference interferometer is usually adjusted so that the white-light
fringe occurs just before the zero-retardation point of the sample
interferogram. In the measurement of single-sided interferograms a
sufficient number of data points can then be sampled before zero retardation for the small double-sided interferogram required for phase
correction but leaves most of the data points beyond zero retardation
so that the desired resolution can be achieved with the minimum scan
length and number of data points [8, 9]. The zero-crossings of the
sinusoidal laser-light interferogram as measured in the reference
interferometer are used to digitize the signal from the main interferometer, since each of these points occurs at equal intervals of

retardation. Thus, when every second zero-crossing of a 632.8 nm helium-neon laser, for example, were used, a bandwidth of 7902 cm^{-1} adequate for work in the mid-infrared region could be handled [8]. With
the laser-fringe referencing system the requirement of a strictly
linear mirror drive is than no longer necessary because velocity fluc-
tuations are compensated.

Phase Correction

When phase errors as a consequence of asymmetries in the optics of the
instrument or sampling procedure and electronic effects occur [8, 9,
13], Eq. (3.7c) no longer gives an accurate representation of the interferogram and corrections to the phase angle have to be applied.
Such a correction enters Eq. (3.7c) as a phase shift which may be a
linear or nonlinear function of the wavenumber $\bar{\nu}$ [8, 28], and the expression for the distorted interferogram function $F_{dist}(s)$ then reads

$$F_{dist}(\bar{\nu}) = \int_{-\infty}^{+\infty} I(\bar{\nu}) \cos(2\pi\bar{\nu}s + \theta_{\bar{\nu}}) \, d\bar{\nu} \qquad (3.16)$$

Since

$$\cos(2\pi\bar{\nu}s + \theta_{\bar{\nu}}) = \cos2\pi\bar{\nu}s\cdot\cos\theta_{\bar{\nu}} - \sin2\pi\bar{\nu}s\cdot\sin\theta_{\bar{\nu}} \qquad (3.16a)$$

the addition of $\theta_{\bar{\nu}}$ to the phase angle has the effect of introducing
sine components to the cosine wave interferogram and the spectrum and
interferogram must be related by the complex Fourier transform [13]:

$$I_{real}(\bar{\nu}) + iI_{imag}(\bar{\nu}) = \int_{-\infty}^{+\infty} F_{dist}(s) \exp(-2i\pi\bar{\nu}s) \, ds \qquad (3.17)$$

with

$$I_{real}(\bar{\nu}) = \int_{-\infty}^{+\infty} F_{dist}(s) \cos2\pi\bar{\nu}s \, ds \qquad (3.17a)$$

and

$$I_{imag}(\bar{\nu}) = \int_{-\infty}^{+\infty} F_{dist}(s) \sin2\pi\bar{\nu}s \, ds \qquad (3.17b)$$

where $I_{real}(\bar{\nu})$ (cosine transform) and $I_{imag}(\bar{\nu})$ (sine transform) are
the real and imaginary parts, respectively, of the complex spectrum.
It can be shown [8, 9, 13] that $I_{real}(\bar{\nu})$ and $I_{imag}(\bar{\nu})$ are related to
the corrected spectrum $I(\bar{\nu})$ by

$$I_{real}(\bar{\nu}) = I(\bar{\nu})\ \cos\theta_{\bar{\nu}} \qquad\qquad (3.18a)$$

and

$$I_{imag}(\bar{\nu}) = I(\bar{\nu})\ \sin\theta_{\bar{\nu}} \qquad\qquad (3.18b)$$

so that the phase shift $\theta_{\bar{\nu}}$ may be obtained from the relation

$$\theta_{\bar{\nu}} = \arctan \frac{I_{imag}(\bar{\nu})}{I_{real}(\bar{\nu})} \qquad\qquad (3.18c)$$

For routine spectroscopic investigation the most economic way of
phase correction is to scan a small double-sided portion of the interferogram around zero retardation and extend it single-sided up to s_{max}
[9]. The cosine and sine transform of the small double-sided interferogram are then used to determine $\theta_{\bar{\nu}}$ according to Eq. (3.18c) and
the undistorted interferogram is obtained by correction of the single-
sided region. In commercial instruments a subroutine for phase correction based on the procedure outlined above is commonly included in the
software provided by the manufacturer.

Instrumentation and Advantages of FTIR Spectroscopy

In practice interferometers are used in two different ways depending
on the scan speed of the moving mirror. For slow-scanning interferometers the velocity of the moving mirror is sufficiently slow that the
modulation frequency f of each of the spectral frequencies is generally less than 1 s^{-1}. This type of system is most commonly used for
FIR spectroscopy. A typical scan speed is of the order of 0.004 mm/s,
so that a wavenumber of 500 cm^{-1}, for example, has a modulation frequency of 0.4 s^{-1} in the interferogram. Frequencies in this subaudio
range are rather difficult to amplify, and amplification is usually
accomplished with a lock-in amplifier upon modulation of the beam with

a mechanical chopper whose frequency is considerably higher than the
modulation frequency of the largest wavenumber retardation to be
measured [8].

Rapid-scanning interferometers differ from the slow-scanning type
in that the mirror velocity of the interferometer is sufficiently
high that the spectral frequencies are modulated in the audio-fre-
quency range. Rapid-scanning is mandatory for the successful measurement of mid-infrared spectra in order to keep the dynamic range of the
interferometer within the limits of the digitization system [8]. A
typical mirror velocity used for the measurement of IR spectra in the
4000 to 400 cm^{-1} region is 3 mm/s. The modulation frequencies for the
upper and lower wavenumber limit are therefore 2400 and 240 s^{-1}, respectively. Since these frequencies are in the audio range, they can
easily be amplified without the necessity for modulating the beam with
a chopper. A block diagram of the basic components of an FTIR instrument is outlined in Fig. 3-7.

In analogy to the conventional dispersive instruments a Nernst or
Globar is used as source for the mid-infrared and a high-pressure mercury arc lamp for the far-infrared region. By far the most common interferometer type used in FTIR spectroscopy is the Michelson interferometer. Naturally, the beam-splitter material has to be chosen according to the wavelength region under examination. While thin, organic self-supporting films [e.g., polyethylene terephthalate)] are
used in the FIR, dielectric layers (for example, Ge or Fe_2O_3) coated
on a transparent substrate (for example, KBr, CsI, or CaF_2) are used
in the MIR and NIR region. For slow-scanning instruments a Golay cell
or He-cooled Ge or Si bolometer may be used as detector. The requirement for faster response for rapid-scanning interferometers is met by

FIGURE 3-7   Block scheme of the basic components of an FTIR spectrom-
              eter system.

the use of a pyroelectric detector (e.g., TGS). Upon amplification of
the signal in which high-frequency contributions have been eliminated
by a band-pass filter, the data are converted to digital form in the
analogue-to-digital (A/D) converter and transferred to the computer
for Fourier transformation. The calculated spectra may then be recorded
or displayed on a screen.

Generally, the interferogram of the background and the sample are
measured separately and after Fourier transformation the corresponding
transmittance or absorbance spectrum of the sample can be calculated.
As example, the intermediate steps of data acquisition for the spectrum
of polystyrene are demonstrated in Fig. 3-8.

Until late 1960 most workers in the field of Fourier transform
spectroscopy used a conventional Fourier transform in which the interferogram is multiplied by a cosine wave of the wavenumber being analyzed. For N input and output points in the interferogram and computed
spectrum the time required for the transform is proportional to N^2.
However, computation time has been cut down considerably with the development of the fast Fourier transform (FFT) algorithm for complex
Fourier transforms which is based on the symmetry and periodicity of
the sine and cosine functions. Using this technique the time required
for the transform is reduced to $N \cdot \log_2 N$. It is not within the scope
of this section to give a description of Fourier transform algorithms,
and for a detailed treatment of this matter the reader is referred to
the literature [8-10, 30, 32].

FTIR spectroscopy possesses several advantages over conventional
dispersive IR spectroscopy. Among these primarily the multiplex or
Fellgett [33] and the throughput or Jacquinot [34] advantages have
contributed to the break through of the FTIR technique in the last
decade. The multiplex advantage becomes obvious from Fig. 3-9. In a
dispersive spectrometer, n spectral increments of the width $\Delta \bar{\nu}$ are
measured successively with a certain signal-to-noise (S/N) ratio in
time T

$$T = n\tau \tag{3.19}$$

where τ is the time required to measure one spectral element $\Delta \bar{\nu}$.

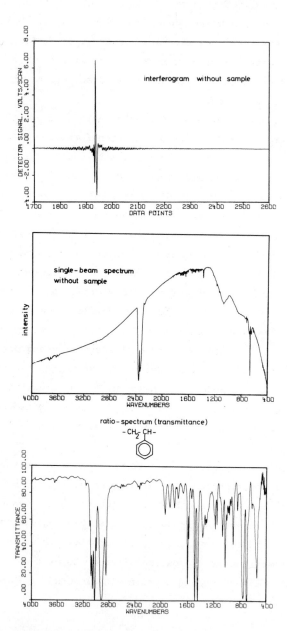

FIGURE 3-8   Intermediate steps of data acquisition for the IR spectrum of polystyrene.

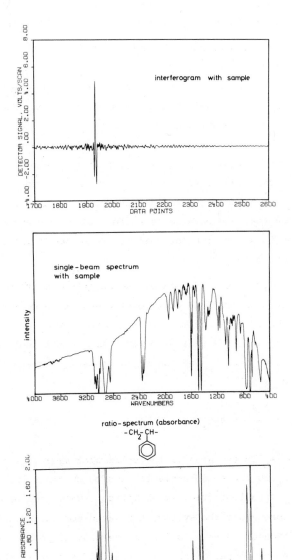

FIGURE 3-8 Continued.

In the FTIR spectrometer all the spectral increments contribute simultaneously to the interferogram and two cases may be distinguished:

1. The time required to scan the interferogram is the same as that
    required to record the complete spectrum with a dispersive instru-
    ment (slow-scanning). For every spectral element the total measur-
    ing time T is then available and a statistical advantage of $\sqrt{n}$
    is gained in the S/N ratio.

2. The interferogram is scanned in time τ (rapid-scanning). Signal
    averaging of n repeated interferogram scans again yields the sta-
    tistical advantage of $\sqrt{n}$ in the S/N ratio. Alternatively, advan-
    tage may be taken of the rapid-scanning capability when changes
    of the chemical and physical structure of a compound have to be
    monitored in short time intervals over the whole mid-infrared
    wavenumber region.

The second principal advantage, the Jacquinot or throughput advantage, means that the radiant power of the source is more effectively
utilized in the interferometer with rotational symmetry about the
radiation direction. The radiant power P is proportional to the product of the source area A_s and the solid angle Ω_c subtended by the collimator mirror (Fig. 3-9):

$$P \sim A_s \Omega_c = \frac{A_s A_c}{f^2} = A_c \Omega_s \qquad (3.20)$$

In Eq. (3.20) A_c is the area of the collimator mirror, f is its focal
length, and Ω_s is the solid angle the source subtends from the collimator mirror. In a loss-free, ideal optical system the product $A\Omega$
is constant throughout the system, and Jacquinot called this quantity
throughput (étendu). When A_c and f are assumed to be the same for the
instruments to be compared, the essential difference is that the true
area of the source $(A_s = r_s^2 \pi)$ is the limiting factor in the interferometer, while the smaller slit area $(A_s = wh)$ is the limiting factor in
the grating spectrometer (Fig. 3-9).

A critical review of the comparative performances of Fourier transform and dispersive IR spectrometers has been recently given by Griffith et al. [35].

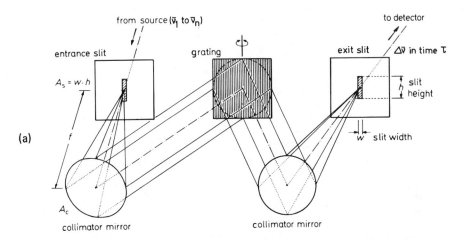

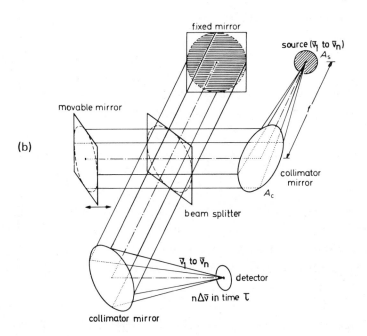

FIGURE 3-9   Schematic representation of (a) a grating monochromator
and (b) a Michelson interferometer demonstrating the factors which determine the multiplex and throughput advantages: A_s, source area; A_c, collimator mirror area; f,
focal length. [Reprinted with permission from R. Geick,
Fresenius Z. Anal. Chem. 288:1 (1977).]

Data Processing

The optical part of modern FTIR spectrometers is completely controlled
by a more or less powerful data system which possesses as basic units
a computer (preferably with 16-, 20-, or 24-bit words), a moving-head
disk drive (single or double platter system), an oscilloscope, and a
teletype. In addition to the software provided by the manufacturers
of FTIR spectrometers, individual software programs can be prepared
for special application and evaluation problems. These programs can
be written in machine language, in BASIC, or in FORTRAN if the respective software compiler is available) and stored together with the
other software and the experimental data on the disk cartridges.

Before outlining the experimental and evaluation possibilities
of such a system let us briefly discuss the storage facilities of the
disk cartridges. We will restrict ourself here to the Nicolet FTIR
system model 7199, but the other manufacturers (Bruker, Digilab) offer
comparable possibilities.

According to Eq. 3.15b a spectrum from 0 to 8000 cm^{-1} (which is
always generated by the Fourier transformation) requires approximately
4000, 8000, or 16000 data points (intensities) to be stored for a resolution of 4, 2, or 1 cm^{-1}, respectively. Each intensity value is
stored in a 20-bit word, where 352 words form one block or sector.
Since the wavenumber difference between the data points is constant
and depends on the frequency of the laser used to digitize the signal
of the main interferometer (see Sec. 3.1.1.2) only the laser frequency has to be stored in the file status block together with the other
parameters necessary to reproduce the spectrum. The file status block
(352 words) and the intensity values [necessary number of words $\hat{=}$
16384/resolution (cm^{-1})] form one file. However, in most cases only
a selected wavenumber region (mainly 400 to 4000 cm^{-1}) is of interest
and the data points representing the other parts of the spectrum can
be deleted. Thus, for the conventional MIR region from 400 to 4000
cm^{-1} the following amounts of 20-bit words or blocks have to be stored:

    1 cm^{-1} resolution: 25 blocks = 8800 words
    2 cm^{-1} resolution: 13 blocks = 4576 words
    4 cm^{-1} resolution: 7 blocks = 2464 words

This binary information of a file representing the experimental data
can be stored on a disk cartridge. Each cartridge has a maximum capacity of 6496 blocks. This amounts to 2286592 words of information or
433 spectra with a resolution of 2 cm^{-1} between 400 and 4000 cm^{-1}.
However, since there are often 2 cartridges on line, one fixed and
one removable, the total capacity is doubled. This storage capacity
for spectra is only insignificantly reduced by the software programs
necessary for data evaluation or manipulation.

Besides these advantages of computer-supported IR spectrometers
in the field of data storage, the computer equipment can be used to
control additional accessories when performing vibrational spectroscopic studies of short-time phenomena in polymers during elongation,
stress relaxation, crystallization, temperature changes, and other
similar experiments [36]. As an illustration Fig. 3-10 shows the experimental arrangement and the possibilities of data manipulation
for FTIR studies of elongation and relaxation phenomena of polymers.
For temperature dependent measurements the stretching device simply
has to be interchanged by a cooling or heating cell. During a stress-
strain experiment the interferograms are recorded in operator-selected
time intervals simultaneous to the stress-strain diagram. Each interferogram is stored together with the respective stress and elongation
(or temperature) values and the time of the measurement on the disk.
Dependent on the applied computer program the polarization direction
of the transmitted light can be changed automatically after a operator
preselected number of spectra has been scanned. A computer-controlled
camera takes pictures from that region of the polymer film which is
irradiated by the infrared beam. The exact position is determined by
means of a small-power helium-neon laser. In this way misinterpretation of the IR spectra because of macroscopic defects in the polymer
film can be avoided. Furthermore, for every spectrum the corresponding
sample geometry is known. This is of extreme importance when performing measurements in the necking region of the deformed polymer.

The powerful computer system supports spectra evaluation by application of special programs developed by the user (e.g., removal of
interference fringes, determination of the film thickness, automatic
detection of intensity changes in selected frequency regions) (Fig.
3-10).

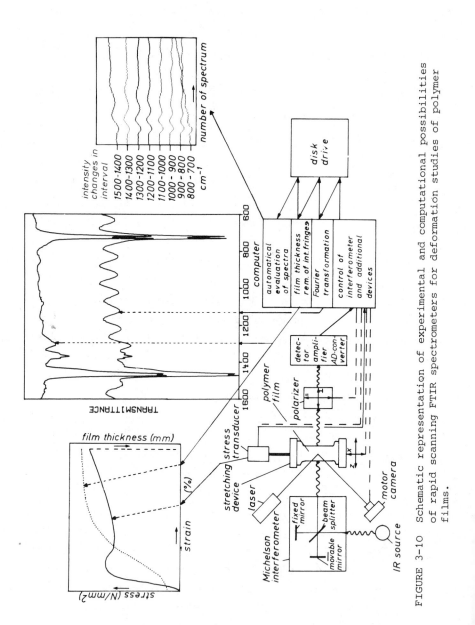

FIGURE 3-10   Schematic representation of experimental and computational possibilities of rapid scanning FTIR spectrometers for deformation studies of polymer films.

3.1.2 Raman Spectrometers

Figure 3-11 shows the block scheme of a usual experimental setup for
a Raman experiment. In modern instruments the sample is generally irradiated by the monochromatic radiation of a gas-laser [37-40]. Some
common gas-laser types with their most powerful excitation lines are
listed in Table 2-1. Spurious laser lines [38] which only prove useful with respect to wavelength calibration have to be removed by a
filter or a premonochromator. The laser irradiates highly plane polarized radiation (polarization purity >98%), and for depolarization
measurements the plane of polarization of the laser beam can easily
be rotated by use of a half-wave plate. In most cases the stray light
is observed under 90° (fibers, powders, bulk materials, liquids, etc.)
or 180° (transparent films, crystals, etc.). A lens focuses the stray
light on the entrance slit of a monochromator. For depolarization
measurements the polarization of the stray light in relation to the
orientation of the incident electric field vector can be analyzed by
means of a polaroid sheet positioned between the entrance slit and
the focusing lens (Fig. 3-12). Since several components (mirrors,
gratings, slits) of a spectrometer influence the polarization character of the light, the analyzed light either has to be scrambled or
the plane of polarization has to be rotated by a half-wave retardation plate before reaching the entrance slit. By appropriate positioning of this plate, the plane of polarization of the light entering the
monochromator can be selected independently of the analyzer position.

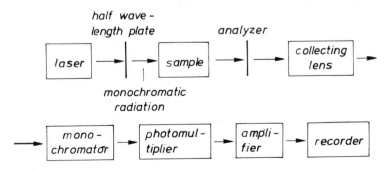

FIGURE 3-11 Block scheme of a Raman spectrometer.

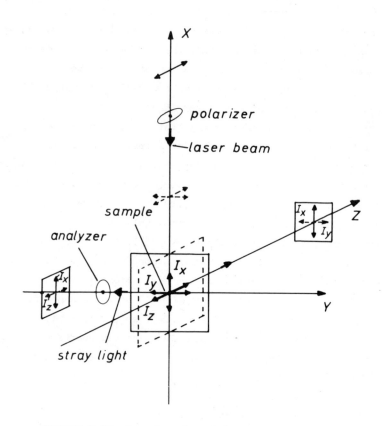

FIGURE 3-12   Geometry for the Raman experiment.

To obtain comparable Raman and IR spectra the monochromator of a
Raman spectrometer must possess a resolving power about an order of
magnitude higher than the monochromator of a corresponding dispersive
IR spectrometer because of the shorter wavelength of the analyzed
light (Table 2-1) in the Raman experiment. The stray light is analyzed
in the monochromator by a system of mirrors and two (double monochromator, Fig. 3-13) or three (triple monochromator) slowly rotating tan-
dem gratings [41-43]. Finally the diffracted light is focused on the
exit slit. A highly sensitive, generally cooled photomultiplier tube
converts the incident photons into electrical pulses which are amplified either by a DC amplifier or a pulse-counting system. A re-

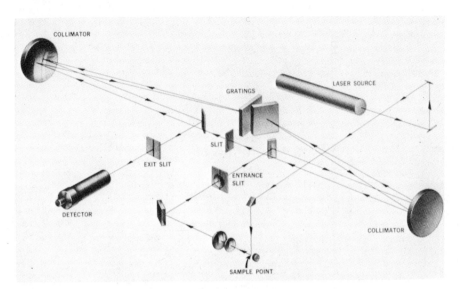

FIGURE 3-13   Double monochromator of the Raman spectrometer Cary 83
(with kind permission of the Varian GmbH, Darmstadt,
West Germany).

corder plots the intensity of the stray light in dependence of the
position of the rotating gratings. To receive a linear wavenumber
scale, the motion of the gratings is generated by a cosecant drive
[21]. The following points have to be kept in mind for the recording
of a Raman spectrum:

1. Since the intensity of the stray light is proportional to v_o^4 the
   lower frequency exitation lines (red, yellow) will yield much
   weaker Raman bands than the higher frequency lines (blue, green).

2. The response of the commonly used S 20 photomultiplier tubes depends on the frequency of the incoming light. From Fig. 3-14 it
   can be seen that the sensitivity of the photomultiplier is about
   two order of magnitudes higher for the blue Ar^+ 488 nm line than
   for the red Kr^+ 647.1 nm line. Within one spectrum which is
   scanned from 0 to 4000 Δcm^{-1}, the sensitivity decreases more
   rapidly when using a krypton laser. Therefore, an optimum signal-
   to-noise ratio will be obtained with higher frequency excitation
   lines. However, sometimes fluorescence and colored samples can

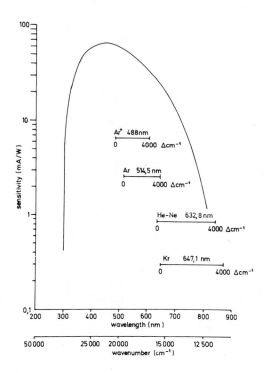

FIGURE 3-14   Sensitivity of an S 20 photomultiplier tube.

cause considerable trouble. Then a change to lower frequency lines
will often be of advantage [45].

With few modifications in the scanning system and the electronic
detection of a concave holographic grating double monochromator it has
been demonstrated that well resolved Raman spectra may be obtained in
short time intervals in the order of seconds [46, 47]. For the study
of transients, unstable samples, or ultra-fast reactions laser Raman
spectrometers were developed [48, 49], which allow for measurement of
Raman spectra within 2×10^{-11} s. These instruments use multichannel
detectors and are able to analyze simultaneously a wide range of wavenumbers during the pulse duration of a pulsed laser beam.

Recently, a double monochromator laser Raman molecular microprobe system (MOLE) has been developed which permits to obtain vibrational spectra of samples handled with a microscope. In the imaging mode an image of the surface from a characteristic line of a selected component can be directly observed on a TV monitor screen with the aid of an image intensifier. This allows to map the distribution of a component in a heterogeneous sample [50].

3.2 SAMPLING TECHNIQUES

The quality of the IR spectrum of a polymer depends largely on a careful and suitable preparation of the sample. Despite correct operation parameters of the spectrometer a considerable amount of information may be lost or misinterpreted as a consequence of incorrect sample thickness, nonuniformity, interference fringes, and impurities or solvent residues. In marked contrast to IR spectroscopy, the Raman spectrum of a polymer can be obtained in favorable cases with no sample preparation whatever. Thus, bulk samples varying greatly in dimensions and shape (powders, pellets, lumps, films, fibers, rods, or plates) may be investigated without further modification. Difficulties are sometimes encountered with polymer samples which are colored or contain highly absorbing fillers because the sample may decompose thermally as a consequence of strong absorption. To overcome these difficulties rotating sample techniques [51, 52] have been proposed. In other cases high backgrounds are encountered owing to the occurrence of fluorescence. Usually, fluorescence is associated with impurities in the polymer and can be removed by preliminary purification such as recrystallization, extraction, or treatment of the polymer with activated charcoal [53-55]. An alternative and most convenient method is to allow the fluorescence to decay by exposure to a laser beam [56, 57]. Improvement can also be obtained by shifting to an exciting line of different wavelength out of the efficient fluorescent excitation [58].

3.2.1   Infrared Spectroscopy

3.2.1.1   Solutions

Despite the advantage of excellent reproducibility of the chemical
and physical state of the sample (especially for quantitative measurements), IR spectroscopy of solutions, widely applied to low-molecular-
weight compounds, is used far less for the investigation of polymers in the MIR (4000 to 400 cm^{-1}) wavelength region. It is commonly
difficult to find solvents which are both transparent to IR radiation
in this wavelength interval and good solvents for polymers as well.
The IR spectra of solvents in common use for IR spectroscopy of polymers have been summarized by various authors [59, 60].

To eliminate the superposition of the polymer spectrum by solvent
absorption bands, these have to be compensated for in the reference
beam. Thus, the concentrations of the solutions should be as high as
possible. However, these often very viscous solutions are difficult
to handle in thin cells normally used for spectroscopic purposes in
the MIR region. Nevertheless, this preparation technique may be valuable for special quantitative purposes where only narrow spectral
ranges are of interest [61] or with unusual solvents [62, 63].

The preceding limitations do not apply to the NIR region (10000
to 4000 cm^{-1}). The overtone and combination bands in this wavenumber
region are of comparably small intensity (see Sec. 4.5), so that they
may be observed at pathlengths of up to 100 mm. In addition, glass
and quartz transmit acceptably in this region, and liquids can be contained in simple cells which are easily cleaned.

With the introduction of the commercially available Fourier transform IR (FTIR) spectrometers the problem of removing interfering absorptions due to solvents has been alleviated. IR analysis of aqueous
solutions, e.g., has been limited by the strong, broad absorption bands
of water. With FTIR, if the aqueous solutions are examined such that
total absorbance does not occur, the water spectrum can be subtracted.
This improvement has been extensively utilized for the investigation
of biological systems where water is the only interesting solvent [64].

3.2.1.2 Preparation as Film

The standard method for the analysis of polymers in the solid state
is their preparation as film with dimensions approximately 15×5 mm,
large enough to occupy the entire cross section of the light beam.
Polymer films intended for qualitative spectroscopic investigation in
the MIR region should be about 0.02 mm thick (greater thicknesses up
to 2 mm are frequently required for the FIR and NIR regions. When the
structure of the polymer under examination contains polar groups (for
example, C-F, C=O, CONH) thicknesses as low as 0.001 mm may be necessary to record the most intense bands satisfactorily. The range of
thicknesses for quantitative measurements depends mainly on the absorption band chosen for quantitative evaluation and varies roughly
between 0.001 and 1mm. Several methods for the preparation of films
are in general use.

Solvent Casting

A solution of the polymer (concentration up to 20% depending on the
required film thickness) is poured onto a leveled glass or metal plate
and spread to uniform thickness; afterward, the solvent is evaporated
under an IR lamp or in a vacuum oven. The choice of solvent is determined not only by its ability to dissolve the sample but also by the
facility of preparing a homogeneous film. Ideally, a microscope slide
with a roughened surface (to avoid interference fringes in the IR spectrum) is used as casting plate. After drying, the sample film can be
stripped from the plate, if necessary, by dipping into warm water.
Casting on a liquid surface, usually mercury, is employed when a highly uniform sample thickness is required. The polymer solution is con-
fined to the required area by a copper ring floating on the mercury
surface. Special attention must be drawn to the complete removal of
solvent from the polymer film, especially with high-boiling solvents
(for example, DMF in polyacrylonitrile, DMSO in polyoxymethylene, and
phenol in polyamides).

Figure 3-15(a) shows the spectrum of an acrylonitrile-methylacrylate
(94:6) copolymer film cast from DMF solution and dried in vacuum at

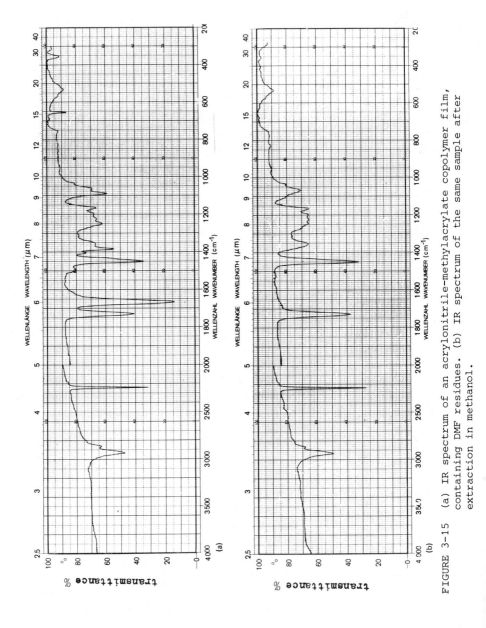

FIGURE 3-15 (a) IR spectrum of an acrylonitrile-methylacrylate copolymer film, containing DMF residues. (b) IR spectrum of the same sample after extraction in methanol.

323 K for 24 hr. The absorption bands at 1667, 1495, 1405, 1385, 1250, 1090, and 655 cm^{-1} belong to residual solvent and may in fact show slight wavenumber shifts owing to solvent-polymer interaction. This solvent-polymer interaction is also reflected in the perpendicular dichroism of the ν(C=O) absorption band of DMF residues in drawn films of this polymer (see Sec. 4.2.3) [65, 66]. Apart from removal by prolonged treatment at high temperatures [67], the DMF residues may be removed by extraction with a suitable solvent (methanol, water) [Fig. 3-15(b)].

Melt Casting and Hot Pressing

Polymers which melt at suitable temperatures without decomposition, oxidation, or degradation can be prepared as film by sandwiching a few milligrams of sample between KBr plates after preheating the sample and the plates to the required temperature on a Kofler bench. Better results can be obtained by hot pressing the sample in a hydraulic press (preferably evacuable) between heated metal plates, which may be coated with thin layers of Teflon as a release agent [68] at pressures up to 30 N/mm^2. Figure 3-16 shows a commercially available hydraulic press which is successfully employed for these purposes in our laboratory. The pressure, temperature of the plates, and press period can be manually adjusted and are automatically controlled in the pressing cycle. Good results have been obtained by positioning the sample between aluminum foils, which can be dissolved in dilute HCl. The pressing temperature ranges of several plastics are summarized in Table 3-1 [69].

3.2.1.3 Pressed Disk Technique and Mulls

A great deal has been written on the application of these methods [70-72], and only the basic principles are outlined here. The principle of the pressed-disk technique is to suspend the powdered sample in an IR transparent matrix by grinding in a mortar or microball mill and pressing the mixture in a special disk press under loads of up to 1000 N/mm^2. The dimensions of the pellets obtainable with about 300 mg of suspension (concentration of the sample approximately 0.5%) in the presses

FIGURE 3-16   Hydraulic press for hot pressing (with kind permission
of Paul Weber AG, Stuttgart, West Germany).

commonly available from accessory manufacturers are usually 13 mm in
diameter and about 1 mm in thickness. The optical properties of some
matrix, window, and internal reflection element materials for IR spectroscopy are listed in Table 3-2. To reduce radiation scattering, the
matrix should have a refractive index close to that of the sample
under investigation. A definite disadvantage of some matrix materials
is the appearance of water absorption peaks near 3450 and 1635 cm^{-1}.

TABLE 3-1 Pressing Temperature Ranges for Some Polymers[†]

Polymer	Pressing temperature range (K)
Polyethylene	377-569
Polypropene	459-555
Polystyrene	401-527
Polyvinyl chloride	453-513
Polytrifluoromonochlorethylene	463-525
Polyamide-6	478-489
Polyamide-6,6	519-527
Polyamide-6,10	478-515
Polyethylene terephthalate	513-543
Poly-n-propylene terephthalate	493-533
Poly-n-butylene terephthalate	493-533
Poly-n-hexylene terephthalate	423-443
Polycarbonate	478-494
Polymethylmethacrylate	483-494
Polyoxymethylene	452-492
Cellulose-acetate	433-454
Cellulose-propionate	417-455
Ethyl-cellulose	405-443
Cellulose-acetate-copolymer	423-443
Styrene-acrylonitrile-copolymer	463-503
Acrylonitrile-butadiene-styrene-terpolymer	463-503

[†]Partly taken from Refs. 69 and 118.

A modification of the pressed-disk technique, preferably applied to water-sensitive samples or samples inclined to interact with the matrix material, employs liquid dispersion materials (paraffin oil, hexachlorobutadiene, perfluorocarbon). Here, the paste of matrix and sample obtained upon grinding is squeezed between IR transparent plates. The inherent disadvantage of this technique is its restriction to limited spectral regions owing to the absorption bands of the dispersion materials which can be only partially compensated in the ref-

TABLE 3-2   Refractive Index and Approximate Transmission Range of
            Some Window, Matrix, and Internal Reflection Element
            Materials

Optical material	Transmission range (cm^{-1})	Refractive index
NaCl	40000 - 650	1.52
KCl	33000 - 500	1.47
KBr	40000 - 4400	1.53
KI	40000 - 300	1.63
AgCl	25000 - 500	2.0
AgBr	20000 - 300	2.2
CaF_2	70000 - 1100	1.40
BaF_2	50000 - 800	1.46
CsBr	30000 - 250	1.67
CsI	40000 - 200	1.74
TlBr	20000 - 300	2.5
TlBr-TlI (KRS-5)	20000 - 3300	2.38
TlCl-TlI (KRS-6)	20000 - 400	2.19
Si	5000 - 800	3.45
Ge	5000 - 450	4.10
ZnS (Irtran-2)	10000 - 700	2.26
ZnSe (Irtran-4)	10000 - 550	2.4
CdTe (Irtran-6)	5000 - 350	2.67
Al_2O_3	50000 - 1500	1.76

erence beam. Important with all dispersion techniques is that the par-
ticle size be smaller than the wavelength of the IR radiation; otherwise band contours will be distorted by the Christiansen effect [71,
73].

3.2.1.4 Fibers

In the preparation of fibers those methods in which the sample morphology is destroyed have to be distinguished from the limited number
of methods where the fiber form is retained. For the qualitative analysis of fibers all preparation techniques so far discussed can be

applied. Thus, hot pressing of fibers to coherent films has been reported for thermoplastic materials [73-77].

In the preparation of fibers by the pressed-disk or mull techniques special attention has to be focused on a sufficient reduction
of partical size. This can be achieved with a Wiley laboratory [78,
79] or a ball mill [80], a hand microtome [81, 82], or simply by cutting the sample with scissors [83, 84]. The fibrous material may also
be ground with the liquid dispersion medium between glass plates [85,
86].

Single monofilaments may be examined destruction-free with the
aid of an IR microscope [22, 87-93] (see also Microsampling in Sec.
3.2.1.6). The use of the microscope is restricted to fiber diameters
between about 0.015 and 0.03 mm. The limiting factors are the lower
limit of the slit width on the one hand and too intense absorption of
polar groups in the polymer structure on the other hand. However, the
thickness may be reduced by microtoming longitudinal sections from the
fiber, imbedded in a suitable wax [94, 95].

In polarization measurements and for quantitative determination,
corrections for the convergence of the light beam have to be applied
[96, 97]. For some problems the preparation of fiber grids has been
applied. In the overtone region where samples of higher thickness are
required, parallelized fiber bundles, uniformly distributed between
transparent plates upon immersion in a matrix of suitable refractive
index, may suffice [98-101]. Investigations in the fundamental vibrational region necessitate the preparation of monolayer fiber grids.
The preparation procedures [102-107] are usually time-consuming and
the obtained spectra often of poor quality. Sources of considerable errors are the nonuniform pathlength of the light beam through the sample
cross section and gaps in the fiber grid. Their effect on the observed
absorbances and their corrections have been considered in detail [89,
97, 108]. Keeping in mind the loss of fiber morphology better results
may be obtained by pressing these fiber grids under load (up to 1000
N/mm^2) into homogeneous films (with or without imbedding in an IR
transparent matrix, for example, AgCl) [109-113]. With the introduc-

tion of the attenuated total reflection (ATR) technique (see below)
another alternative for the destruction-free IR spectroscopic investigation of fibers has become available.

Comprehensive articles on the preparation of fibrous materials for
IR spectroscopic studies can be found in literature [108, 114-118].

3.2.1.5 Microtoming

The use of a microtome has been suggested if sample thickness has to
be reduced without changing the physical state by solution, melting,
or high pressure [69, 94, 95, 118-120]. Alternatively, this sample
preparation may be utilized when polymeric materials can not be dissolved or molten without degradation (e.g., cross-linked resins). A
necessary condition for the successful use of the microtome is the
choice of the suitable cutting parameters. The cutting speed, form of
the knife, and angle relative to the cut as well as the cutting temperature depend mainly on the hardness of the sample. In general, soft
materials require higher cutting speeds than hard materials. Within
certain limits the suitable hardness of the sample can be adjusted by
working at low or high temperatures or swelling the polymeric material
in a suitable solvent [73]. For investigations of the state of order
it should be kept in mind that the microtoming procedure may sometimes
induce slight crystallization or orientation in the sample under examination [118].

3.2.1.6 Microsampling

To obtain IR spectra from small specimens (e.g., biological samples,
microtomed sections, single fibers) optical attachments are required
which reduce the image of the monochromator slit at the sample position.
For most purposes microilluminators [4] with slit-image reductions in
the range from 1:3 to 1:6 will satisfy the optical requirements. Extremely small samples may be investigated in a reflecting microscope
[22, 92, 94]. The optical scheme of an IR microscope is shown in Fig.
3-17.

To prevent undesireable heating of the sample by the focused polychromatic radiation the microscope is placed behind the monochromator
[92]. The dispersed light leaves the exit slit of the monochromator,

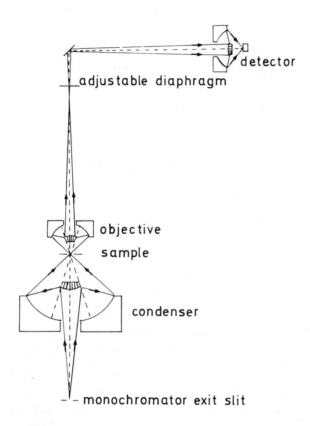

detector

adjustable diaphragm

objective

sample

condenser

monochromator exit slit

FIGURE 3-17   Optical scheme of IR microscope.

passes through the condensor and forms a reduced image (approximately
1:10) of the slit at the sample position. A new image of the sample
and slit is then formed by the objective at the plane of the diaphragm
and finally the light is focused onto the detector. To eliminate interference by atmospheric absorptions in single-beam operations, the spectrometer should be purged with dry and CO_2-free air. In double-beam
operation the atmospheric absorptions may be compensated with the aid
of a long path (1 m) gas cell in the reference beam [121]. Some authors
have proposed an arrangement where the light beam alternately passes
through the sample and a reference area [122, 123]. By combination of
the microscope attachment with a polarizer, orientation measurements

may be performed on small anisotropic specimens such as monofilaments
or single crystals. Owing to the dimensions of the investigated samples
some sources of errors have to be kept in mind in obtaining their IR
spectra. Thus, false radiation occurring as a consequence of diffraction phenomena or imperfections in the optical system will cause the
apparent intensity of absorption bands to be too small. For quantitative examinations corrections have to be applied which depend on the
convergence of the light beam, the refractive index of the sample, and
the measured absorbance values [22, 94, 97, 124]. The effects of convergence in IR dichroism measurements have been considered by Fraser
and Vettegren [96, 124].

3.2.1.7 Reflection Spectroscopy

ATR Spectroscopy

The first applications of attenuated total reflection (ATR) or internal reflection spectroscopy were reported independently by Fahrenfort
and Harrick [125-128]. Many macromolecular materials (rubbers, fibers,
fabrics, coatings, laminates) which often fail to yield useful transmission spectra are accessible to IR spectroscopic investigation by
the ATR technique [129-134]. The principles of the method are based
on the phenomenon of total internal reflection. For the experiment the
sample (refractive index n_2) is brought into direct contact with the
reflecting surface of a prism of high refractive index n_1 ($n_1 > n_2$)
(see Fig. 3-18). When the angle of incidence of the light beam at the
prism sample interface exceeds the critical angle α_c, total internal
reflection takes place. At the critical angle of total internal reflection the following relation holds:

$$\sin\alpha_c = \frac{n_2}{n_1} = n_{21} \qquad (n_1 > n_2) \qquad (3.21)$$

By the superposition of the incident and reflected light beam, a
standing wave perpendicular to the totally reflecting interface is
established. In the optical denser medium the electric field component
of this standing wave shows the dependence of a sine function but de-

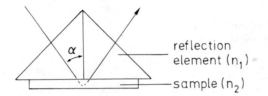

FIGURE 3-18 Principal optical scheme of ATR spectroscopy. n_1 and n_2
are refractive indexes.

creases exponentially in the optical rarer medium (Fig. 3-19) [135].
In contrast to metallic reflection, a large electric field component
can be localized in the interface in the case of total internal reflection. The penetration depth d_p for which the electric field component has decreased to 1/e of its value in the interface, has been

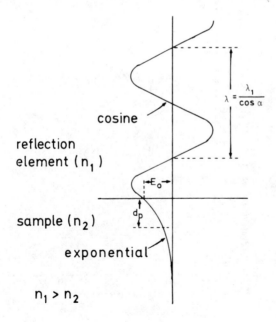

FIGURE 3-19 Standing wave pattern established near totally reflecting
interface: α, angle of incidence; λ_1, wavelength of incident radiation; λ, wavelength of standing wave; E_o, electric field component in the interface; d_p, penetration
depth; n_1 and n_2, refractive indexes.

derived theoretically [135]:

$$d_p = \frac{\lambda_1}{2\pi[\sin^2\alpha - (n_2/n_1)^2]^{1/2}} \tag{3.22}$$

Here α is the angle of incidence, λ_1 the wavelength of radiation in the optical denser medium, and n_1 and n_2 are the refractive indexes. This equation demonstrates the dependence of the penetration depth d_p on three parameters:

1. The penetration depth increases with increasing wavelength.
2. For a given value of n_2/n_1 the penetration depth approaches a maximum value toward the critical angle of total internal reflection α_c (Fig. 3-20).
3. d_p increases the closer the refractive indexes are to each other.

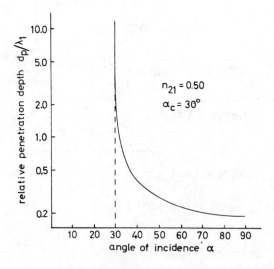

FIGURE 3-20   Dependence of the relative penetration depth on the angle of incidence: λ_1, wavelength of the incident radiation; n_{21}, refractive indexes ratio.

The penetration depth of the radiation into the optical rarer medium (sample) is usually of the order of a few micrometers, which is sufficient to constitute a short absorbing path. Therefore, total internal reflection will be attenuated in the wavelength regions of sample absorption. The spectrum recorded is thus very similar to the transmission spectrum of the sample surface with the only difference being the higher intensities of absorption bands at longer wavelengths.

The propagation of electromagnetic radiation in an absorbing material is characterized by the complex refractive index n_2^*:

$$n_2^* = n_2(1 - i\kappa_2) \tag{3.23}$$

which is directly related to molecular properties of this medium by

$$k_2 = \frac{4\pi n_2 \kappa_2}{\lambda_2} \tag{3.24a}$$

and

$$k_2 = a_2 c_2 \tag{3.24b}$$

where κ is the absorption index, c is the concentration, and k_2 and a_2 are the absorption coefficient and absorptivity, respectively. For $\kappa_1 = 0$ (nonabsorbing optical denser medium) and $n_1 > n_2$, two physical phenomena may be distinguished [136]:

1. Regular external reflection: $\sin\alpha < n_2/n_1$.
   When $\kappa_2 < 0.2$, the influence of absorption on the reflectivity can be neglected; only the intense bands of a substance can be observed.
2. Total internal reflection: $\sin\alpha \geq n_2/n_1$.
   It has been shown [125] that the attenuation of light totally reflected at the interface of a dielectricum of high refractive index n_1 and the investigated sample (refractive index n_2) becomes measurable in the vicinity of the critical angle α_c for κ_2 values as low as 0.1 to 0.0001.

In analogy to transmission measurements the total reflectivity R for samples of weak absorptions may be written [137]

$$R \simeq 1 - kd_e \qquad (3.25)$$

where k is the absorption coefficient and d_e is the so-called effective thickness which defines the sample thickness required to obtain the same absorbance in a transmission measurement as with a single total reflection and which is characteristic of the interaction of the above-mentioned standing wave with the absorbing optically rarer medium. For light polarized parallel or perpendicular to the plane of reflection, the following effective thicknesses were derived [137, 138]:

$$d_{e\parallel} = \frac{\lambda_1 n_{21}(2\sin^2\alpha - n_{21}^2)\cdot\cos\alpha}{\pi(1 - n_{21}^2)[(1 + n_{21}^2)\cdot\sin^2\alpha - n_{21}^2](\sin^2\alpha - n_{21}^2)^{1/2}} \qquad (3.26a)$$

$$d_{e\perp} = \frac{\lambda_1 n_{21}\cos\alpha}{\pi(1 - n_{21}^2)(\sin^2\alpha - n_{21}^2)^{1/2}} \qquad (3.26b)$$

The term $d_{e\parallel}$ is always larger than $d_{e\perp}$, a fact which has to be kept in mind for orientation measurements.

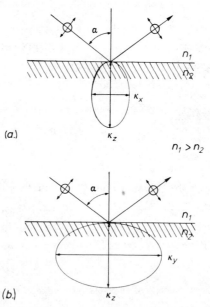

FIGURE 3-21   Sample-reflection element-polarization direction geometry
              for the determination of main absorption indexes by ATR
              spectroscopy.

The absorption of anisotropic materials is characterized by the three axes of the absorption index ellipsoid. Experimentally four different sample-reflection element-polarization direction geometries are available for the determination of the main absorption indexes κ_x, κ_y, and κ_z by ATR spectroscopy (Fig. 3-21) [139].

For ideal contact and single reflection the following relations have been derived for the reflectivity R and the absorption indexes [140]:

$$\log \frac{1}{R_{\perp 2}} = A_{\perp 2} = k_1 \kappa_y \qquad (3.27a)$$

$$\log \frac{1}{R_{\parallel 2}} = A_{\parallel 2} = k_2 [\kappa_z \sin^2\alpha + (\sin^2\alpha - n_{21}^2)\kappa_x] \qquad (3.27b)$$

$$\log \frac{1}{R_{\perp 1}} = A_{\perp 1} = k_1 \kappa_x \qquad (3.27c)$$

$$\log \frac{1}{R_{\parallel 1}} = A_{\parallel 1} = k_2 [\kappa_z \sin^2\alpha + (\sin^2\alpha - n_{21}^2)\kappa_y] \qquad (3.27d)$$

where

$$k_1 = \frac{4n_{21}^2 \cos\alpha}{(\sin^2\alpha - n_{21}^2)^{1/2}(1 - n_{21}^2)}$$

$$k_2 = \frac{4n_{21}^2 \cos\alpha}{(\sin^2\alpha - n_{21}^2)^{1/2}(\sin^2\alpha - n_{21}^2 + n_{21}^4 \cos^2\alpha)}$$

and A is the apparent absorbance. The numbers 1 and 2 designate parallel and perpendicular alignment of the drawing direction toward the plane of reflection. The symbols $\parallel$ and $\perp$ characterize the direction of polarization with reference to the plane of reflection. With multiple reflection elements where the number of reflections N is known R has to be substituted by R^N. Thus the absorption indexes may be evaluated from Eqs. (3.27a) through (3.27d).

In addition to an appropriate choice of the area of efficient optical contact - the larger the area of contact, the greater the absorption intensity - success in obtaining good ATR spectra depends mainly on two variables:

1. The refractive index of the reflection element n_1.
2. The angle of incidence α of light on the interface reflection
   element-sample.

The refractive index of the sample and thus the critical angle
of total internal reflection varies anomalously in the region of absorption bands. As a consequence there is some distortion of the band
contours which can be reduced by using a reflection element of higher
refractive index or by increasing the angle of incidence. In either
event the depth of penetration is reduced [Eq. (3.22)], and hence the
spectrum is weaker, but the intensity loss may be compensated for by
application of a multiple reflection element. The optical scheme of a
commercially available multiple reflection ATR attachment is shown in
Fig. 3-22. To record ATR spectra which correspond to conventional
transmission spectra, absorption-free, high-refractive-index reflection elements at an angle of incidence 2 to 8° above the critical
angle of total internal reflection should be used. The requirements
which have to be met by the material of the reflection element (refractive index, transmission, toughness, chemical inertness, surface
polish) and its geometrical form (precision of the dimensions and
angles) are high. Suitable materials for the IR region are (see also
Table 3-2) AgCl (n = 2.0), KRS-5 (n = 2.4), Si (n = 3.4), and Ge (n = 4.1).

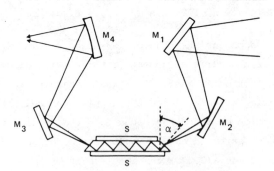

FIGURE 3-22   Optical scheme of a multiple-reflection ATR attachment:
M_1, M_2, M_3, and M_4 are adjustable mirrors; α, angle of
incidence; S, sample.

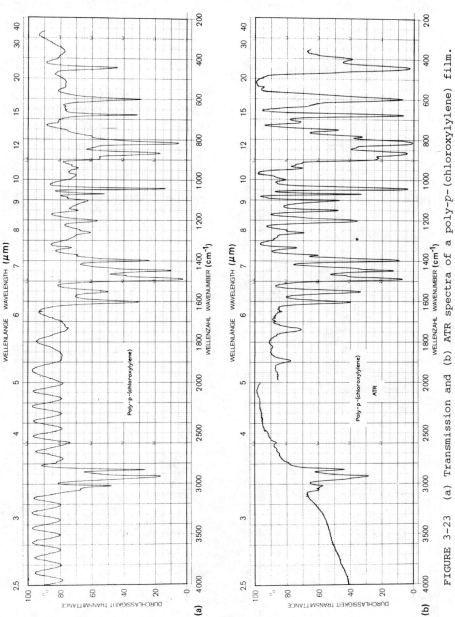

FIGURE 3-23   (a) Transmission and (b) ATR spectra of a poly-p-(chloroxylylene) film.

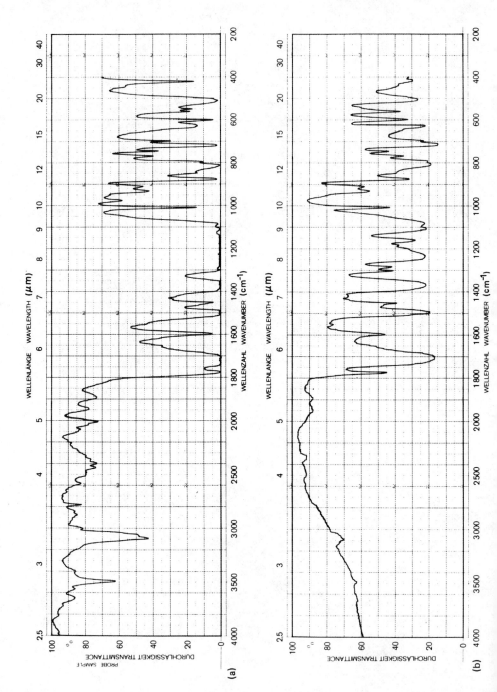

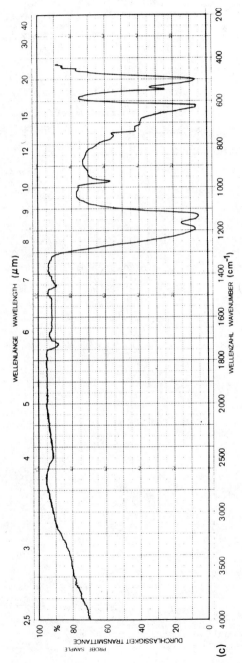

FIGURE 3-24   Transmission spectrum of the original film, (b) and (c) ATR spectra
of the two surfaces of the polymer film (see text).

For the investigation of organic samples a trapezoidal KRS-5 crystal with an angle of incidence of 45° and 25 reflections yields the most satisfactory results.

Fibers and fabrics - among the most difficult to handle by transmission spectroscopy - have proved to be quite amenable to study by multiple internal reflection spectroscopy. For routine spectra pieces of fabric are brought into contact with either side of the reflection element or fiber bundles are wound around the reflection prism [134].

A further advantage of ATR spectroscopy is the absence of troublesome interference fringes, which can be a source of error, e.g., in the spectra of polymer films (Fig. 3-23).

Apart from its use as an alternative to transmission spectroscopy to mitigate sample preparation problems, ATR spectroscopy has been applied successfully in the examination of surface coatings and laminates. The information of transmission and ATR spectra of the sample can often be pieced together to give a satisfactory picture of the surface and bulk composition. In Fig. 3-24 the transmission and ATR spectra of a polymer film are shown. Reference to standard spectra reveals the main component of the transmission spectrum to be the polyimide from pyromellitic dianhydride and 4,4'-diaminodiphenylether.

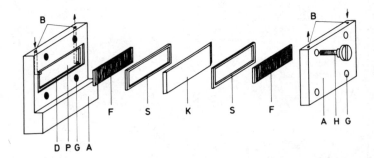

FIGURE 3-25   ATR sample holder for the deuteration of fibrous materials: A, sample holder block; F, fiber sample; S, gold-coated silicone rubber gasket; K, reflection element; B, inlet and outlet ducts; G, thread for sample holder screw; P, slot for fiber sample; D, slot for gasket; H, sample holder screw.

Strong bands are superimposed at about 1200, 620, and 500 cm^{-1}. Inspection of the ATR spectra of both sample surfaces reveals that the sample is a polypyromellitimide film coated on one side with Teflon FEP.

Generally, information on the surface structure of anisotropic materials is available from polarization measurements with the ATR technique. To obtain data on the existence of skin/core structures, the degree of order and dichroism of polyamide-6 fibers has been measured in relation to the experimentally determined penetration depth [139-141]. Quite recently the surface orientation in aromatic polyamide fibers has been characterized by ATR spectroscopy with polarized radiation [142]. Construction of a suitable sample holder (Fig. 3-25) has offered the possibility of characterizing the state of order in cellulosic [143] and aromatic [142] fibers by the determination of the extent of deuterium exchange (see also Sec. 4.2.4). Other fields of application in which ATR spectroscopy has proved valuable are studies of polymer surfaces upon irradiation. The ATR technique may reveal degradation or oxidation well before the bulk properties of the sample show significant changes [144-147].

Reflection-Absorption Spectroscopy

In contrast to ATR spectroscopy the less frequently applied reflection-absorption (RA) spectroscopy [148-150] involves external reflections. The sample is located on a metal surface, and in the process of reflection on this surface the IR beam interacts with the species of interest (Fig. 3-26). Similarly to ATR spectroscopy a standing wave perpendicular to the reflecting surface is established (Fig. 3-19). In RA spectroscopy, on the other hand, with metal surfaces and normal incidence a node exists very near the surface [151]. Therefore, no interaction with the sample, being located on the same side of the interface as the incoming beam, can be expected and no IR spectrum will be obtained. The node is not at the surface for high angles of incidence, and the field strength at the surface is large enough to obtain reasonable IR spectra [152]. As a consequence, in RA spectroscopy the angle of incidence is usually chosen between 70 and 89°.

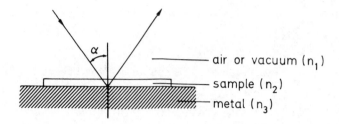

FIGURE 3-26   Principal optical scheme of reflection-absorption spec-
              troscopy.

Unlike the ATR technique, reflections in RA spectroscopy are not loss-
less, and it has been shown [153, 154] that there is an optimum number
of reflections after which the signal-to-noise ratio degenerates.
While ATR spectroscopy is generally applied to the IR spectroscopic
characterization of the first micrometers of a bulk sample, the thickness of the material of interest in RA spectroscopy commonly ranges
from 50 nm down to monolayers. For monolayers however, often only the
most intense absorptions originating from species such as C-F or C=O
groups are readily observed.

Apart from the detection of thin polymer films on metal substrates
[149], RA spectroscopy has also been successfully applied in the investigations of the adhesion between polymers and metals [155] and reaction at metal polymer interfaces. Chan and Allara [156, 157] have
combined ATR and RA spectroscopy for the studies of oxidation products
formed at a copper polyethylene interface.

3.2.2   Raman Spectroscopy

3.2.2.1   Solutions

Provided ample purification of the solvent and polymer, the solution
of the polymer is simply contained in a glass capillary tube and the
scattered light viewed at right angle to the exciting radiation passing axially through the tube [Fig. 3-27(a)] [158, 159]. Larger multi-
pass cells may be utilized when greater volumes of sample are avail-

able [54]. Water has an extremely weak Raman spectrum and, in consequence, is an excellent solvent for Raman spectroscopy. Thus, solution studies on water-soluble polymers (e.g., adhesives, biopolymers)
are becoming of increasing importance [160].

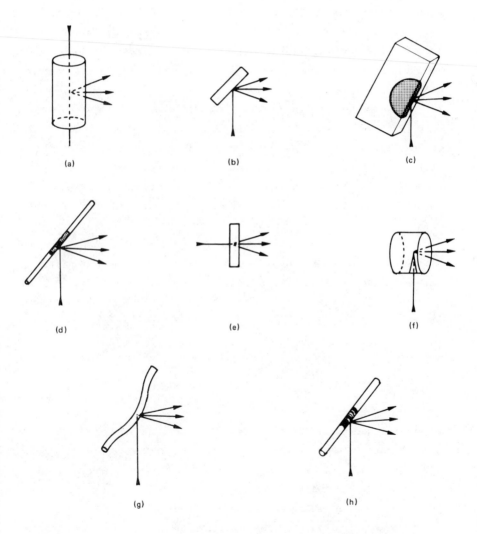

FIGURE 3-27   Sample preparation for Raman spectroscopy (see text).

3.2.2.2 Solids

The geometry of the sample arrangement relative to the exciting beam
and collecting lens can be varied by the design of the sample holder
and depends particularly on the nature of the sample under investigation. Thus, highly scattering or turbid samples are usually studied
by front surface reflection [Fig. 3-27(b)]. Powdered samples are
either simply tamped into a conical cavity for front surface illumi-
nation or into a transparent glass tube for transverse or 90° excitation [Fig. 3-27(c) and (d)]. With transparent samples it is possible
to illuminate from the backside and collect the radiation generated
within the sample [Fig. 3-27(e)]. For translucent samples the transmission technique with a hole drilled into the sample has proved valuable [Fig. 3-27(f)].

Extremly useful data can be derived from Raman spectra of single
crystals. Spectra obtained by excitation and observation parallel to
axes of the polarizability ellipsoid with polarized radiation allow
an unambigous classification of fundamental and lattice modes into
the various symmetry classes.

Fibers can be mounted according to Fig. 3-27(g) and (h). The experimental arrangement for observing oriented fiber Raman spectra has
been described by Hendra [161]. The depolarization ratio found in the
spectra recorded with the fiber bundle vertically, horizontally, and
end on enable one to distinguish modes where the vectors are parallel
or perpendicular to the fiber axis, respectively.

3.2.3 Special Techniques

3.2.3.1 Preparation of Deuterated Samples

A powerful technique for structure-spectra correlation and investigation of the state of order in polymers involves isotope exchange [162,
163] (see also Sec. 4.3.5). Pronounced frequency shifts can be expected upon substitution of hydrogen for deuterium. Apart from the
polymerization in deuterated media, hydrogen atoms attached to nitrogen or oxygen (polyamides, cellulose, polyvinyl alcohol) can be re-

placed by deuterium through direct contact of the sample with excess D_2O vapor or liquid. To avoid rehydrogenation by the atmosphere, the sample has to be mounted in a sealed cell. The design of a deuteration cell used in our laboratory is shown in Fig. 3-28. The sample is deuterated by bubbling a stream of dry nitrogen through liquid D_2O prior to flushing the cell. The cell can also be conveniently used for drying or rehydrogenating the sample under investigation. A modified ATR sample holder constructed for deuteration experiments on fibers is shown in Fig. 3-25 [105].

3.2.3.2 Oriented Specimens

For the investigation of deformation processes in plastic materials orientation of polymer chains has to be achieved along one or more preferential axes. Furthermore, polarization measurements of polymer specimens exhibiting directional properties have often contributed to the correct assignment of absorption bands. Orientation in polymeric materials may be produced by mechanical drawing or rolling. Several

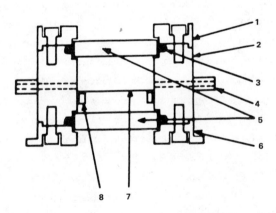

FIGURE 3-28   Deuteration cell for IR transmission spectroscopy:
(1) front cover, (2) cell body, (3) O-ring seal, (4) gas
inlet, (5) KBr windows, (6) back cover, (7) sample, and
(8) annular sampler holder.

thermoplastics can be oriented at room temperature as a result of
cold drawing, usually with a concomitant uniaxial orientation of the
molecular chains. Sometimes the application of higher temperatures
or swelling of the sample with a plasticizing agent can assist the
drawing operation [164, 165]. To avoid troublesome handling of the
oriented specimen and for elastomers with a reversible elastic deformation behavior a combination of a stretching jig with a sample
holder is highly recommended. Such a combination of a stretching unit
with a variable temperature cell for IR and Raman spectroscopic investigations are shown in the Figs. 3-30 and 3-31 [166, 167].

Quite recently more sophisticated devices for the simultaneous
measurement of FTIR spectra and stress-strain diagrams during deformation, recovery or relaxation processes have been reported [26, 36,
168-170]. In such a film stretching machine the polymer film under
examination can be uniaxially drawn while mounted in the sample com-
partment of the FTIR spectrometer (Fig. 3-29). The specimen to be
tested is held between two clamps which are movable by means of a
spindle drive with variable velocity. The clamps are attached to displacement and force transducers and two voltages proportional to the
two mechanical quantities - displacement and force - are recorded and
the stress-strain diagram of the sample is obtained taking into account the initial cross section of the sample under investigation.
For orientation measurements the polarization direction of the incident radiation can be alternately adjusted parallel or perpendicular
to the stretching direction by a pneumatically rotatable polarizer
unit. The change in polarization direction is automatically operated
by the FTIR system after a preselected number of interferogram scans
(see also Fig. 3-10 in Sec. 3.1.1.2). Specific values of the dichroic
ratio $R = A_{\parallel}/A_{\perp}$ of any absorption band at small strain intervals may
then be obtained by relating the mean absorbance value of the two
subsequent polarization spectra to the absorbance value of the corresponding perpendicular polarization spectrum and vice versa. The
exact position of the illuminated sample area can be monitored with
the built-in laser system. Installation of a heating device further
offers the possibility to study the deformation mechanism as a func-

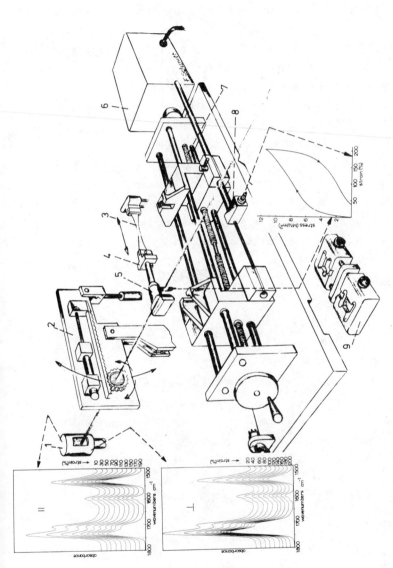

FIGURE 3-29   Film stretching machine: (1) detector, (2) pneumatically rotatable polarizer unit, (3) sample, (4) clamp, (5) force transducer, (6) driving motor, (7) threaded spindle, (8) displacement transducer, and (9) specimen preparation device. (The authors gratefully acknowledge the assistance of F. Schmitt, Bayer AG, Leverkusen, West Germany, in the design and construction of this stretching machine.)

tion of temperature. To illustrate this technique polarization spectra of the $\nu(C=O)$ and $\delta(NH)$ vibration region of a polyesterurethane recorded during elongation in a loading-unloading cycle have been inserted in Fig. 3-29 along with the corresponding stress-strain diagram. Further improvements of this technique have been obtained by exploiting the data processing capability of the dedicated computer for the evaluation and representation of the FTIR data [36, 169].

On the basis of such investigations the mechanical behavior of the material under examination may be related to the structural deformation on a molecular level (see Sec. 4.3.4.5).

Biaxial orientation is effected by successive drawing in perpendicular directions [171], bulging of circular shaped specimens [172], or mechanical rolling [173]. The operation of rolling of thin films required for IR spectroscopy is simplified if the film is laminated on a deformable IR transparent matrix, for example, AgCl [174].

3.2.3.3 Special Cells

For the measurement of IR spectra at low and high temperatures, it must be taken into account that radiation emitted by cells, sample, and detector [175] can lead to incomparable spectra with shifts in the absorption background. To overcome these problems either a spectrometer with a double-chopper system or two completely identical (in terms of geometry and temperature) cooling or heating cells have to be used. In the second case, only the effect of sample emission will not be eliminated, but for many studies this method is sufficient and far more economic. In the literature various types of cooling or heating cells built for special applications are reported [14, 118, 167, 168, 175-178]. To obtain a well-defined thermal pretreatment, a cell for annealing, cooling, or heating a polymer within the cell with an adjustable cooling or heating rate between 80 and 650 K has been developed [166] (Fig. 3-30). To avoid asymmetry in the sample and reference beam either two identical cells or one cell with comparable optical paths can be used. The latter consists of two parts: the cell body with cooling chamber, heater, interchangeble

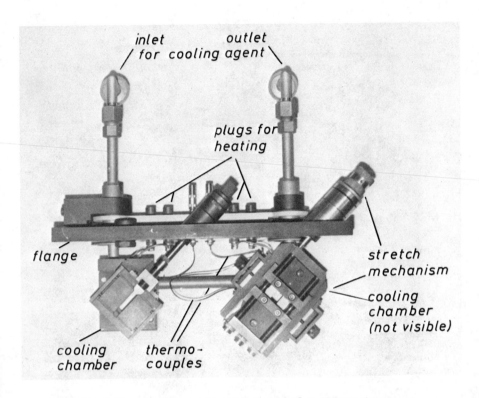

FIGURE 3-30 Cooling and heating cell for IR spectroscopy.

sample holder, inlet pipe for cooling agent, and feedwire for current
and thermocouple. A built-in stretching unit enables the drawing of
samples at different temperatures within the cell.

For Raman spectroscopy, which generally uses the single-beam
technique, only one cell is required [167, 179, 180] (Fig. 3-31). Besides an alignment mechanism for the X, Y, and Z directions, the most
important parts are the cell body and the flange with cooling chamber,
heater, etc. (Fig. 3-31). For orientation measurements a suitable
sample holder allows one to stretch the sample at the desired temperatures within the cell. After fixing the sample to the sample holder
the cell body is moved to the flange and sealed. To eliminate condensation water or ice, the cells can be evacuated or flushed with inert

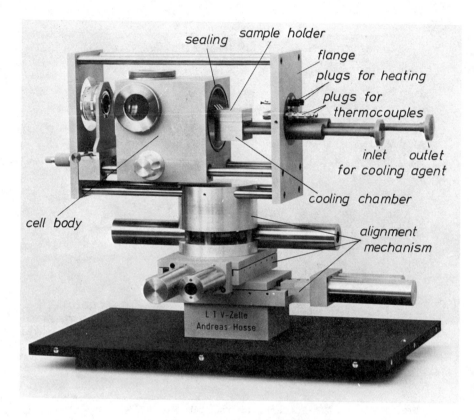

FIGURE 3-31   Cooling and heating cell for Raman spectroscopy

gas. Cooled methanol from a cryomat, liquid nitrogen, cooled gaseous
nitrogen, or a cryocooler can be used for cooling the sample. Since
the cooling agent can flow through the cooling chamber, a cycle may
be used and long-time studies with polymers are possible. For examination above room temperature, only the heater is used. In comparing
the thermoelectromotive force of the two thermocouples with an adjustable set-point, the temperature of the sample can be controlled. Different programs for the set-point adjustment enable a well-defined
heating, cooling, or annealing process to occur. During annealing
operations, the variation in temperature is less than ±0.1 K over the
entire temperature range.

REFERENCES

1. R. P. Baumann, *Absorption Spectroscopy*, Wiley, New York, 1962.

2. W. J. Potts, *Chemical Infrared Spectroscopy*, *vol. 1, Techniques*, Wiley, New York 1962.

3. L. W. Herscher, in *Applied Spectroscopy*, ed. D. N. Kendall, Reinhold, New York 1963, p. 88.

4. W. Brügel, *Einführung in die Ultrarotspektroskopie*, Steinkopff, Darmstadt 1969.

5. D. P. Thackeray, in *Laboratory Methods in Infrared Spectroscopy*, eds. P. G. J. Miller and B. C. Stace, Heyden, London 1972, p. 1.

6. H. Volkmann, in *Handbuch der Infrarot-Spektroskopie*, ed. H. Volkmann, Verlag Chemie, Weinheim 1972, p. 23.

7. K. Kiss-Eröss, *Comprehensive Analytical Chemistry VI*, Elsevier, Amsterdam 1976.

8. P. R. Griffiths, *Chemical Infrared Fourier Transform Spectroscopy*, Wiley, Interscience, New York 1975.

9. R. Geick, in *Topics in Current Chemistry*, Springer, Berlin, 58: 73 (1975).

10. R. J. Bell, *Introductory Fourier Transform Spectroscopy*, Academic Press, New York, 1975.

11. P. Grosse, Beckman Report 1:3 (1970).

12. R. C. Milward, Polytec Technical Bulletin 1 (1972).

13. S. T. Dunn, C. T. Foskett, R. Curbelo, and P.R. Griffith, in *Computers in Chemical Research*, eds. C. E. Klopfenstein and C.L. Wilkins, Academic Press, New York 1972

14. J. Derkosch, *Absorptionsspektralanalyse im ultravioletten, sichtbaren und infraroten Gebiet*, Akad. Verlagsges., Frankfurt 1967.

15. E. Knözinger, Angew. Chem. 88:1 (1976).

16. J. Brandmüller and H. Moser, *Einführung in die Ramanspektroskopie*, Steinkopff, Darmstadt 1962.

17. J. R. Ferraro, in *Raman Spectroscopy*, *vol. 1*, ed. H. A. Szymanski, Plenum Press, New York 1967, p. 44.

18. H. W. Schrötter, in *Raman Spectroscopy*, *vol. 2*, ed. H. A. Szymanski, Plenum Press, New York 1970, p. 90.

19. T. R. Gilson and P. J. Hendra, *Laser-Raman Spectroscopy*, Wiley, New York 1970.

20. M. Tobin, *Laser-Raman Spectroscopy*, Wiley, New York 1971.

21. P. J. Hendra, in *Laboratory Methods in Infrared Spectroscopy*, eds. P. G. J. Miller and B. C. Stace, Heyden, London 1972, p. 230.

22. R. Zbinden, *Infrared Spectroscopy of High Polymers*, Academic Press, New York 1964.

23. P. Krömer, Analysentechnische Berichte, Perkin Elmer 14 (1968).

24. Spectroscopy Datasheet, D-459, Instrument Division, Perkin Elmer Corp., Norwalk, Conn.

25. R. S. Stein, J. Appl. Polym. Sci. 5:96 (1961).

26. B. E. Read, D. A. Hughes, D. C. Barnes, and F. W. Drury, Polymer 13:485 (1972).

27. A. A. Michelson, Phil. Mag. 31:256 (1891).

28. R. Geick, Fresenius Z. Anal. Chem. 288:1 (1977).

29. L. Genzel, Z. Anal Chem. 273:391 (1975).

30. D. Ziessow, *On-line Rechner in der Chemie - Grundlagen und Anwendungen in der Fourierspektroskopie*, Walter de Gruyter, Berlin 1972.

31. J. R. Ferraro and L. J. Basile, *Fourier Transform Infrared Spectroscopy - Applications to Chemical Systems, vol. 1*, Academic Press, New York 1978.

32. J. W. Cooley and J. W. Tukey, Math. Comput. 19:297 (1965).

33. P. B. Fellgett, J. Phys. Radium 19, 187:237 (1958).

34. P. Jacquinot, Rep. Progr. Phys. 13:267 (1960).

35. P. R. Griffith, H. J. Sloane, and R. W. Hannah, Appl. Spectrosc. 31:485 (1977).

36. K. Holland-Moritz, W. Stach, and I. Holland-Moritz, J. Mol. Spectrosc. (in press).

37. R. Pressley, *Handbook of Lasers*, CRC Press, Cleveland 1971.

38. W. Demtröder, *Grundlagen und Techniken der Laserspektroskopie*, Springer, Berlin 1977.

39. H. Walther (ed.), in *Topics in Applied Physics, vol.2*, Springer, Berlin 1975.

40. U. Köpf, *Laser in der Chemie*, Salle und Sauerländer, Frankfurt 1979.

41. D. Landon and S. Porto, Appl. Optics 4:762 (1965).

42. Jarrell-Ash, Spectrum Scanner 22:5 (1967).

43. A. Weber, The Spex Speaker 11 (1966).

44. D. O. Landon and P. R. Reed, The Spex Speaker 17 (1972).

45. P. J. Hendra and C. J. Vear, Analyst 95:321 (1970).

46. M. Delhaye, Appl. Optics 7:2195 (1968).

47. J. M. Beny, B. Sombret, F. Wallart, and M. Leclercq, J. Mol. Struct. 45:349 (1978).

48. M. Bridoux, A. Deffontaine, M. Delhaye, F. Grase, and C. Reiss, in *Proc. 5th Int. Conf. on Raman Spectroscopy*, Freiburg 1976, p. 760.

49. P. Dhamelincourt, F. Wallart, M. Leclerq, A. T. N'Guyen, and D. O. Landon, Anal. Chem. 51, 3:414A (1979).

50. M. Delhaye, in *Spectroscopy in Chemistry and Physics - Modern Trends*, eds. F. Comes, A. Müller, and W. J. Orville Thomas, Elsevier, Amsterdam 1980.

51. W. Kiefer and H. J. Bernstein, Appl. Spectrosc. 25:60 (1971).

52. W. Kiefer, W. J. Schmid, and J. A. Topp, Appl. Spectrosc. 29: 434 (1975).

53. H. J. Sloane, Appl. Spectrosc. 27:217 (1973).

54. H. J. Sloane, Appl. Spectrosc. 25:430 (1971).

55. J. L. Koenig, Appl. Spectrosc. Rev. 4, 2:233 (1971).

56. Laser-Raman Notes, The Spex Speaker 15 (1970).

57. J. R. Allkins, The Spex Speaker 14 (1969).

58. P. J. Hendra, in *Polymer Spectroscopy*, ed. D. O. Hummel, Verlag Chemie, Weinheim 1974, p. 151.

59. D. O. Hummel and F. Scholl, *Atlas der Kunststoffanalyse, vol. I/2*, Carl Hanser Verlag, München 1968.

60. J. E. Stanfield, D. E. Sheppard, and H. S. Harrison, in *Laboratory Methods in Infrared Spectroscopy*, ed. R. G. J. Miller, Heyden, London 1965, p. 31.

61. M. Tyron and E. Horowitz, in *Analytical Chemistry of Polymers, vol. 2*, ed. G. M. Kline, Wiley, Interscience, New York 1962, p. 302.

62. H. A. Szymanski, K. Broda, J. May, W. Collins, and D. Bakalik, Anal. Chem. 37:617 (1965).

63. I. Lindquist, Acta Chem. Scand. 9:73 (1955).

64. J. L. Koenig, J. Polym. Sci., Part D59 (1972).

65. H. W. Siesler, Colloid Polym. Sci. 255:321 (1977).

66. R. Schmolke, Faserforsch. Textiltechn. 16:514 (1965).

67. R. Schmolke, H. Herma, and V. Gröbe, Faserforsch. Textiltechn. 16:589 (1965).

68. P. Arnold and H. A. Willis, in *Polymer Science*, ed. A. D. Jenkins, North Holland, Amsterdam 1972, p. 1587.

69. J. C. Henniker, *Infrared Spectroscopy of Industrial Polymers*, Academic Press, London 1967, p. 60.

70. H. Röpke and W. Neudert, Z. Anal. Chem. 170:78 (1959).

71. G. Duyckaerts, Analyst, 84:201 (1959).

72. M. M. Stimson, in *Progress in Infrared Spectroscopy, vol. 1*, ed. H. A. Szymanski, Plenum Press, New York 1962, p. 143.

73. D. O. Hummel and F. Scholl, *Atlas der Kunststoffanalyse*, *vol. I/1*, Carl Hanser Verlag, München 1968, p. 62.

74. C. C. Fagot, Appl. Spectrosc. 19:30 (1965).

75. J. A. Coakley and H. H. Berry, Appl. Spectrosc. 20:418 (1966).

76. H. Zimmer, Melliand Textilber. 50:1141 (1969).

77. D. Minkwitz and H. Zimmer, Melliand Textilber. 11:909 (1974).

78. R. T. O'Connor, E. F. Du Pre, and E. R. McCall, Anal. Chem. 29: 998 (1957).

79. H. G. Higgins, Austral. J. Chem. 9:496 (1957).

80. F. H. Forziati, W. K. Stone, J. W. Rowden, and W. D. Appell, J. Res. Nat. Bur. Stand. 45:109 (1950).

81. M. K. Wharton and F. M. Forziati, American Dyestuff Reporter 50:515 (1961).

82. M. R. Harvey, J. E. Stuart, and B. G. Achhammer, J. Res. Nat. Bur. Stand. 56:225 (1956).

83. B. Cleverly and R. Herrman, J. Appl. Chem. 11:344 (1961).

84. H. Spedding, Unicam Spectrovision 8:7 (1960).

85. F. H. Forziati, J. W. Rowden, and E. K. Plyler, J. Res. Nat. Bur. Stand. 46:288 (1951).

86. A. Crook and P. J. Taylor, Chem. Ind. 1958:95.

87. G. Caroti and J. H. Dusenbury, Nature 178:162 (1956).

88. R. G. Quynn and R. Steele, Nature 173:1240 (1954).

89. N. V. Michajlov and M. V. Shablygin, Vysokomol. Soedin. 4:1155 (1962).

90. C. G. Cannon, Chem. Ind. 1957:29.

91. U. A. Schwair, Thesis No. 197, Technical University, Stuttgart 1966.

92. V. J. Coates, A. Offner, and E. H. Siegler, J. Opt. Soc. Amer. 43:984 (1953).

93. P. Bouriot and A. Parisot, J. Polym. Sci. C16:1393 (1967).

94. E. R. Blout, G. R. Bird, and D. S. Grey, J. Opt. Soc. Amer. 40: 306 (1950).

95. A. Elliott, E. J. Ambrose, and R. B. Temple, J. Sci. Instrum. 27:21 (1950).

96. R. D. B. Fraser, J. Chem. Phys. 21:1511 (1953).

97. R. D. B. Fraser, J. Opt. Soc. Amer. 48:1017 (1958).

98. R. D. B. Fraser, J. Chem. Phys. 28:1120 (1958).

99. R. G. J. Miller, Chem. Ind. 1957:190.

100. K. H. Bassett, C. Y. Liang, and R. H. Marchessault, J. Polym. Sci. A1:1687 (1963).

101. A. Elliott, in *Symposium on Techniques in Polymer Science*, Royal Institut of Chemistry, London 1956, p. 36.

102. P. Holliday, Nature 163:602 (1949).

103. Y. Nishimo, Analyst (Jap.) 4:174 (1955).

104. H. Sobue and S. Fukuhara, J. Chem. Soc. Jap. 60:86 (1957).

105. F. Grass, H. Siesler, and H. Krässig, Melliand Textilber. 52: 1001 (1971).

106. G. A. Tirpak and J. P. Sibilia, J. Appl. Polym. Sci. 17:643 (1973).

107. D. J. Carlson, F. R. S. Clark, and D. M. Wiles, Text. Res. J. 46:318 (1976).

108. B. C. Stace, in *Laboratory Methods in Infrared Spectroscopy*, eds. R. G. J. Miller and B. C. Stace, Heyden, London 1972.

109. R. J. E. Cumberbirch and H. Spedding, J. Appl. Chem. 12:83 (1962).

110. J. K. Smith, W. J. Kitchen, and D. B. Mutton, J. Polym. Sci. C2:499 (1963).

111. R. G. Zbankov, *Infrared Spectra of Cellulose and Its Derivatives*, Consultants Bureau, New York, 1966.

112. S. Burgess and H. Spedding, Chem. Ind. 1961: 1166

113. J. A. Knight, M. P. Smoack, R. A. Porter, and W. E. Kirkland, Text. Res. J. 37:924 (1967).

114. C. Ruscher and R. Schmolke, Faserforsch. Textiltechn. 11:383 (1960).

115. R. G. Zbankov, Zavod. Lab. 29:1438 (1963).

116. J. H. Rau, Melliand Textilber. 44:1098 (1963).

117. D. O. Hummel, Melliand Textilchem. 1:139 (1965).

118. J. Dechant, *Ultrarotspektroskopische Untersuchungen an Polymeren*, Akademie Verlag, Berlin 1972.

119. N. Iwanow and R. Schneider, Bull. Inst. Textile Fr. 55 (1958).

120. H. Tadokoro, S. Seki, I. Nitta, and R. Yamadera, J. Polym. Sci. 28:244 (1958).

121. E. R. Blout and M. J. Abbate, J. Opt. Soc. Amer. 45:1028 (1955).

122. G. H. Haggis, J. Sci. Instrum. 33:491 (1956).

123. M. A. Ford, W. E. Seeds, and G. R. Wilkinson, J. Opt. Soc. Amer. 48:249 (1958).

124. V. I. Vettegren and I. I. Novak, Opt. Spektrosk. 14:545 (1963).

144   Experimental Techniques

125. J. Fahrenfort, Spectrochim. Acta 17:698 (1961).

126. J. Fahrenfort, Spectrochim. Acta 18:1103 (1962).

127. N. J. Harrick, J. Chem. Phys. Solids 8:106 (1959).

128. N. J. Harrick, Ann. N. Y. Acad. Sci. 101:928 (1963).

129. P. A. Wilks and T. Hirschfeld, Appl. Spectrosc. Rev. 1:99 (1967).

130. G. Jayme and G. Traser, Angew. Makromol. Chem. 21:87 (1972).

131. R. L. Harris and G. R. Svoboda, Anal. Chem. 34:1655 (1962).

132. S. E. Polchlopek, in *Applied Infrared Spectroscopy*, ed. D. N. Kendall, Reinhold, New York 1966, p. 462.

133. K. H. Reichert, Farbe und Lack, 72:13 (1966).

134. D. O. Hummel, H. Siesler, E. Zoschke, I. Vierling, U. Morlock, and T. Stadtländer, Melliand Textilber. 12:134 (1973).

135. N. J. Harrick, *Internal Reflection Spectroscopy*, Interscience, New York 1967.

136. K. Gottlieb, Z. Instrumentenkunde 75:125 (1967).

137. N. J. Harrick and F. K. Du Pre, Appl. Opt. 5:1739 (1966).

138. N. J. Harrick and A. I. Carlson, Appl. Opt. 10:19 (1971).

139. J. Dechant, Faserforsch. Textiltechn. 25:24 (1974).

140. P. A. Flournoy and W. J. Schaffers, Spectrochim. Acta 22:5 (1966).

141. G. Heidemann, Chemiefasern 20:204 (1970).

142. F. Druschke, H. W. Siesler, G. Spilgies, and H. Tengler, Polym. Eng. Sci. 17:93 (1977).

143. H. W. Siesler, H. Krässig, F. Grass, K. Kratzl, and J. Derkosch, Angew. Makromol. Chem. 42:139 (1975).

144. D. M. Wiles and D. J. Carllson, Macromolecules 4:179 (1971).

145. M. G. Chan and W. L. Hawkins, Polym. Prep. Am. Chem. Soc. Div. Polym. Chem. 9:1638 (1968).

146. P. Blais, D. J. Carllson, and D. M. Wiles, J. Polym. Sci. A1 10:1077 (1972).

147. H. Wagner and L. Wuckel, Plaste Kautsch. 18:426 (1971).

148. R. G. Greenler, R. R. Rahn, and J. P. Schwartz, J. Catalysis 23:42 (1971).

149. H. G. Tompkins, in *Methods of Surface Analysis*, ed. A. W. Czanderna, Elsevier, Amsterdam 1975.

150. H. G. Tompkins, Appl. Spectrosc. 30:377 (1976).

151. M. Born and E. Wolf, in *Principles of Optics*, Pergamon, New York 1969, p. 279.

152. R. G. Greenler, J. Chem. Phys. 44:310 (1960).

153. H. G. Tomkins and R. G. Greenler, Surf. Sci. 28:194 (1971).

154. R. G. Greenler, J. Chem. Phys. 50:1963 (1969).

155. R. J. Jakobsen, Amer. Chem. Soc. Div. Colloid Surface Chem.,
     April 1976, No. 103.

156. M. G. Chan and D. C. Allara, Polym. Eng. Sci. 14:12 (1974).

157. M. G. Chan and D. C. Allara, J. Colloid Interfac. Sci. 47:697
     (1974).

158. B. Schrader, W. Meier, E. Steigner, and F. Zöhrer, Z. Anal.
     Chem. 254:257 (1971).

159. S. K. Freeman and D. O. Landon, Anal. Chem. 41:398 (1969).

160. W. L. Peticolas, Biochimie 57:417 (1970).

161. P. J. Hendra, Adv. Polym. Sci. 6:151 (1969).

162. C. Y. Liang, in *Newer Methods of Polymer Characterization*, ed.
     Bacon Ke, Interscience, New York 1964, p. 33.

163. A. Koshimo, J. Appl. Polym. Sci. 9:55 (1965).

164. W. J. Dulmage and A. L. Geddes, J. Polym. Sci. 31:499 (1958).

165. W. Haule, H. Kleinpoppren, and A. Scharmann, Z. Naturforsch.
     13a:64 (1958).

166. K. Holland-Moritz and I. Modrič, Progr. Colloid Polymer Sci.
     57:212 (1975).

167. K. Holland-Moritz, European Spectroscopy News 1:4 (1976).

168. K. Holland-Moritz, W. Stach, and I. Holland-Moritz, Progr.
     Colloid Polym. Sci. 67 (1980) (in press).

169. H. W. Siesler, in *Spectroscopy in Chemistry and Physics -
     Modern Trends*, eds. F. Comes, A. Müller, and W. J. Orville
     Thomas, Elsevier, Amsterdam 1980.

170. R. P. Wool and W. O. Statton, J. Polym. Sci. 12:1575 (1974).

171. J. Dechant and C. Ruscher, Faserforsch. Textiltechn. 16:180 (1965).

172. A. J. De Vries and C. Bonnebat, Polym. Eng. Sci. 16:93 (1976).

173. A. Elliott, E. Ambrose, and R. Temple, Nature 163:567 (1949).

174. E. Ambrose, A. Elliot, and R. Temple, Proc. Roy. Soc. A199:183
     (1949).

175. J. Stokr, Z. Ruzicka, and S. Ekwal, Appl. Spectrosc. 28:479 (1974).

176. N. F. J. Brockmeier, J. Appl. Polym. Sci. 12:2129 (1968).

177. M. J. Hannon and J. L. Koenig, J. Polym. Sci. A2, 7:1085 (1969).

178. R. Danz and J. Dechant, Faserforsch. Textiltechn. 23:199 (1972).

179. F. A. Miller and B. M. Harney, Appl. Spectrosc. 24:291 (1970).

180. J. Stokr and B. Schneider, Appl. Spectrosc. 24:461 (1970).

Chapter 4

APPLIED SPECTROSCOPY

4.1 QUANTITATIVE ANALYSIS

In many applications of vibrational spectroscopy the analyst is interested not only in the qualitative chemical nature of the investigated sample but also in the quantitative composition (e.g., degree
of purity, amount of additives, copolymer composition). However,
quantitative absorption spectroscopy is not limited to purely analytical aspects but is also widely applied to basic problems of
polymeric structure: state of order, configurational and conformational regularity, sequence distribution, orientation measurements,
etc.. Owing to the dependence of IR and Raman intensities on the
change of dipole moment and polarizability of the vibrating molecules, respectively, additional structural information (calculation of bond moments and polarizabilities, correlation of the intensities of characteristic group frequencies with structural parameters) may be derived from absolute intensity data [1].

4.1.1 Infrared Spectroscopy

Detailed accounts of quantitative IR spectroscopy are available in
the literature [2-7], and in what follows an attempt is made to provide a summary of theoretical and practical aspects.

Light incident on a system may suffer transmission, reflection,
scattering, and absorption. For the portion of radiation effectively entering the sample, the Lambert-Bouguer law of absorption holds
under the assumption of a homogeneous medium and monochromatic

146

radiation:

$$T = \frac{\Phi}{\Phi_o} = 10^{-\alpha b} \tag{4.1}$$

or

$$A = \log\frac{\Phi_o}{\Phi} = \alpha b \tag{4.2}$$

where T = transmittance

A = absorbance

Φ_o = intensity of radiation effectively entering the sample

Φ = intensity of radiation after passing through the sample

α = absorption coefficient (cm^{-1})

b = path length of radiation within the sample (cm)

The absorption coefficient α is related to the absorption index κ by the equation:

$$\alpha = \frac{4\pi n \kappa}{\lambda} \tag{4.3}$$

where n is the refractive index of the sample at wavelength λ. Beer could show that the absorption coefficient α is proportional to the concentration of the absorbing medium:

$$\alpha = ac \tag{4.4}$$

When c is expressed in moles per dm^3 then a is called the *molar absorptivity* $(cm^2/mmol)$.

Thus, the combined law of Lambert-Bouguer-Beer, the basis of quantitative absorption spectroscopy reads

$$A = \log\frac{\Phi_o}{\Phi} = acb \tag{4.5}$$

Deviations from strict linearity between absorbance A and concentration c may be

1. Sample specific: e.g., association or dissociation phenomena in in the system under investigation

2. Due to instrumental inadequacies: lack of monochromatic radiation, stray light, etc.

3. Related to the technique of sample preparation: nonuniform dis-
    tribution of absorbing material

Experimentally measured radiation intensities I_o and I deviate
from the theoretical values Φ_o and Φ. These differences are mainly
due to reflection and scattering losses, stray light, and finite
spectral slit width of the spectrometer [8, 9]. The errors associated
with the cell optics (reflection and interference phenomena) and
their suppression and correction have been discussed by various
authors [10-12].

Radiation losses owing to scattering depend mainly on the size
of the scattering particles in relation to the radiation wavelength
and may occur in solid polymer samples containing additives, such as
fillers, pigments, etc. On the other hand, short-wavelength stray
light may increase the observed transmittance by superposition with
radiation of the analytical wavelength. A correction procedure to

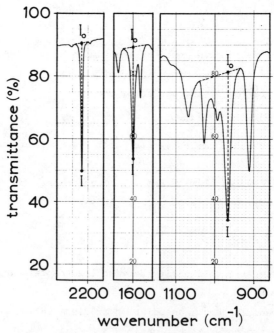

FIGURE 4-1   Various base-line constructions for the compositional
             analysis of an acrylonitrile-butadiene-styrene copolymer.

account for this error has been described by Brügel [3].

An important instrumental factor for correct quantitative analysis is the ratio of the spectral slit width and the half width of the analytical absorption band, which should be smaller than 0.2 to keep the maximum error of the absorbance measured at the peak maximum A_{max} in a 3% limit [13, 14]. Notwithstanding these aberrations from Beer's law, empirical relationships between absorbance and concentration may serve as valuable reference for special analytical problems.

In practical quantitative analysis the absorbance A_{max} determined at the peak maximum, applying an appropriate base line (Fig. 4-1) [15-17] is generally adequate. For more accurate purposes the integrated absorbance $\int \log(I_o/I)_{\bar{\nu}} \, d\bar{\nu}$, the area under the analytical absorption band recorded linearly in absorbance versus wavenumber $\bar{\nu}$, should be given preference. This quantity is less sensitive to errors due to radiation scattering and finite slit width of the spectrometer. The primary advantage of this intensity parameter, however, is its sensitivity towards alterations in the form and shape of absorption bands as a consequence of changes in the intermolecular forces or state of order (see, for example, Figs. 4-83 and 4-85). The value of the integrated absorbance $\int \log(I_o/I)_{\bar{\nu}} \, d\bar{\nu}$ can be determined by graphical integration or weighing of the area obtained by cutting along the curves, drawn on paper of uniform thickness. Most spectrometers record linearly in percent transmittance %T ($T = I/I_o$) versus wavenumber $\bar{\nu}$ (cm^{-1}), and for quantitative evaluation the spectra have to be converted to absorbance [18]. In modern instruments the procedures of transformation from percent transmittance to absorbance and integration between operator-selected wavenumber limits can be performed automatically. Broad absorption bands usually overlap with their neighbors and a weak band will thus often appear as a shoulder on a stronger band. In such a case reliable results as to wavenumber position and absorbance can only be derived from the separated band complex. This can be achieved either graphically with the aid of a curve resolver [19] or by the use of a computer [20] (see Fig. 4-2).

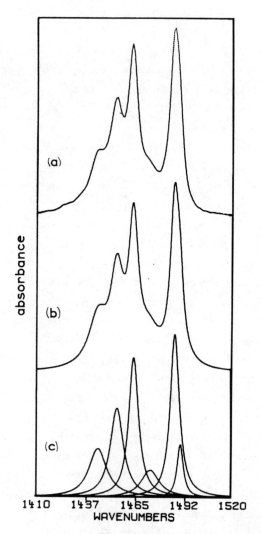

FIGURE 4-2 Computer separation of absorption band complex: (a) experimental absorption profile, (b) calculated absorption profile, (c) calculated single components.

The absolute integrated absorption intensity of a band is defined as

$$A_{abs} = \int_{-\infty}^{+\infty} a_{\bar{\nu}} \, d\bar{\nu} = \frac{1}{cb} \int_{-\infty}^{+\infty} \log\left(\frac{\Phi_o}{\Phi}\right)_{\bar{\nu}} d\bar{\nu} \qquad (4.6)$$

On the basis of a triangular energy distribution of the slit system and a Lorentzian absorbance curve, the calculation of A_{abs} from the experimental value of $\log(I_o/I)_{\bar{\nu}_o}$ (A_{max}), the observed half width of the absorption band $\Delta\bar{\nu}_{1/2}$, and the spectral slit width s has been discussed by Ramsay [13]. Furthermore, A_{abs} may be determined by extrapolation of the measured integral absorption B

$$B = \frac{1}{cb} \int \log\left(\frac{I_o}{I}\right)_{\bar{\nu}} d\bar{\nu} \qquad (4.7)$$

toward cb = 0 or A_{max} = 0 [4, 21, 22].

For the general case of multicomponent systems where all components absorb at the analytical wavenumbers of the other components, Beer's law reads

$$A(\bar{\nu}_i) = a_1(\bar{\nu}_i)c_1 b + \ldots + a_i(\bar{\nu}_i)c_i b + \ldots + a_n(\bar{\nu}_i)c_n b \qquad (4.8)$$

Here $a_1(\bar{\nu}_i)$ is the absorptivity of component 1 at wavenumber $\bar{\nu}_i$. The resulting system of n linear equations can only be solved when the absorptivities of all components are known from studies on pure standards. In case the absorptivities cannot be determined from measurements on pure standards the system can still be analyzed if an isolated analytical band for each component can be found and at least the same number of mixtures of linear independent quantitative compositions are available as components have to be determined [23, 24].

The difficulty to obtain accurate values for the thickness of solid samples (polymer films, KBr pellets) necessitates the use of comparative methods. The absorbance A of the analytical band is compared to the absorbance A' of an internal standard which is either present or has been added to the sample in known concentration:

$$\frac{A}{A'} = \frac{acb}{a'c'b} = kc \qquad\qquad (4.9)$$

In analogy the composition of a binary system can be determined from the ratio of isolated absorption bands characteristic of each component {for example, $\nu(C\equiv N)$ and $\nu(C=O)$ stretching vibrations in copolymers of acrylonitrile and vinylacetate or methylacrylate [25]} :

$$\frac{A_1}{A_2} = \frac{a_1 c_1 b}{a_2 c_2 b} = k\frac{c_1}{c_2} \qquad\qquad (4.10)$$

Empirical calibration of the absorbance ratios with standards of known composition is the most generally applied method in quantitative IR spectroscopy of polymers. However, care has to be taken

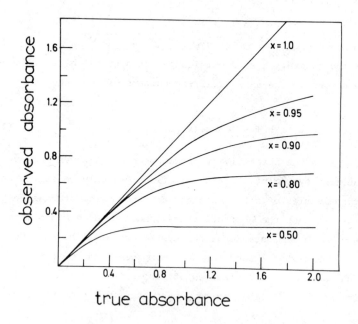

FIGURE 4-3   The effect of nonuniform distribution of absorbing material on the observed absorbance: x, fraction of incident radiation passing through the absorbing material. (Reprinted with permission from Ref. 26, Copyright by the American Chemical Society.)

in the choice of analytical and reference bands because their intensity should be independent of molecular order phenomena (e.g., crystallinity, sequence distribution in copolymers).

The effect of inhomogeneous sample distribution (e.g., gaps in
fiber grids or holes in polymer films) on the measured absorbance
has been studied by Jones [26]. The results are shown in Fig. 4-3
where the value of 1 - x would represent the fraction of incident
radiation passing through clear spaces of the absorbing material.
The errors incurred when oriented polymer samples are investigated
with unpolarized light have been discussed by Stace [27]. A thorough
account on the accuracy of absorption measurements has been given
by Martin [28]. Principally, it can be shown that the relative error
of absorbance is minimal if concentration and sample thickness are
chosen so that the observed percent transmittance %T lies in the
range from 60 to 20% (absorbance 0.2 to 0.7) (Fig. 4-4).

Numerous references to applications of quantitative IR spectroscopy in the polymer field have been assembled in various books
[29-33].

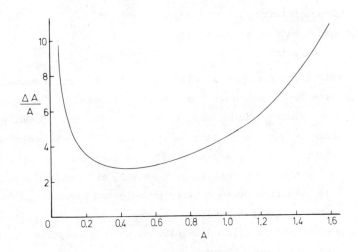

FIGURE 4-4   Relative error ΔA/A at constant error of T = 1% as a
function of absorbance.

4.1.2 Raman Spectroscopy

In Raman spectroscopy the relation between signal intensity and concentration is linear rather than logarithmic as in IR spectroscopy.
However, despite several accounts on the application of Raman scattering to quantitative analysis [34-42], Raman spectroscopy has not
yet gained a comparable widespread acceptance as an analytical tool.
This is largely a consequence of the problems encountered with intensity measurements in a single-beam emission technique. Thus, the intensity of the observed Raman signals depends on a number of sample
properties:

1. Refractive index
2. Molecular environment
3. Fluorescence
4. Color

and instrumental factors [40, 43]:

1. Intensity of Raman source
2. Polarization geometry of exciting radiation
3. Properties of the cell
4. Spectral sensitivity of the spectrometer
5. Background emission and straylight
6. Detector sensitivity

A general scheme for quantitative analysis has been developed
[43] which eliminates to a large degree the influence of almost all
(except sample color and fluorescence) factors mentioned above. However, measurements have to be made in sufficiently dilute solutions
and a Raman band of the solvent is used as an internal standard.
Background corrections are applied by measuring the band intensities
above a linear base line drawn between preselected wavenumbers on
each side of the analytical bands.

Local overheating of colored samples owing to absorption can be
obviated by relative motion between the sample and the focused laser
beam. In the rotating-sample techniques, the samples are rotated at
up to 50 revolutions/s [44, 45], and much better signal-to-noise

ratios are obtained than from static samples. On the other hand, the laser beam, while remaining focused on the sample, can be rapidly scanned over its surface in either a linear [46] or circular [42] manner. This can be accomplished with a rotating reflector plate or a rotating lens, respectively.

Alternatively to the internal standard method where a solvent band or the band of an added standard is used for obtaining quantitative relative intensities [47, 48], the Raman spectra of the sample and the standard can be measured separately (external standard) by a cell replacement technique [49, 50]. Recently, a new technique has been described for obtaining quantitative relative intensity measurements with a rotating cylindrical cell with separate compartments for the sample and reference [42, 44, 51, 52]. The quantitative determination of component ratios in solid polymeric samples (e.g., copolymer analysis) is based on the relative band ratio method. With Raman lines which are characteristic of the individual components and which are chosen to be as nearly free from interference as possible, the actual relative composition can be determined from calibration measurements on standard samples of known composition [53, 54]. In analogy to IR spectroscopy the intensity of a Raman line can be defined and measured by the maximum peak intensity i_o or the absolute integrated intensity I which is expressed mathematically by the integral

$$I = \int_{-\infty}^{+\infty} i(\bar{\nu}) \, d\bar{\nu} \qquad\qquad (4.11)$$

and eliminates the light-distributing influences of the apparatus [1, 40].

4.2 IDENTIFICATION AND ANALYTICAL APPLICATIONS

Of all the physical techniques applied to the identification of polymeric materials IR spectroscopy is certainly the most widely used. However, maximum information about the composition of the system under examination will be gained only from appropriate combination with other chemical and/or physical techniques. Although it is not within

the scope of this book to give a detailed account of polymer analysis,
the discussion of some relevant problems may demonstrate the utility
of vibrational spectroscopy for identification purposes. A complete
analysis (in the sense of accounting for 100% of the material as
chemical compounds) will require a preliminary separation of any additives from the polymer prior to the determination of the single
components in the original material. Various analytical separation
procedures and schemes have been thoroughly treated in the literature
[55-61].

For the rapid qualitative identification of the most frequently
encountered polymers from their IR spectra upon separation of any
low-molecular-weight additives, systematic schemes have been proposed
[32, 58, 61]. Thus, the general nature of the polymer under examination may readily be derived from certain characteristic group frequencies, and in favorable cases reference to one of the published
collections of polymer spectra will finally reveal the exact chemical
structure of the polymer.

4.2.1 Selected Analytical Problems

Principally, additives such as plasticizers, antioxidants, emulsifiers,
fire retardant materials, etc., can be separated from the polymer
under investigation by extraction with suitable solvents or solution
and reprecipitation of the polymer. The isolation of additives in
commercial polychloroprene by the last-mentioned procedure is demonstrated in the IR spectra of Fig. 4-5(a) to (e). In the IR spectrum
of the original material [Fig. 4-5(a)] the absorption bands at 1690,
1490, 1270, and 1190 cm^{-1} which are superimposed on the spectrum of
polychloroprene indicate the presence of certain additives. These
absorption bands are no longer observable in the spectrum of Fig. 4-5
(b) upon repeated precipitation of the polymer from methylene chloride solution. The spectrum of the residue obtained from the combined
filtrates of the precipitation procedures is shown in Fig. 4-5(c).
Reference to spectra collections of additives commonly encountered
in polychloroprene reveals the spectrum to be primarily composed from

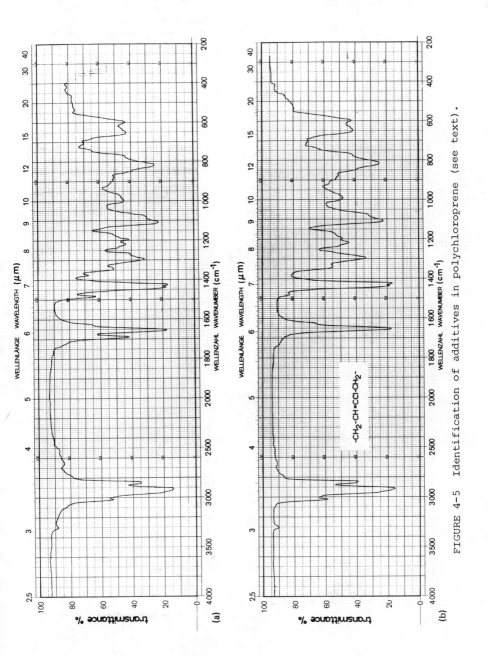

FIGURE 4-5 Identification of additives in polychloroprene (see text).

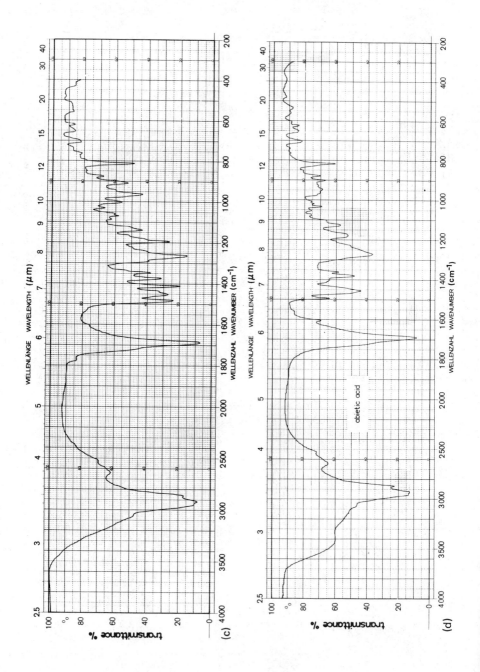

(c)

(d)

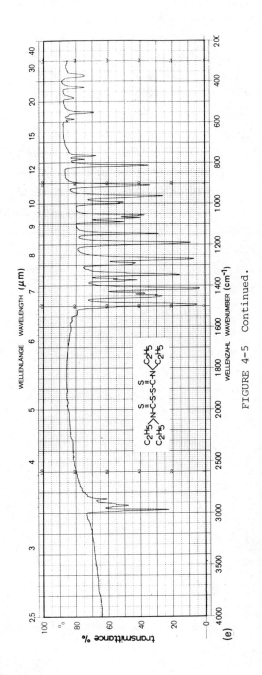

FIGURE 4-5 Continued.

the absorptions of abietic acid [Fig. 4-5(d)] and tetraethylthiuram-
disulfide [Fig. 4-5(e)].

In textile analysis the combination of electron microscopy and
vibrational spectroscopy alongside chemical separation procedures
have proved powerful tools for the qualitative and quantitative
identification of the components in fiber blends and fabrics. In
Fig. 4-6 the IR spectrum of an unknown fabric which had been previously extracted with a mixture of methanol and benzene is shown. The intense absorption bands at 1650 and 1530 cm^{-1} belong to characteristic
vibrations of the amide group for example in proteins. The absorption
bands at 1435, 1335, 1250 cm^{-1} and in the 600 to 700 cm^{-1} region can be
assigned to polyvinyl chloride. Additionally, broad absorption bands
in the 900 to 1000 cm^{-1} region and at about 450 cm^{-1} can be observed.
The scanning electron micrograph of the same sample (Fig. 4-7) demonstrates the fabric to be composed of three components:

1. Natural wool, detectable by its characteristic scale morphology
   (diameter about 20 μm)

2. Polymer filaments (diameter between 15 and 20 μm)

3. Glass fibers (diameter about 5 μm)

Thus, the IR spectrum in Fig. 4-6 is a superposition of the characteristic absorptions of the protein structure of wool, polyvinyl

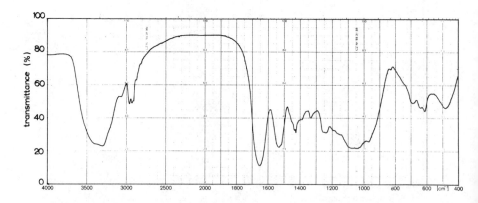

FIGURE 4-6   IR spectrum (KBr pellet) of unknown fabric (see text).

FIGURE 4-7   Scanning electron micrograph of unknown fabric (see text).

chloride, and the broad intense absorption bands in the spectrum of
silica.

Similarly the composition of the fiber fleece shown in the scanning electron micrograph of Fig. 4-8 could be determined by IR spectroscopic investigation of the material separated by differential
solution. Upon extraction of the adhesive formulation on polyacrylate
basis (observable as "sails" between the single fibers in Fig. 4-8),
three fiber components with different morphology could be distinguished by electron microscopy [Fig. 4-9(a) to (c)]. The IR spectrum
(on KBr pellet) of the extracted sample, shown in Fig. 4-10, indicates the presence of polyvinyl chloride (600-700 cm^{-1}), an aromatic
polyester (1730, 720 cm^{-1}) and possibly a cellulose derivative (broad

FIGURE 4-8 Scanning electron micrograph of original, unknown fiber
fleece (see text).

absorption at about 1050 cm^{-1}). The fiber components were separated
by treatment of the sample with tetrahydrofurane, hexafluoroisopropan-
ol, and sulfuric acid and identified as polyvinyl chloride [round
cross section, Fig. 4-9(a)], rayon [polylobal cross section, Fig.4-
9(b)], and polyethylene terephthalate [trilobal cross section, Fig.
4-9(c)], respectively.

Very often NMR spectroscopy will be helpful in establishing the
exact nature of a polymer which has been assigned to a certain class
of compounds by IR spectroscopy. In Fig. 4-11(a) to (c) the IR spectra
of three different aliphatic polyesters are shown. From the ratio of
the proton signal intensities in the corresponding NMR spectra of
their $CDCl_3$ solutions [Fig. 4-12(a) to (c)], the polymers can be
readily identified as adipic acid-ethylene glycol

$$-\underset{O}{\underset{\|}{C}}-\underline{CH_2-CH_2-CH_2-CH_2}-\underset{O}{\underset{\|}{C}}-O-\underline{CH_2-CH_2}-O-$$

$$\quad\quad\quad 1.7 \quad\quad\quad 2.3 \quad\quad\quad 4.3$$
$$\quad\quad\quad ppm \quad\quad\quad ppm \quad\quad\quad ppm$$

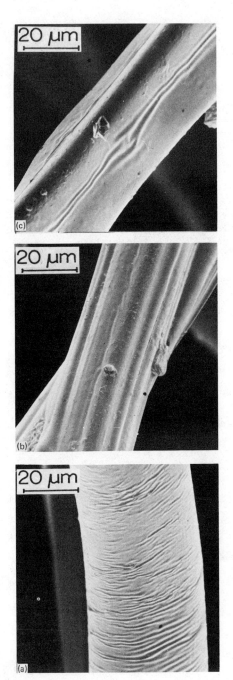

FIGURE 4-9   Scanning electron micrographs of the individual fiber
components (see text).

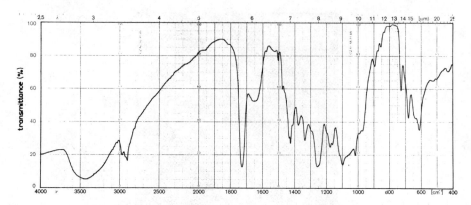

FIGURE 4-10   IR spectrum (KBr pellet) of fiber fleece upon extraction
with a mixture of methanol and benzene (see text).

adipic acid-butane diol

$$-\underset{\underset{O}{\|}}{C}-CH_2-CH_2-CH_2-CH_2-\underset{\underset{O}{\|}}{C}-O-CH_2-CH_2-CH_2-CH_2-O-$$

| | 1.7 | 2.3 | | 4.1 | 1.7 |
| | ppm | ppm | | ppm | ppm |

and adipic acid-hexane diol polyesters,

$$-\underset{\underset{O}{\|}}{C}-CH_2-CH_2-CH_2-CH_2-\underset{\underset{O}{\|}}{C}-O-CH_2-CH_2-CH_2-CH_2-CH_2-CH_2-O-$$

| | 1.7 | 2.3 | | 4.1 | 1.4 | 1.7 |
| | ppm | ppm | | ppm | ppm | ppm |

respectively.

Infrared spectroscopy can also be very helpful and time-saving
for the analysis of lacquers and their composites [62, 63]. However,
often other analytical methods (e.g., gas chromatography, polarography,
potentiometry or chemical analysis) have to be taken into account to
avoid misinterpretation and errors.

These difficulties are caused by the complex chemical composition
of lacquers and their components. Most lacquers are multicomponent
systems, and often some of these components exhibit less characteristic absorption bands or band combinations and do not allow an exact
identification when certain other components are present. Additionally, characteristic band features can be obscured by absorptions of

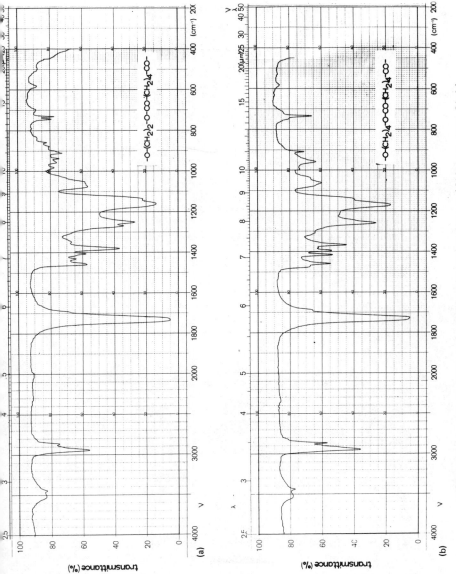

FIGURE 4-11 IR spectra of aliphatic polyesters (film on KBr disk).

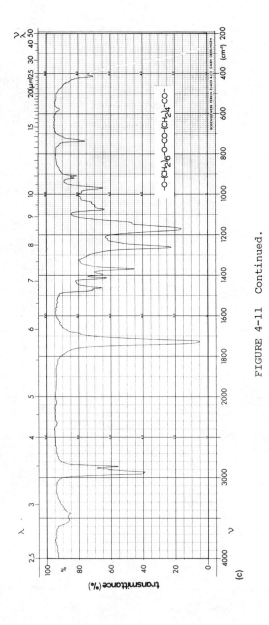

FIGURE 4-11 Continued.

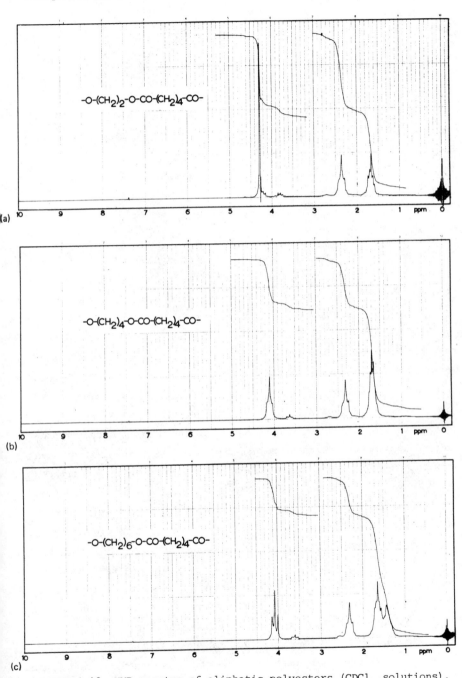

FIGURE 4-12   NMR spectra of aliphatic polyesters (CDCl$_3$ solutions).

other components. This situation may be illustrated by the analysis
of alkyd resins. The formulations of these resins are primarily based
on the following components:

1. Acidic component: e.g., adipic acid, azelaic acid, phthalic acid,
    trimellitic acid, pyromellitic acid

2. Alcoholic component: e.g., ethylene glycol, butylene glycol,
    pentaerythritol

3. Fatty acids and oils for modification: e.g., linseed oil, tung
    oil, soybean oil

4. Other components for modification: e.g., silicone, epoxy resin,
    benzoic acid, styrene

Because of the absence of very characteristic absorption bands in
their IR spectra, it is difficult to identify fatty acids, oils, or
the alcoholic components of alkyd resins. However, components such as
phthalic acid, isophthalic acid, and styrene do not cause any serious
problems.

The analysis of a resin requires a detailed interpretation and
discussion of the spectra, but the situation becomes even more complex
for lacquers, which are generally composed of various resins, oils,
and esters. Nevertheless, an experienced spectroscopist can usually
elaborate a quantitative determination of the most important formulation components. Furthermore, the reliability of the analysis can
now be considerably enhanced by application of computer-supported
IR spectroscopy (see also Sec. 4.2.2), which allows a fast addition
or subtraction of selected spectra. Figures 4-13 and 4-14 illustrate
the application of conventional dispersive IR spectroscopy to the
analysis of unknown composites.

Figure 4-13(a) shows the spectrum of a varnish of unknown composition. By inspection of the spectrum we can immediately identify a
melamine resin because of its characteristic bands at 813 and 1560 cm^{-1}.
The bands between 950 and 840 cm^{-1} are specific for the component the
melamine resin is modified with. The bands at 3010 and 700 cm^{-1} indicate a styrene modified resin. Although the most characteristic bands
at 1605 and 1580 cm^{-1} of phthalic acid resin are overlapped by the

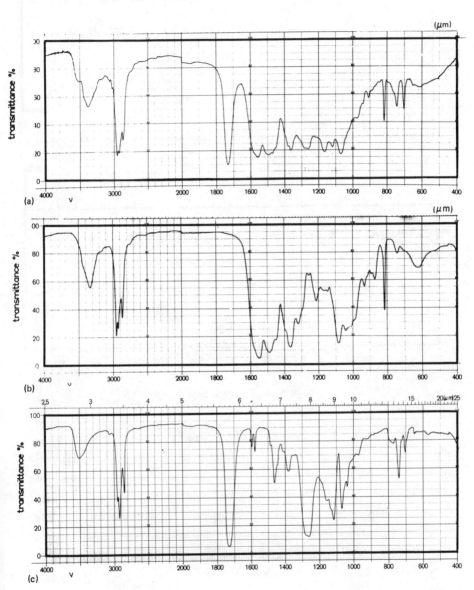

FIGURE 4-13   IR spectroscopic analysis of unknown varnish (see text).

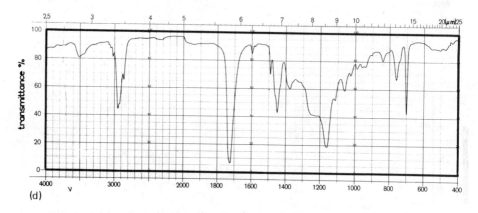

(d)

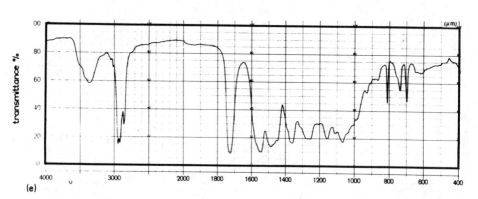

(e)

FIGURE 4-13   Continued.

melamine resin band at 1560 cm^{-1} the asymmetry of this band and the
shoulders at 1605 and 1580 cm^{-1} give a first indication for the presence of this component. The other characteristic bands at 3500, 1072,
1041 (shoulder), and 745 cm^{-1} help to identify this resin. Although,
there are some other useful bands to ensure the composition of this
resin we will not discuss the assignment in more detail. An extended
discussion of the IR spectroscopic analysis of lacquers is given in
[84]. Figure 4-13(b) to (d) shows the spectra of the individual components.

   Upon identification of the most significant components a quantitative determination of the composition can be carried out (see
Sec. 4.1). The result is shown in Fig. 4-13(e). The spectrum was

scanned from a mixture of 40% melamine resin [Fig. 4-13(b)], 30%
phthalic acid resin [Fig. 4-13(c)], and 30% styrene modified alkyd
resin [Fig. 4-13(d)]. The spectra of Fig. 4-13(a) and (e) are almost
identical and show only small differences which are mainly caused by
the melamine resin (differences between 950 and 850 cm^{-1}).

Another example for the application of IR spectroscopy to the
qualitative and quantitative analysis of lacquers is shown in Fig.4-14.
The spectrum of the clear varnish [Fig. 4-14(a)] can be interpreted
quite easily. The occurrence of the bands at 1605 and 1580 cm^{-1} indicates a phthalic acid resin. However, of these two bands generally
observable in the spectra of such a resin the lower frequency absorption is slightly more intense. The opposite can be observed in Fig.
4-14(a). This effect can be attributed to the superposition of the
1605 cm^{-1} band by the ν(C=C) stretching vibration of an acrylate
resin. This assignment is in agreement with the occurrence of the
sharp band at 1490 cm^{-1}. Figure 4-14(b) shows the identified resin
which is mainly composed of hydroxypropylacrylate and styrene. From
test mixtures it could be derived [Fig. 4-14(d)] that the varnish
under examination is composed approximately of 20% phthalic acid
resin [Fig. 4-14(c)] and 80% styrene-hydroxypropylacrylate resin
[Fig. 4-14(b)].

To illustrate the application of IR and Raman spectroscopy to
quantitative analysis, the determination of acrylonitrile-butadiene-
styrene (ABS) terpolymer composition may serve as an example here.

In Fig. 4-15 the IR spectrum of a hot-pressed film of an ABS
copolymer is shown. The absorbance of the absorption bands at 2240,
1601, and 965 cm^{-1} is representative of the acrylonitrile, styrene,
and 1,4-*trans*-butadiene content, respectively [64-66]. Upon calibration of the system with ABS samples of known composition, determined
by independent methods (NMR, elemental analysis) and under the assumption of an approximately constant 1,4-*trans*-, 1,4-*cis*-, and
1,2-butadiene ratio, the IR method may be conveniently applied for
a rapid determination of the copolymer composition. The content of
the 1,4-*trans*-, 1,4-*cis*-, and 1,2-butadiene units in the polybuta-

Applied Spectroscopy

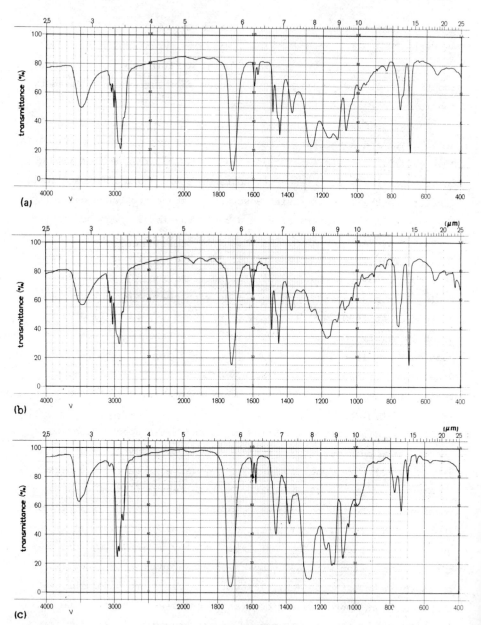

(b)

(c)

FIGURE 4-14   IR spectroscopic analysis of unknown varnish (see text).

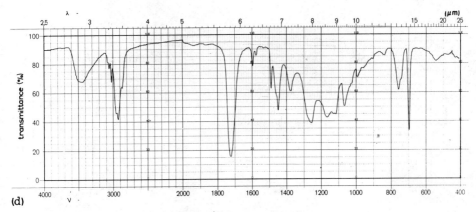

FIGURE 4-14   Continued.

diene blocks can be derived from studies of the ν(C=C) Raman band
complex at about 1650 cm^{-1} (see also Sec. 2.3) with polybutadienes
of different stereoregularity.

In silicone resins the ratio of methyl to phenyl groups which
may be correlated with their coating behavior can be determined by
IR spectroscopy [67-70]. In the IR spectrum of a silicone polymer
[Fig. 4-16(a)] the intense absorption bands at 1430 and 1260 cm^{-1}
can be assigned to the ν(C-C) of the aromatic ring and the δ_s(CH$_3$)

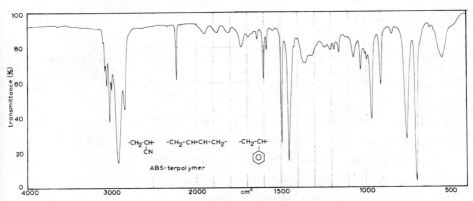

FIGURE 4-15   IR spectrum of acrylonitrile-butadiene-styrene copolymer (hot-pressed film).

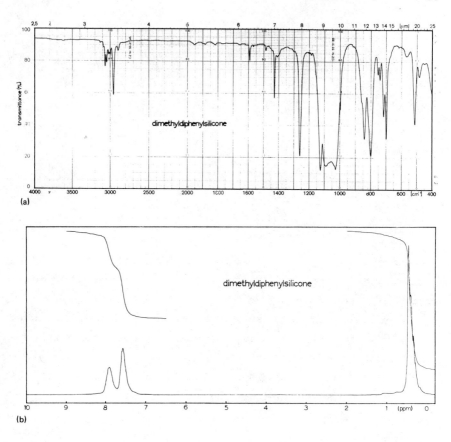

FIGURE 4-16   (a) IR spectrum of dimethyldiphenylsilicone (capillary
               film between KBr disks). (b) NMR spectrum of dimethyl-
               diphenylsilicone (CCl$_4$ solution).

vibrations, respectively [71-73]. The absorbance ratio of these bands
is directly related to the ratio of phenyl and methyl groups present
in the resin. By using standard silicone polymers for which the exact
amount of methyl and phenyl content has been previously obtained from
the intensity of the NMR signals between 0 and 1 ppm and 7 and 8 ppm,
respectively [Fig. 4-16(b)], a calibration curve can be set up.

    Analogous IR band ratio techniques have been reported for numer-
ous other copolymers (see Sec. 4.8).

4.2.2 Computer-Supported Infrared Spectroscopy

In recent years, with the introduction of commercial FTIR spectrometers and dispersive instruments associated with computer facilities,
many applications of IR analysis that were extremely difficult with
the conventional technique are now readily accomplished. Because the
data can be stored digitally in a computer, mathematical manipulations
can be performed with the stored spectra.

4.2.2.1 Difference Spectroscopy

The procedure of difference spectroscopy by absorbance subtraction
which has been popularized by Koenig [74, 75] provides an elegant
method to effect spectral separation of mixtures or to detect small
compositional changes in the sample under examination. In the IR spectrum of a polymer containing, for example, two additives the total
absorbance at any wavenumber $A_{T_1}^{\bar{\nu}}$ is simply the algebraic sum of the
absorbance of each component in the mixture:

$$A_{T_1}^{\bar{\nu}} = A_{P_1}^{\bar{\nu}} + A_X^{\bar{\nu}} + A_Y^{\bar{\nu}} \qquad (4.12)$$

where $A_{P_1}^{\bar{\nu}}$, $A_X^{\bar{\nu}}$, and $A_Y^{\bar{\nu}}$ are the absorbances of the pure polymer and the
additional components X and Y, respectively. In analogy it can be
written for the spectrum of a pure standard polymer sample

$$A_{T_2}^{\bar{\nu}} = A_{P_2}^{\bar{\nu}} \qquad (4.13)$$

In order to eliminate the interfering absorbance of the unmodified
polymer, it is desired to subtract $A_{T_2}^{\bar{\nu}}$ from $A_{T_1}^{\bar{\nu}}$:

$$A_S^{\bar{\nu}} = A_{T_1}^{\bar{\nu}} - kA_{T_2}^{\bar{\nu}} \qquad (4.14)$$

Selecting a wavenumber region where only the polymer absorbs, the
removal of the absorptions due to the pure polymer at all frequencies
can be accomplished using the subtraction criterion

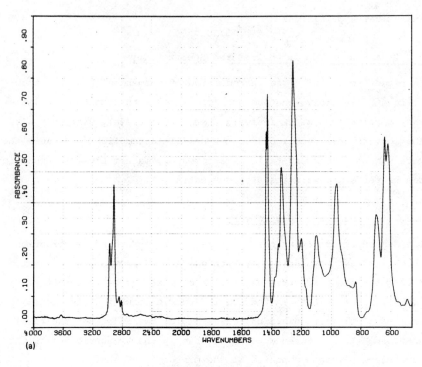

(a)

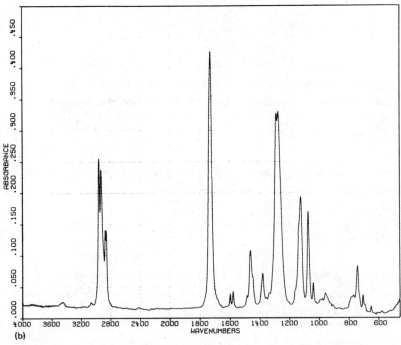

(b)

176

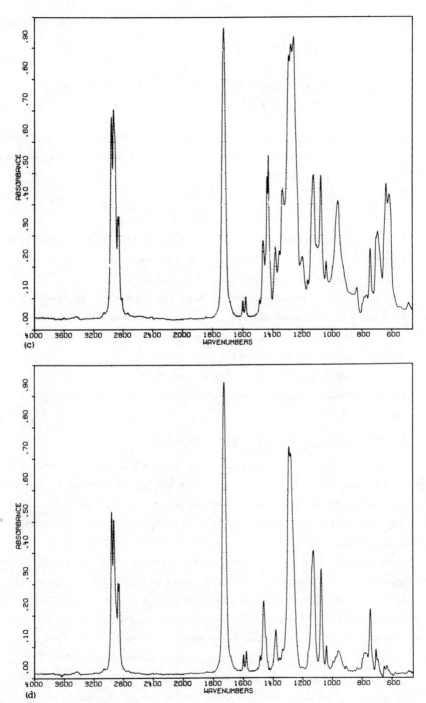

FIGURE 4-17 FTIR difference spectroscopy of plasticized polyvinyl chloride:
(a) IR spectrum of plasticized polyvinyl chloride (film on KBr disk), (b)
unplasticized polyvinyl chloride (film on KBr disk), (c) dioctylphthalate
(capillary film between KBr disks); and (d) difference spectrum: (a) - (b).

$$A_{P_1} - k A_{P_2} = 0 \qquad \text{with } k = \frac{A_{P_1}^{\bar{\nu}}}{A_{P_2}^{\bar{\nu}}} \qquad \text{(scaling factor)} \qquad (4.15)$$

In this way, nonanalytical differences associated with the amount of sample being viewed by the beam are automatically eliminated. The difference spectrum then reads

$$A_S^{\bar{\nu}} = A_X^{\bar{\nu}} + A_Y^{\bar{\nu}} \qquad (4.16)$$

A similar second difference spectrum could then be computed upon identification of X or Y to isolate the spectrum of the residual unknown component.

Recently, the procedure has been put on a more theoretical basis by the development of a spectra ratio technique and a least-squares curve-fitting algorithm for the determination of the appropriate scaling factor in quantitative measurements [76, 77].

The spectral processing operation of absorbance subtraction can be successfully employed in a wide field of polymer analysis and polymer physics:

Detection and Identification of Additives

Under favorable conditions the time-consuming chemical separation procedures of multicomponent systems may be successfully replaced by the spectroscopic absorbance subtraction technique. This also applies to systems where one or more components occur in small concentrations such as in the detection and identification of additives. Once the spectrum of a standard pure polymer has been obtained and stored in the computer, it can be used at any time to remove the base polymer spectrum from the spectrum of a modified polymer sample. Thus, additives such as plasticizers, antioxidants, vulcanizing agents, fillers, etc., can be readily detected, identified, and determined quantitatively by a proper calibration procedure [74]. The isolation of the spectrum of dioctylphthalate (DOP) from the spectrum of plasticized polyvinyl chloride is demonstrated in Fig. 4-17 [78].

End Group Analysis

The detection and identification of end groups is of considerable importance for the determination of chain length and degree of branching.
If a standard polymer of very high molecular weight and low degree of
branching is available, it can be used to subtract the base polymer
IR spectrum from the spectrum of any low-molecular-weight analogue,
thereby accentuating the absorptions of the end groups. The typical
results of a digital subtraction employed for the determination of
methyl groups in polyethylene is shown in Fig. 4-18. Appropriately
adjusted subtraction of the spectrum of high-density polyethylene from
a low-density polyethylene sample yields an isolated $\delta_s(CH_3)$ band at
1378 cm^{-1}, while the $\gamma_w(CH_2)$ doublet at 1368 and 1352 cm^{-1} has been
canceled. Having obtained the difference spectrum, the peak height
can be estimated and the concentration of methyl groups calculated.
The digital method is substantially faster than the manual method
applying standard polymethylene or high-molecular-weight polyethylene
wedges in the reference beam [79].

Study of Oxidation and Degradation Reactions

Valuable information regarding the various stages of oxidation or
degradation may be gained by subtracting the spectrum of the pure
polymer from the oxidized or degraded sample [80-82]. The kinetic
aspects of polymer reactions will be treated in Sec. 4.7.

Molecular Interactions

As a consequence of molecular interactions arising from chemical or
physical effects (e.g., hydrogen bonding or dipole association) shifts
in the absorption frequency and/or changes in the absorptivity of particular absorption bands will be observed. Either of these spectral
differences can be isolated using the difference spectra technique.
The frequency shift due to physical interactions makes it impossible
to subtract the affected absorption bands, while the unaffected bands
cancel. The shape of such an absorption band in the difference spectrum will then be very similar to the first derivative of a band with
a maximal and minimal value. The frequency shift of the $\nu(C=O)$

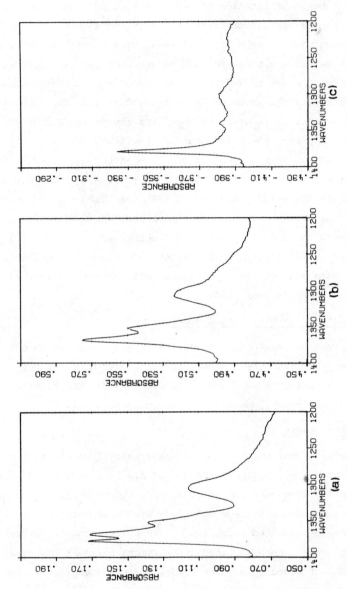

FIGURE 4-18   Determination of methyl group content in polyethylene by FTIR difference spectroscopy: (a) IR spectrum of low-density polyethylene film, (b) IR spectrum of high-density polyethylene film, and (c) difference spectrum: (a) - (b).

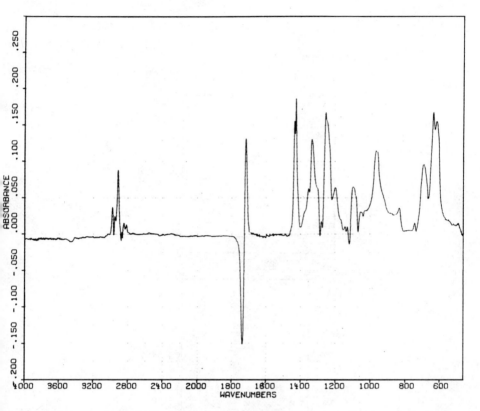

FIGURE 4-19   FTIR difference spectrum: (dioctylphthalate-plasticized
polyvinyl chloride) - (dioctylphthalate).

stretching vibration in the IR spectrum of dioctylphthalate upon

plasticization of polyvinyl chloride (Fig. 4-19) suggests the for-

mation of complexes between the carbonyl groups of DOP and C-Cl groups

in the polymer chain segments [78]. When changes in absorptivity oc-

cur, residual positive or negative absorbance will appear in the dif-

ference spectra for the affected modes.

Coleman and Painter [81, 82] have utilized the potential of digi-

tal subtraction and addition techniques to study the compatibility of

polymer blends. They have demonstrated that as a consequence of phase

separation the spectra of incompatible blends can be duplicated almost

perfectly by weighted addition of the individual component polymers.

Conversely, interaction of compatible polymers caused considerable

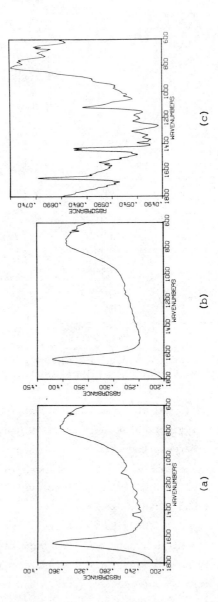

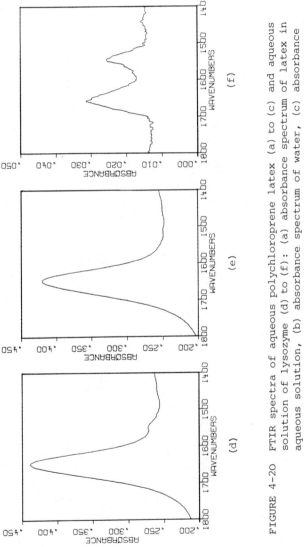

FIGURE 4-20 FTIR spectra of aqueous polychloroprene latex (a) to (c) and aqueous solution of lysozyme (d) to (f): (a) absorbance spectrum of latex in aqueous solution, (b) absorbance spectrum of water, (c) absorbance spectrum of latex with water absorptions removed by subtraction, (d) absorbance spectrum of aqueous lysozyme solution, (e) absorbance spectrum of water, and (f) difference spectrum (d) - (e).

differences between the IR spectrum of the blend and the spectrum synthesized from the absorbance spectra of the pure components.

Aqueous Solutions

Over many years the application of IR spectroscopy to the study of aqueous solutions has been hampered by the strong interfering absorptions of water. However, if the pathlength of light through the solution is limited such that total absorbance does not occur, the water spectrum may be subtracted to isolate the spectral features of the solute. The digital compensation of the broad water absorption from the spectrum of an aqueous polychloroprene latex is demonstrated in Fig. 4-20 (a) - (c).

The practicality of digital subtraction of the water absorptions has also widened the scope of IR spectroscopy to include the investigation of biological macromolecules where water is the most interesting solvent. Primarily conformational effects as a function of pH or temperature have become accessible to spectroscopic characterization by utilizing the structure of the amide I band region as a specific probe [75, 83]. As an example, the spectra of an aqueous solution of lysozyme and pure water are shown in Fig. 4-20 (d) - (f) along with the difference spectrum in which the absorbance contributions of water have been eliminated.

4.2.2.2 Infrared Studies of Optically Dense Materials

IR spectra of optically dense materials, such as carbon-black-filled polymers, coals, degraded polymers, etc., obtained on conventional dispersive IR spectrometers are generally of poor quality. However, taking advantage of the high energy throughput of FTIR systems coupled with the signal-averaging capabilities, excellent quality IR spectra of optically dense materials may be obtained. Figure 4-21 shows the IR spectra of a carbon-black-filled polybutadiene recorded on an (a) conventional and (b) FTIR spectrometer. No meaningful spectrum could be obtained with the dispersive spectrometer, whereas it is evident that the good quality spectrum obtained using FTIR readily permits identification of the sample [81].

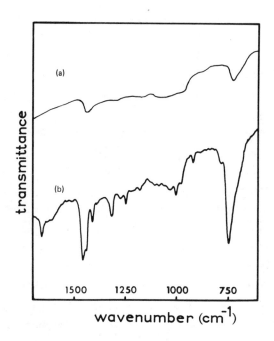

FIGURE 4-21 ATR spectra of carbon black-filled 1,4-*cis*-rich polybutadiene run on a (a) dispersive instrument and (b) FTIR
spectrometer. [Reprinted with permission from W. W. Hart,
P. C. Painter, J. L. Koenig, and M. M. Coleman, Appl.
Spectrosc. 31:220 (1977).]

4.2.3 Pyrolysis, Combustion, and Degradation of Polymers

The difficulties associated with the identification of intractable,
cross-linked rubbers and similar insoluble, filled materials can be
sometimes overcome by spectroscopic examination of the collected
liquid and gaseous pyrolysis products as well as any involatile residues. If the thermal degradation can be performed in a reproducible
manner, the IR spectra of the products will be characteristic of the
investigated polymer [31, 84-91]. In recent years the pyrolysis-gas
chromatography method (preferably coupled with mass spectrometry)
has gained wide recognition as a means of polymer identification

[92-95]. Furthermore, pyrolysis-thin-layer chromatography combinations
have been successfully applied for the identification of a wide variety of polymeric materials [96-99].

There are various types of pyrolysis units in use to achieve the
required temperature of about 823 to 1223 K [57, 92, 100-102]. In
furnace-type pyrolyzers, the pyrolysis chamber, through which a car-
rier gas flows to the chromatographic column, is preheated to a selected temperature before the sample is placed in the hot zone.

In the filament pyrolysis unit the sample is placed either on the
surface of the filament or inside a small container that is held within the filament coil. Pyrolysis is achieved by passing a current
through the filament, which, through resistive heating, raises the
temperature of the system to a value which is determined by the applied voltage. The filament pyrolysis unit is particularly useful as
it permits additional information to be obtained by stepwise pyrolysis
[103, 104].

In Curie-point pyrolyzers the filament wire is a ferromagnetic
material which undergoes induction heating when exposed to a radio-
frequency field and rapidly reaches a specific temperature known as
the *Curie temperature*. A range of pyrolysis temperatures from 493 to
1273 K may be obtained by using wires of different Curie point [105,
106]. Laser pyrolyzers can achieve high temperatures in extremely
short times [107, 108].

If the decomposition of the polymer under investigation is con-
ducted under controlled chemical and thermal conditions certain
polymer systems will yield their basic constituents as degradation
products. Stahl has shown [97-99] that certain copolycondensates can
be readily identified by thin-layer-chromatographic investigation of
the degradation products obtained from controlled thermal treatment
of their alkali melt. Polyester urethanes, for example, will be
cleaved according to the following scheme:

$$
- O - \underset{\substack{\text{diamine}}}{\overset{\substack{O \\ \parallel}}{C}} - \underset{H}{\overset{}{N}} - R_1 - \underset{H}{\overset{}{N}} - \underset{\substack{\text{diol}}}{\overset{\substack{O \\ \parallel}}{C}} - O - R_2 - O - \underset{\substack{\text{diacid}}}{\overset{\substack{O \\ \parallel}}{C}} - R_3 - \overset{\substack{O \\ \parallel}}{C} - O -
$$

The diamine of the corresponding diisocyanate can be identified by
thin-layer chromatography upon thermal treatment of the alkali melt
up to about 523 K. In a successive step the diols and diacids are
obtained from the acidified reaction mixture at temperatures up to
423 K and above 423 K, respectively. As an alternative to thin-layer
chromatography, the diols and diacids can be trapped and identified
by IR spectroscopy. In this manner the composition of the polyester
urethane whose IR spectrum is shown in Fig. 4-22 has been shown to
be based on 2,4-2,6-toluylene diisocyanate, hexanediol, and adipic
acid.

The most powerful method at present is probably to follow
pyrolysis by gas chroamtography and to apply IR investigation alongside retention time and mass spectrometry to characterize the separated volatile products. The ideal solution is to combine the gas
chromatograph with analytical instruments for "on the fly" analysis,
thereby eliminating the need for trapping [109]. In this respect
considerable effort has gone into gas chromatography-mass spectrometer combinations [110, 111]. The interfacial systems involved in
pyrolyzer-gas chromatograph-mass spectrometer combinations are all
eliminated if pyrolysis is performed directly in the vacuum chamber
of the mass spectrometer [112].

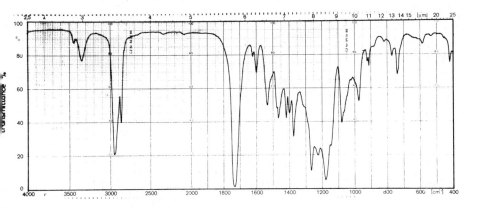

FIGURE 4-22   IR spectrum of polyester urethane on the basis of
              2,4-2,6-toluylene diisocyanate, adipic acid, and
              hexanediol (film on KBr disk).

The IR spectra of gas chromatographic effluents may be taken either from liquid (trapping technique) or gaseous samples. Analysis in the gas phase is often desirable, since this approach eliminates the difficult process of condensing the sample from a vapor stream. In addition, in recent years, the generation of the gaseous combustion products of synthetic polymers, especially poisonous gases due

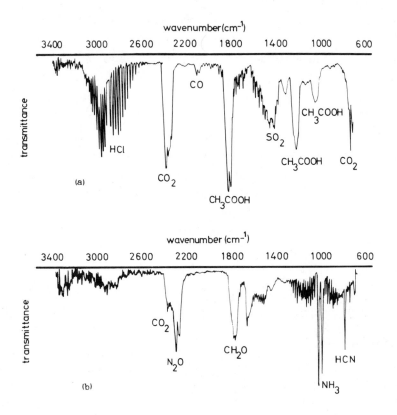

FIGURE 4-23   FTIR spectra of gaseous combustion products of
(a) polyvinyl chloride composite programmed at 40 K/min
at 648 K in air and (b) urea formaldehyde composite
pulsed at 573 K in air. [Reprinted with permission from
S. A. Liebman, D. H. Ahlstrom, and P. R. Griffith,
Appl. Spectrosc. 30:355 (1976).]

to fires has been increasingly drawing attention as a serious problem. With the use of data obtained from pyrolysis and combustion studies, details in polymer microstructure and degradation mechanisms can be deduced [113]. Fast-scanning FTIR spectrometers interfaced to a flow-through heated light-pipe gas cell permit the continuous measurement of the IR spectra of any volatile effluents and establish a general technique for the analysis of the gaseous products of polymer degradation.

The FTIR spectrum of the volatile products obtained by a programmed heat treatment (40 K/min) of a polyvinyl chloride composite at 648 K in air is shown in Fig. 4-23(a) [114]. It can be seen that SO_2, HCl, CH_3COOH, CO, CO_2, and hydrocarbons were evolved during combustive degradation. Similarly, the volatile products of a urea-formaldehyde composite formed by a direct heat pulse (573 K) in a combustive atmosphere were studied by on-line FTIR spectroscopy [Fig. 4-23(b)]. As a consequence of the limited time available only four scans were signal averaged for a total measurement time of 8 s. Nevertheless, the direct identification and relative quantitation of CH_2O, NH_3, CO_2, N_2O, HCN, and hydrocarbons can be performed from successive spectra.

Generally, the analytical potential for the identification of gaseous mixtures is drastically increased when the separating power of a gas chromatograph is combined with the selectivity of vibrational spectroscopy. Thus, in a typical GC-FTIR experiment (Fig. 4-24) the effluent from the GC column is continuously passed through the GC detector (preferably a nondestructive thermal conductivity detector) down a heated transfer line into a light-pipe which has been designed as an IR gas cell [115]. When a more sensitive GC detector such as the destructive flame ionization detector (FID) is required, an effluent stream splitter has to be used and the majority of the sample is passed through the IR cell with only a small portion passing to the FID. The optimum dimensions of the light-pipe construction which are critical to the success of GC-FTIR experiments have been discussed in detail [115, 116]. A further improvement in spectroscopic detection sensitivity has been achieved with the use of a liquid-nitrogen-cooled

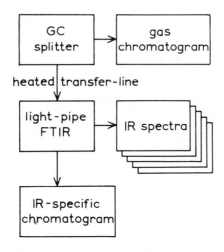

FIGURE 4-24   Block diagram of a GC-FTIR system.

mercury-cadmium telluride (MCT) mid-infrared photoconductive detector [115]. To avoid the storage of unnecessary data during a gas-
chromatographic run, a system has been developed [116] which allows
to monitor the absorbance in several operator-selected frequency
ranges simultaneously as a function of time to yield an IR-specific
chromatogram in an analogous way that the total ion current is used
in GC-mass spectrometry experiments. When the absorbance in any selected frequency range exceeds a certain threshold value the system
automatically stores full resolution interferograms for later examination. Interpretation and assignment of the spectral data to the
corresponding peaks in the spectrally separated chromatogram and
comparison to the chromatogram as recorded by the GC detector will
then provide a more complete analytical picture of the mixture under
consideration.

4.2.4   Gel Permeation Chromatography - FTIR Spectroscopy

Gel permeation chromatography (GPC) is a powerful tool for the separation of polymeric materials in solution in terms of the molecular
weight distribution of the sample under investigation. The fast scan,
sensitivity, and digital processing make FTIR spectroscopy an ideal
detector for the analysis of the eluted fractions. Preliminary applications of the on-line combination of GPC and FTIR spectroscopy by a
flow-cell technique have been reported [117]. The difficult situation
of solvent opacity and peak spreading in flow-cell IR detection is
alleviated by the high sensitivity of FTIR systems. Thus, thin flow-
cells with very small internal volume (10 µl) (allowing more solvent
transmission and suppressing peak broadening) can be used for detection and fractions of interest can be readily collected for further
analysis. Further to storing spectra characteristic of distinct points
in the chromatographic separation the FTIR system transforms and dis-
plays spectra in real time during the chromatographic procedure. The
software which is similar to that developed for GC-FTIR provides for
the integration of absorbances in preselected frequency regions and
the plot of IR-specific chromatograms for the peak location. The
absorbance subtraction technique has been shown not only to provide
a means for removing not too intense solvent absorptions but also
to elucidate the composition of chromatographically unresolved components.

        As an example Fig. 4-25 presents a series of spectra collected
as a function of retention time during elution of an unresolved GPC
peak containing mineral oil and phenyl silicone [117]. Despite the
chromatographically incomplete separation the successively collected
spectra allow a straightforward identification of the individual components by functional group analysis. The blank region in the spectra
between 700 and 800 cm^{-1} corresponds to the region of CCl_4 solvent
opacity.

        Alternatively to the flow-cell technique attempts to automate
solvent removal in these experiments may contribute to a more efficient exploitation of the frequency range available for spectroscopic
interpretation purposes [126].

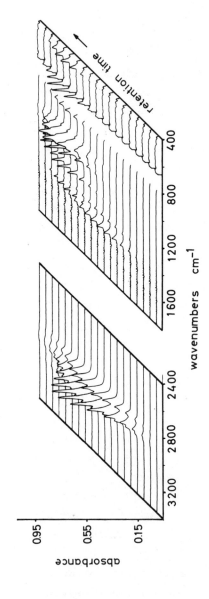

FIGURE 4-25   FTIR spectra obtained during GPC separation of mineral oil and phenyl silicone. (Reproduced with kind permission of Nicolet GmbH, Offenbach, West Germany.)

4.3 STATE OF ORDER IN POLYMERS

In Sec. 2.4 the nomenclature and origin of IR and Raman bands characteristic of the structural order in polymers has been discussed in
some detail. Real crystallinity bands owing to intermolecular interactions of adjacent chains in the crystallographic unit cell have
been unambiguously observed and assigned only in very few polymers
[118-121]. For some polymers, however, absorption bands originating
from regular intramolecular interactions in polymer chains of a certain conformational regularity can be used for the determination of
crystallinity [122-125]. This procedure is justified because conformational regularity is a necessary condition for a regular three-dimensional arrangement of polymer chains. However, some polymers may
occur in different crystalline modifications whose individual absorption bands cannot be completely separated in their vibrational spectra. Furthermore, conformations which are correlated with certain absorption bands may occur in different phases. Polymers with methylene
sequences, for example, frequently crystallize in the *trans* conformation of the fully extended chain segments. This conformation, however, may also occur in the amorphous regions alongside other possible conformations and the assumption that the *trans/gauche* ratio is
equivalent to the crystalline/amorphous ratio is not applicable.

     Generally, with increasing crystallinity or state of order, ab-
sorption bands originating from interactions between conformationally regular units show a decreasing bandwidth, while bands due to vibrations of amorphous regions with random conformations decrease in
intensity.

     All vibrational spectroscopic methods for studying order phenom-
ena are based on the evaluation of changes in wavenumber and intensity of the conformational regularity and crystallinity bands in dependence of the experimental conditions (e.g., polarization of the
radiation, temperature, pretreatment of the sample etc.).

     As shown in Sec. 2.4 the inter- and intramolecular interactions
between structural units in a crystallographic cell cause optically

active in-phase and out-of-phase vibrations for which certain phase
relations hold [127-129]. The application of these relations can be
extremely useful for the determination of the state of order in many
conformationally regular polymers, especially, polymers exhibiting a
center of inversion within the repeat unit. Thus, besides the well-
known interpretation of the band splitting observable in the spectra
of polyethylene and other polymers with long methylene sequences,
these relations were successfully applied to the interpretation of
the wavenumber positions of the amide I and amide II bands in polypeptides [130, 131], the absorption bands of the monoclinic α-crystal
structure of polyamide-6 in the 1000 cm^{-1} wavenumber region [132],
and in a modified version to account for Raman data and the results
of the normal coordinate analysis of polyglycine [133].

For the qualitative and quantitative determination of certain
conformations of crystalline modifications and the state of order in
partially crystalline polymers the following approaches are in common use:

1. Separation of the spectra of amorphous and crystalline phases
   or different crystalline phases by computer-supported vibrational
   spectroscopy

2. Studies at high and low temperatures of samples with defined
   thermal history (e.g., quenched, annealed, slowly cooled)

3. Measurements of the dichroic and depolarization ratio

4. Studies of hydrogen bonding

5. Deuterium exchange

Points 4 and 5 are applicable only to a limited number of polymers.

4.3.1 Quantitative Determination of the State of Order

Under the assumption that the polymer under investigation can be
treated as a two-phase system, the fraction x of the crystalline
phase or a single rotational isomer (e.g., *trans/gauche*) can be
evaluated from the absorbances of absorption bands which are peculiar
to the respective modification:

$$A_{cr} = a_{cr}xb \tag{4.17}$$

$$A_{am} = a_{am}(1 - x)b \tag{4.18}$$

Elimination of x from Eqs. (4.17) and (4.18) yields

$$\frac{A_{am}}{A_{cr}} = \frac{a_{am}b}{A_{cr}} - \frac{a_{am}}{a_{cr}} \tag{4.19}$$

and a_{am}/a_{cr} can be obtained from a calibration plot of A_{am}/A_{cr} versus $1/A_{cr}$ derived from polymer samples of different states of order. Alternatively, the absorptivities of the crystalline and/or amorphous absorption bands (a_{cr}, a_{am}) may be determined from the measured absorbances of two samples with different degrees of crystallinity [23, 24] or the molten polymer (x = 0) [134-136]. With the value of a_{cr}/a_{am} the percentage x(%) of crystalline regions can then be evaluated from the equation:

$$x(\%) = \frac{100}{(A_{am}/A_{cr})(a_{cr}/a_{am}) + 1} \tag{4.20}$$

Several authors have correlated the absorbances of crystallinity-sensitive absorption bands with values derived from other techniques which characterize the state of order (e.g., density or specific volume, x-ray diffraction) [137-142]. Once a linear relationship has been established, the degree of crystallinity for any sample can then be derived from the corresponding calibration curve (see, for example, Fig. 4-26). When only absorption bands characteristic of a single phase are available, the combination of vibrational spectroscopy with an independent technique is a necessary condition in order to obtain absolute values of crystallinity. Differences in sample thickness can be eliminated by reference to an absorption band which is insensitive to the state of order of the polymer under examination.

The Raman spectrum of partially crystalline polyethylene [Fig. 4-27(c)] in the frequency ranges of the C-C-stretching (I), CH_2-twisting (II), and CH_2-bending (III) vibrations represents the super-

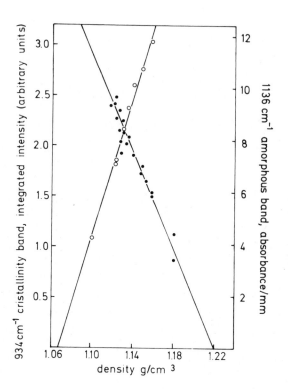

FIGURE 4-26   Intensity of "crystalline" (o) and "amorphous" (●) IR
absorption bands of polyamide-6,6 as a function of density. [Reproduced with permission from H. W. Starkweather
and R. E. Moynihan, J. Polym. Sci. 22:363 (1956).]

position of sharp bands characteristic for the all-*trans* conformation
of the crystalline phase [Fig. 4-27(a)] and broad scattering phenomena which reflect a meltlike coiled chain conformation [Fig. 4-27(b)].
From close inspection of the CH_2-bending region, however, Strobl and
Hagedorn [143] have derived that three different components contribute to a spectrum. These have been related to the orthorhombic crystalline phase, a meltlike amorphous phase and a disordered phase of
anisotropic nature, where chains are stretched but do not show lat-

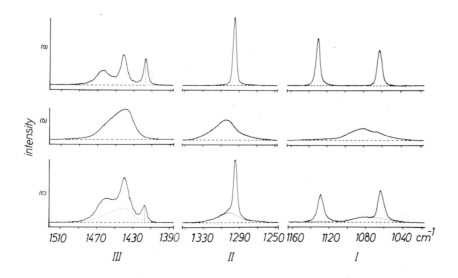

FIGURE 4-27   Raman spectra of the (I) C-C-stretching vibration, (II)
CH$_2$-twisting vibration, and (III) CH$_2$-bending vibration:
(a) extended chain polyethylene at 298 K, (b) the melt
at 423 K, (c) branched, partially crystalline polyethylene
at 298 K. The amorphous component in the spectrum is indicated. (Courtesy of Prof. G. R. Strobl, University of
Mainz, West Germany.)

eral order. The mass fractions involved in the three phases can be
determined directly from the integral intensities of characteristic
bands by a graphical separation procedure and show agreement with the
crystallinities derived from the density and small- and wide-angle
x-ray diagrams. Furthermore, the analysis of the temperature-dependent measurements permits a detailed examination of the process of
partial melting.

   The difficulty assigning absorption bands to a certain modification of a semicrystalline polymer has been alleviated by the recent
introduction of FTIR spectrometers and computer-equipped dispersive
spectrometers. One of the major advantages of these instruments is
that spectral information can be stored in the computer and mathemat-

ical manipulations may be readily performed. With the aid of an absorbance subtraction technique absorption bands associated with certain crystalline modifications or rotational isomers in various semi-crystalline polymers have been successfully isolated from the composite spectrum [74, 81, 144-148].

Let us suppose that the IR spectrum of the polymer under examination can be treated as a superposition of the spectral characteristics of the corresponding *gauche* and *trans* isomers (e.g., in polyethylene terephthalate [149]. In the spectrum of any specimen the total absorbance A_1 can then be expressed by

$$A_1 = A_{tr} + A_g \tag{4.21}$$

where A_{tr} and A_g are the absorbances of the *trans* and *gauche* components, respectively. Upon thermal treatment of this sample, a certain fraction x of the *gauche* structures is transformed to the *trans* isomer. The total absorbance then reads

$$A_2 = (1 + x)A_{tr} + (1 - x)A_g \tag{4.22}$$

In order to remove the *gauche* absorptions in spectrum 2, spectrum 1 is subtracted:

$$A_S = A_2 - kA_1 \tag{4.23}$$

where the adjustable scaling parameter k is chosen k = 1 - x. A_S then reads

$$A_S = 2xA_{tr} \tag{4.24}$$

i.e., the remaining spectrum is due to the *trans* component only.

The subtraction technique can be extended to the quantitative determination of a component by adjustment of the scaling factor until the known interferent peaks just disappear (preferably in combination with a calibration procedure by a series of samples whose state of order has been determined by an independent technique).

IR spectroscopic investigations of polyethylene terephthalate have contributed to a detailed knowledge of its molecular structure and the changes occurring as a consequence of thermal and mechanical pretreatment. Several absorption bands in the IR spectrum of polyethylene terephthalate vary in intensity as a function of specimen crystallinity and have been associated with the amorphous and crystalline phases, respectively [150-153]. However, it has been pointed out [120, 121, 154, 155] that many differences in the spectra of amorphous and partially crystalline samples can be interpreted in terms of conformational isomerism of the -O-CH$_2$-O- units. While the crystalline phase is limited to the extended *trans* conformation [156], *gauche* and *trans* conformations may occur in the amorphous region [157, 158].

Absorption bands at 1453, 1370, 1040, and 895 cm^{-1} and at 1470, 1340, 973, and 846 cm^{-1} have been assigned to vibrations of the *gauche* and *trans* ethylene glycol segments, respectively [151, 153, 155]. The weak intensity of *trans* bands in melt-quenched samples indicates that the conformation of the aliphatic segments of the amorphous phase is predominantly *gauche*. Crystallization upon thermal treatment is accompanied by an increase of *trans* content at the expense of *gauche* structures. This transformation is reflected in the IR spectrum by the intensity increase and decrease of absorption bands associated with the respective rotational isomers. The crystalline component spectrum of annealed, melt-quenched polyethylene terephthalate was first obtained by D´Esposito and Koenig [146] by digitally subtracting the spectrum of melt-quenched polyethylene terephthalate annealed at 323 K for 5 hr from the spectrum of a melt-quenched polyethylene terephthalate annealed at 513 K for 5 hr until bands associated with the *gauche* conformation (see above) were reduced to the base line. In the difference spectrum the *trans* absorption bands mentioned above and some additional bands (1685, 1368, shoulder at about 1225, 1125, 1109, 1024, and 988 cm^{-1}) which have been partially assigned to true crystallinity bands [146, 159] are brought out clearly by the subtraction procedure.

The spectroscopically determined total *trans/gauche* ratio $1 = (t_{cr} + t_{am})/g$, the *trans/gauche* ratio of the amorphous regions

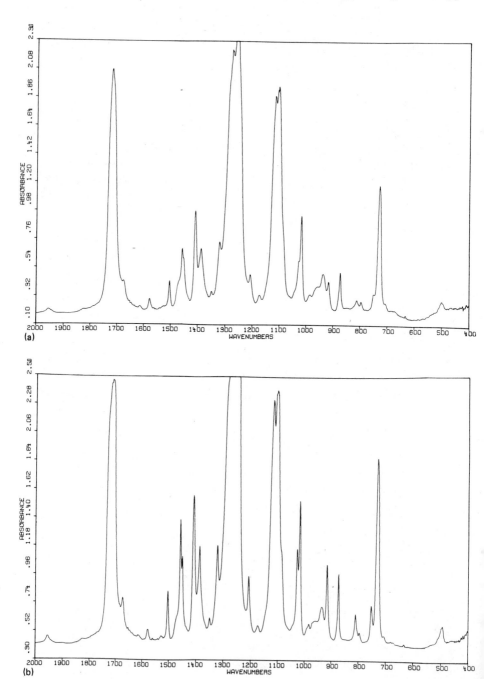

(a)

(b)

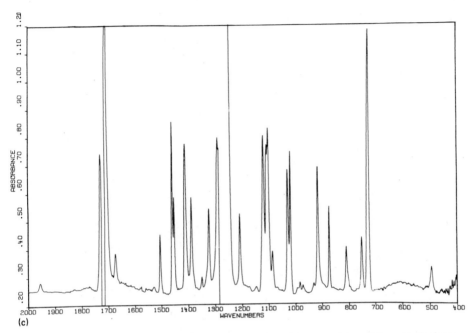

(c)

FIGURE 4-28   Application of the FTIR absorbance subtraction technique
to polybutylene terephthalate films of different state
of order: (a) FTIR spectrum of melt-quenched polybutylene
terephthalate film, (b) FTIR spectrum of melt-quenched
polybutylene terephthalate film annealed at 483 K for
15 hr, (c) difference spectrum (b) - (a).

$p = t_{am}/g$, and the degree of crystallinity x are related by [147]

$$x = \frac{1 - p}{1 + 1} \qquad (4.25)$$

Thus, combination of the IR spectroscopic determination of l with an
independent technique for the determination of x allows the calculation of the parameter p [158].

Significant changes can also be observed in the IR spectrum of a
primarily amorphous polybutylene terephthalate film upon thermal
treatment [Fig. 4-28(a) and (b)]. Several absorption bands decrease
(1470, 1393, 960, and 850 cm^{-1}) and increase (1460, 1450, 1388, 1324,
1210, 1030, 917, 810, and 750 cm^{-1}) in intensity as a consequence of
crystallization. From x-ray investigation it has been established

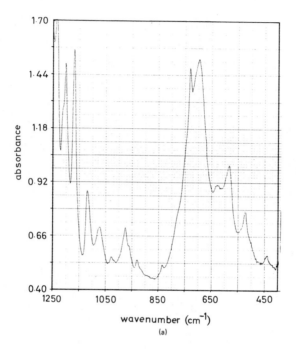

wavenumber (cm⁻¹)

(a)

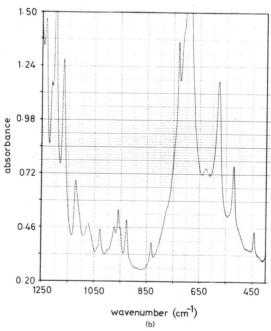

wavenumber (cm⁻¹)

(b)

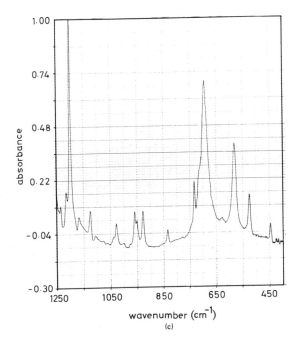

absorbance

wavenumber (cm⁻¹)

(c)

FIGURE 4-29   Application of the FTIR absorbance subtraction technique
              to polyamide-6 films of different states of order: (a)
              FTIR spectrum of melt-quenched polyamide-6 film, (b) FTIR
              spectrum of melt-quenched polyamide-6 film annealed at
              473 K for 14 hr, (c) difference spectrum: (b) - (a).

that in the relaxed crystalline modification the aliphatic segments
of this polymer occur in the *gauche-trans-gauche* conformation [160]
(see also Sec. 4.3.4.3). The absorption bands characteristic of this
crystalline modification have been isolated by subtraction of the
spectrum Fig. 4-28(a) from the spectrum Fig. 4-28(b) so that the broad
absorption band at 1470 cm^{-1} of the $\delta(CH_2)$ vibration of the amorphous
phase is reduced to the base line [Fig. 4-28(c)]. The absorption bands
isolated in the 1450 to 1460 cm^{-1}, 1350 to 1390 cm^{-1}, and 750 to 920
920 cm^{-1} regions have been assigned to CH_2-bending, -wagging, and
-rocking vibrations, respectively, of the *gauche-trans-gauche* conformation [161]. Additionally, the subtraction spectrum reflects any
shifts in wavenumber position (1730 cm^{-1}) and changes in absorptivity
(for example 875 cm^{-1}) upon crystallization. The consequences of the

reversible, stress induced crystalline phase transition of polybutylene terephthalate on its vibrational spectrum will be discussed
in Sec. 4.3.4.5.

Dependent upon thermal and mechanical pretreatment the structure
of polyamide-6 is composed of various fractions of the amorphous,
mesomorphous, and monoclinic crystalline modifications [162]. Here
too, the two types of spectra encountered for this polymer [125,
162, 163] can be attributed to different conformational isomers. In
the amorphous and γ^* mesomorphous form the molecular planes of the
amide group and the aliphatic segments are distorted and the IR
spectra of these modifications are very similar. In contrast, the
crystalline α-modification only occurs in the fully extended chain
conformation with coplanar amide and aliphatic segments. Several absorption bands in the IR spectrum have been assigned to the α-modi-
fication [125, 164, 165]. The isolation of these absorption bands
in the 1250 to 400 cm^{-1} wavenumber region by the subtraction technique (with the disappearance of the 975 cm^{-1} band as criterion for
adjusting the scaling parameter) is demonstrated in Fig. 4-29(a) to
(c). Especially the absorption bands at 1030, 960, and 930 cm^{-1},
which have been assigned to in-plane skeletal vibrations of the amide
group and the $\gamma_r(CH_2)$ absorption band at 833 cm^{-1} have been frequently used for the determination of the so-called α-crystallinity.

In conclusion, this type of subtraction procedure represents an
excellent method to separate spectral features of single modifications, and the obtained data can be used not only for quantitative
determinations of the state of order but also as an experimental basis
for theoretical normal-coordinate analysis studies.

4.3.2 Investigations at High and Low Temperatures

For the characterization of polymer structure the study of vibrational spectra recorded at various temperatures has become of increasing importance. Any changes observed in absorption intensity,
wavenumber position, and band shape directly reflect the temperature
dependence of the vibrational behavior of polymers as a consequence

of changes in the inter- and intramolecular interactions and the state of order [166, 167]. The vibrational spectra of polymers recorded in certain temperature ranges are of special value in studies of melting and recrystallization processes, thermal degradation, hydrogen bonding, and polymorphism and greatly facilitate the assignment of conformational regularity and crystallinity bands. As an illustration Fig. 4-30 shows the IR spectra of the polyester-12,12 (dodecanediol-dodecanedioic acid polyester) at (a) 400 K, (b) 320 K,

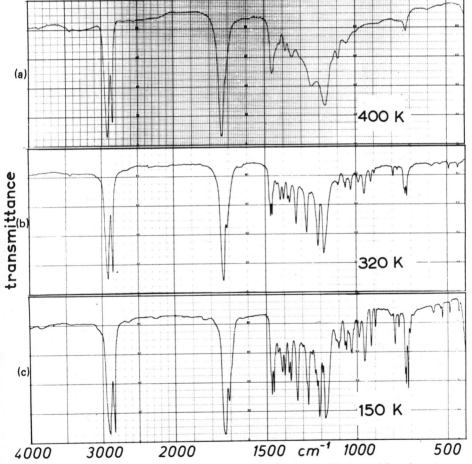

FIGURE 4-30 IR spectra of dodecanediol/dodecanedioic acid polyester (PE-12,12) (a) at 400 K, (b) at 320 K, and (c) at 150 K.

and (c) 150 K, respectively. The spectrum in Fig. 4-30(a) is representative of the molten polyester and spectrum (b) is typical for the
partially recrystallized sample. While some bands vanish upon recrystallization, other new bands appear. According to the nomenclature of
Sec. 2.4 the bands increasing in intensity should be classified as
conformational regularity bands. Upon cooling to 150 K these bands
become sharper and partially exhibit correlation splitting.

Furthermore, studies of polymer and polymerization reactions at
various temperatures provide the necessary data to establish the
corresponding reaction kinetics. The instrumental aspects of variable temperature measurements have been discussed in Sec. 3.2.3.3.

The following section will treat in some detail the results of
studies on the conformational regularity, state of order and crystallization of the comblike isotactic poly(1-alkylethylene)s:

$$- CH - CH_2 - CH - CH_2 -$$
$$\underset{\underset{CH_3}{|}}{(CH_2)_n} \qquad \underset{\underset{CH_3}{|}}{(CH_2)_n} \qquad (n = 0, 1, 2, \ldots 19)$$

For this class of polymers the results of vibrational spectroscopic
investigations at low and high temperatures in combination with
thermoanalytical and x-ray diffraction data have contributed to a
more detailed knowledge of their chemical and physical properties
which depend in a characteristic manner on the length of the side-
chain methylene sequences and their inter- and intramolecular interactions.

A simple although theoretically not very informative physical
constant is the melting point of the crystalline regions (Table 4-1).
The state of order in the first homologues of this series is strongly
influenced by intramolecular forces within the main chain. Thus, isotactic polypropene crystallizes in dependence on the pretreatment in
a 3:1 helix or a planar zigzag conformation [169, 170]. Isotactic
poly(1-ethylethylene) [171-177] and isotactic poly(1-propylethylene)
[178-180] can crystallize in three modifications.

Luongo [172, 181] studied the change of modification II of poly-

TABLE 4-1 Nomenclature, Molecular Weight, and Crystalline Melting Point of Investigated Poly(1-alkyl-ethylene)s

New name	Old name	Abbreviation	Number of CH_2 groups in the side chains	Molecular weight	Melting point (K)
(Polymethylethylene)	Polypropene	P-α-3	0	3.3×10^6	435
Poly(1-ethylethylene)	Polybutene-1	P-α-4	1	1.8×10^6	393
Poly(1-propylethylene)	Polypentene-1	P-α-5	2	1.0×10^6	346
Poly(1-butylethylene)	Polyhexene-1	P-α-6	3	3.0×10^5	?
Poly(1-pentylethylene)	Polyheptene-1	P-α-7	4	6.0×10^4	?
Poly(1-hexylethylene)	Polyoctene-1	P-α-8	5	5.5×10^5	283
Poly(1-heptylethylene)	Polynonene-1	P-α-9	6	6.3×10^5	295
Poly(1-octylethylene)	Polydecene-1	P-α-10	7	8.0×10^5	302
Poly(1-nonylethylene)	Polyundecene-1	P-α-11	8	8.0×10^5	309
Poly(1-decylethylene)	Polydodecene-1	P-α-12	9	5.4×10^5	322
Poly(1-undecylethylene)	Polytridecene-1	P-α-13	10	5.0×10^5	321
Poly(1-dodecylethylene)	Polytetradecene-1	P-α-14	11	1.1×10^5	330
Poly(1-tridecylethylene)	Polypentadecene-1	P-α-15	12	3.0×10^5	327
Poly(1-tetradecylethylene)	Polyhexadecene-1	P-α-16	13	1.7×10^5	340
Poly(1-hexadecylethylene)	Polyoctadecene-1	P-α-18	15	2.8×10^5	344
Poly(1-eicosylethylene)	Polydocosene-1	P-α-22	19	4.5×10^5	364

(1-ethylethylene) to modification I in dependence on time and temperature. He considered the IR band at 1150 cm^{-1} which occurs in all
three modifications (Fig. 2-15) as internal standard for the studies
of the quantitative transition from modification II into modification I. Figure 4-31 shows the time dependence of the absorbance ratio
of the band at 846 cm^{-1} (modification I) and the band at 1150 cm^{-1}.
Since the time for a transition between modification II and I was
shorter for thin films, it was concluded [181] that the change in the
structure begins at the film surface because of tensions within it.

    A similar behavior was observed for the temperature dependence
of this absorbance ratio (Fig. 4-32). The spectra were recorded in
time intervals of 30 min. The most obvious changes in the spectra
upon cooling the sample occur between 293 K and 273 K. Thus, the
typical absorptions of modification I increase drastically, while
those characteristic of modification II decrease.

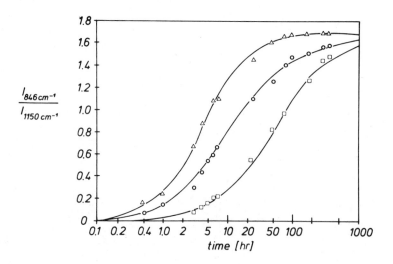

FIGURE 4-31   Absorbance ratio of the IR bands of poly(1-ethylethylene)
               at 1150 cm^{-1}(internal standard) and 846 cm^{-1} (modifica-
               tion I) as function of time: (Δ) 50 μm sample thickness,
               (o) 125 μm sample thickness, (□) 200 μm sample thickness.

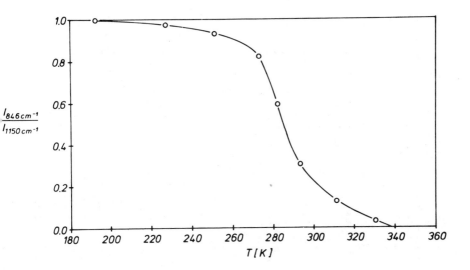

FIGURE 4-32   Absorbance ratio of the IR bands of poly(1-ethylethylene)
at 1150 cm^{-1} (internal standard) and 846 cm^{-1} (modification I) as function of temperature.

Another helpful band combination to study the change from modification II to modification I in dependence on time or temperature
is the doublet at 905 and 925 cm^{-1}. The first band occurs in modification II, the second in modification I (Figs. 2-15 and 4-33). According to recent recalculations of the vibrational frequencies of poly-
(1-ethylethylene) in connection with the normal coordinate analysis
of poly(1-propylethylene) and its deuteroderivatives [182], three
selectively deuterated poly(1-ethylethylene)s [183] and higher poly-
(1-alkylethylene)s [184], these bands can be predominantly assigned
to coupled vibrations of the helical backbone. Therefore, they are
extremely sensitive to the conformation of the investigated polymer.

With increasing side chain length, the melting points of isotactic
poly(1-alkylethylene)s decrease. Poly(1-butylethylene), poly(1-pentyl-
ethylene), and poly(1-hexylethylene) can only crystallize on stretching
or annealing the polymer samples at ambient temperatures. Also with
increasing number of methylene groups in the side chain the polymers

$$-CH_2-CH-$$
$$\overset{|}{C}H_2-CH_3$$

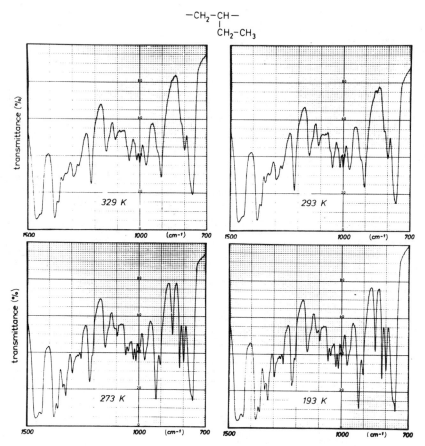

FIGURE 4-33   IR spectra of poly(1-ethylethylene) at different temperatures.

crystallize more easily and the melting points gradually approach
the melting point of polyethylene. This behavior may be explained
by the increasing influence of regular and defined intra- and intermolecular interactions between the methylene sequences of the side
chains. Thus, in poly(1-butylethylene), poly(1-pentylethylene), and
poly(1-hexylethylene) the side chains are too short for regular intra-
and intermolecular interactions which would lead to a conformationally
regular arrangement of the side chains, necessary for a regular three-
dimensional order of the chains. On the other hand, the side chains
are too long and prevent a regular conformation of the main chains.

With increasing number of the methylene groups within the side chains, a zigzag conformation, as in other samples with longer methylene sequences, is favored.

P-α-10 and the following members of the series can crystallize in two modifications. X-ray studies on isotactic poly(1-alkylethylene)s indicate differences in the state of order depending on the preparation method [185, 186], but a definite determination of the structure could not be given. Extended differential thermal analysis (DTA) studies [186-188] of isotactic poly(1-alkylethylene)s with long side chains revealed that in modification II[+] below the endothermic peak indicating the melting of the polymer another significant endothermic peak appears whose enthalpy is considerably enhanced by slowly cooling or annealing the polymer at ambient temperatures, while the enthalpy of the second peak remains nearly constant. The DTA curve of modification I shows only the melt peak. Figure 4-34 shows as example the DTA curve of three isotactic poly(1-eicosylethylene) samples. Curve a and b were measured after melting the sample followed by slow cooling to 328 K and annealing at this temperature for 30 min and 64 hr, respectively. Preceding DTA studies revealed 338 K as optimum annealing temperature [193]. Curve c is due to a solution-crystallized sample (modification I) and does not show a comparable first endothermic peak, a phenomenon which is however not representative for all homologues [187]. The significance of the endothermic peaks (especially that of the first peak, since the second peak indicates the complete melting of the sample) cannot be explained by DTA studies, alone. It proves to be very useful to combine the DTA measurements with x-ray or vibrational spectroscopic studies. To obtain reproducible, comparable IR and Raman spectroscopic data a defined thermal pretreatment of the samples is required. Such a pretreatment can be achieved in the cooling and heating cells described in Sec. 3.2.3.3.

Before looking at more specific results, let us shortly discuss the Raman spectra between 1500 and 800 cm^{-1} of some poly(1-alkylethylene)s scanned at low temperatures (Fig. 4-35). All samples with exception of poly(1-hexylethylene) which was annealed for two weeks

---

[+] Designation according to Ref. 185.

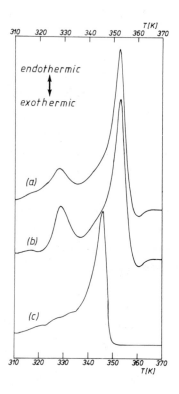

FIGURE 4-34   DTA curves of poly(1-eicosylethylene): (a) slowly cooled
from the melt and annealed at 328 K for 30 min, (b) annealed at 328 K for 64 hr, and (c) solution crystallized
sample.

at 258 K [188] were molten in the cells and subsequently cooled slowly to the indicated temperatures. With increasing number of methylene
groups in the side chains conformational regularity and crystallinity
seem to become more pronounced. The influence of the main-chain vibrations which are observable in the band at 1331 cm^{-1}, the band complex between 1100 and 1150 cm^{-1} and the bands between 800 and 880 cm^{-1}
gradually decreases, while the bands due to a regular *trans* conformation of the side-chain methylene groups arise. Apart from the half
bandwidth and some small bands especially that at 891 cm^{-1} the Raman spectra of poly(1-hexadecylethylene) and poly(1-eicosylethylene)
resemble the spectrum of low density polyethylene [189, 190]. This

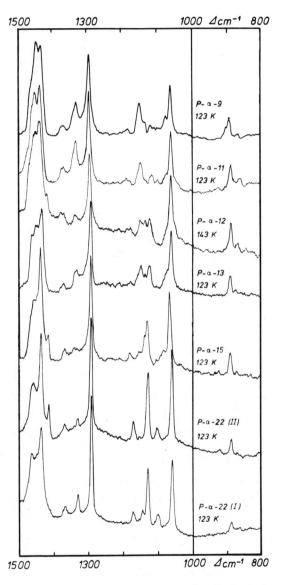

FIGURE 4-35   Raman spectra of various poly(1-alkylethylene)s at 123 K.

effect is also evident from the infrared spectra [190, 192]. However, these spectra are more complex than the Raman spectra and can give us in this case more detailed information of the conformational regularity and crystallinity of these polymers.

The following most characteristic bands in the Raman and IR spectra of modification II do not occur in modification I:

1. Splitting of the CH_2-bending vibration in the Raman and IR spectra.

   Raman: $a_g \rightarrow A_g + B_{1g}$ 1416, 1442 cm^{-1}

   IR: $b_{2u} \rightarrow B_{2u} + B_{3u}$ 1452, 1463 cm^{-1}

2. Splitting of the CH_2-rocking vibration in the IR spectra.

   IR: $b_{2u} \rightarrow B_{2u} + B_{3u}$ 720, 731 cm^{-1}

   The 731 cm^{-1} band is superimposed by the second band of the band progression series of the CH_2-rocking mode in the spectra of those polyalkylethylenes with $n \geq 9$. The bands of the progression series are split into doublets.

3. The IR-active combination band of the IR-active CH_2-rocking mode at 731 cm^{-1} (B_{3u}) and the Raman-active CH_2-rocking mode at 1170 cm^{-1} (B_{1g}) arises in the spectra of modification II at 1893 cm^{-1}.

With increasing temperature spectral changes can be observed in the spectra at temperatures which coincide with the first endothermic DTA peak of analogously prepared samples. Figure 4-36(a) shows a plot of $d[\log(I_o/I)]/dT$ of the IR-active CH_3-rocking band near 891 cm^{-1} in dependence on the temperature of some poly(1-alkylethylene)s with long side chains. The temperatures of the peak maxima are comparable with those of the first endothermic peak [191]. The same qualitative behavior is shown by the band progression of the CH_2-rocking and wagging vibrations of the side-chain methylene sequences, but because of the rather complex band features an analogous quantitative evaluation is not possible. Furthermore, the temperature dependence of the correlation splitting of the CH_2-rocking vibration in a B_{3u} and B_{2u} mode with bands at 731 and 720 cm^{-1}, respectively, proves to be inconclusive since the most intense progression band arises at about the same wavenumber. However, poly(1-decylethylene) and the following homologous polymers show a band at 1893 cm^{-1} which is not overlapped by other bands. This band is assigned to a combination band of the Raman-active CH_2-rocking fundamental (B_{1g}) at 1170 cm^{-1} and the IR-active CH_2-rocking fundamental (B_{3u}) at 731 cm^{-1} [192]. The band always occurs in polymers with methylene sequences or alkanes which possess a similar arrangement of the methylene groups within the unit cell as orthorhombic polyethylene. Figure 4-36(b) shows the temperature dependence of the band intensity for poly(1-hexadecylethylene) and poly(1-eicosylethylene). The change in the intensity decrease coincides with the beginning of the first endothermic DTA peak [193].

Inspection of the corresponding Raman spectra of poly(1-hexadecylethylene) and the following homologues shows a correlation splitting of the CH_2-bending vibration into the B_{1g} and A_g modes, which is generally not obscured by overlapping with other bands, as in the corresponding IR spectra. Wavenumber position and intensity of the low

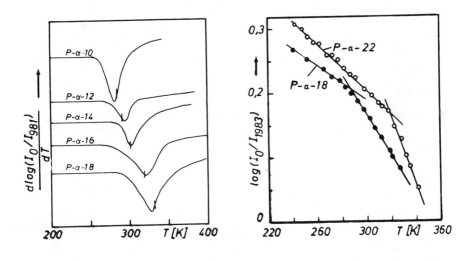

FIGURE 4-36 (a) Plot of $d[log(I_0/I_{891})]/dT$ as function of temperature (the marks indicate the peak maximum of the corresponding DTA curves. (b) Absorbance of the 1893 cm^{-1} combination band in relation to temperature.

frequency band (A_g) depends more on temperature and side-chain length
than the high frequency band (B_{1g}) [191, 193]. Thus from these studies
we can conclude that the first endothermic peak of the poly(1-alkyl-
ethylene)s indicate a change in the polyethylene-like three-dimensional arrangement of the side chains with two methylene sequences running
through a unit cell. The IR and Raman spectra of the other modification do not indicate intermolecular interactions between two adjacent
methylene sequences with a polyethylene-like structure. The correlation splitting cannot be observed and the combination band at 1893
cm^{-1} is absent. Nevertheless, the spectra indicate that in this modification the side chain must possess almost the same extended *trans*
conformation of the methylene groups. Thus, the characteristic Raman
bands and the progression bands appear at nearly the same wavenumber.
The bands due to main-chain vibrations are more pronounced [193].

Let us finally discuss the temperature-dependent changes of the
state of order in a quenched poly(1-octylethylene) film (Fig. 4-37)
[188]. The spectrum at 321 K was recorded from the melt and the spectrum at 131 K after quenching the sample. There do not arise essentially new bands and only a sharpening of bands can be observed. On heating the quenched sample to 244 K a loss in intensity and broadening
of bands occur. When a spectrum is scanned again within an hour at
about the same temperature new bands arise and furthermore again a
decrease in the bandwidth can be observed. The new bands can be seen
more pronounced in the spectrum of the sample after slow cooling from
the melt to 129 K. These characteristic changes in the spectra at
245 K occur at the beginning of the cold crystallization shown in the
corresponding DTA curve.

The reversible, stress-induced transformation of the so-called
α-form of polybutylene terephthalate (PBT) with an approximately
gauche-trans-gauche conformation of the aliphatic segments to the extended, all-*trans* β-form [160, 408] will be discussed in detail in
Sec. 4.3.4.5. Here, attention is primarily focused on the structural
changes observed in the IR and Raman spectra of PBT as a function of
the thermal pretreatment. A scheme of the combined thermal and mechanical pretreatments of the investigated PBT fibers, previously charac-

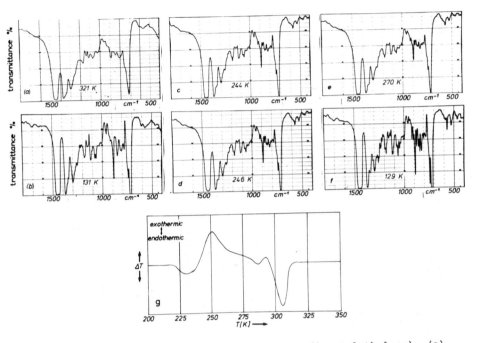

Figure 4-37   IR spectra and DTA curve of poly(1-octylethylene): (a)
melt, (b) quenched from the melt, (c), (d), and (e)
heated from (b) to the indicated temperature, (f) slowly
cooled from the melt, and (g) DTA trace of the quenched
sample.

terized by x-ray diffraction [408], is outlined in Fig. 4-38 which
in addition shows the corresponding Raman spectra. The most significant spectral changes are observed for the 885 cm^{-1} band. This
band can be assigned to a coupled rocking vibration of a *gauche* CH_2
group [512]. It appears in the melt-quenched sample A and increases
in intensity during annealing at 418 K (B) corresponding to the grow-
ing amount of *gauche-trans-gauche* conformations. As expected, this
band decreases upon stretching at 300 K (A → D) and subsequent annealing at 458 K under tension (D → E). In analogy to the IR spectra annealing of the sample D at 458 K without tension or recovery of the
sample E at 300 K results in equivalent spectra (F). The predominating
α-form is indicated by the largely increased band at 885 cm^{-1}. The IR
spectra of Fig. 3-39 demonstrate the change of the orientated α-form
(F) to the less orientated α-form (B) during heating from 338 K to
468 K.

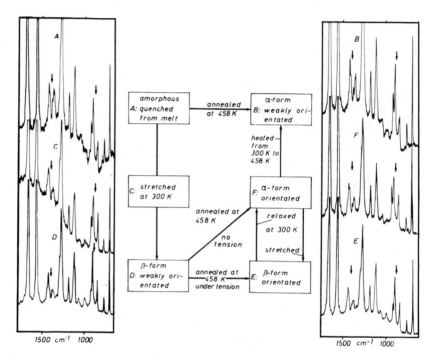

FIGURE 4-38   Sample pretreatment and corresponding Raman spectra
of a polybutylene terephthalate fiber.

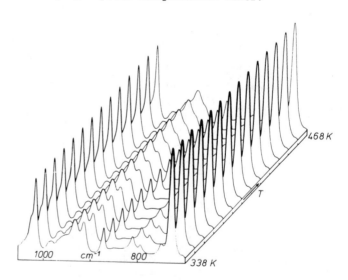

FIGURE 4-39   FTIR spectra obtained during heating of a polybutylene
terephthalate film (sample F).

As mentioned previously vibrational spectroscopic investigations at high and low temperatures are of interest for many other aspects of polymer structure elucidation. In Sec. 4.3.3 IR spectroscopic investigations of the stability of hydrogen bonding in polyamides at elevated temperatures will be discussed in some detail. The importance of variable temperature measurements with reference to reaction kinetics will be treated in Sec. 4.7. Further references to high- and low-temperature investigations of selected polymers by IR and Raman spectroscopy may be found in the pertinent literature [30, 32, 194, 195].

4.3.3 Hydrogen Bonding

The elucidation of the role played by hydrogen bonds in the structure and properties of polymeric solids has been the subject of numerous investigations [32, 196-219]. As far as the mechanical properties are concerned the contributions from hydrogen bonding are still open to question. While some investigators stress the important part of hydrogen bonding for these aspects [203, 211, 220], others conclude that too much emphasis has been placed on their role in determining mechanical properties [210, 212]. Although the energies of hydrogen bonds are weak (20-50 kJ/mol) in comparison to covalent bonds (of the order of 400 kJ/mol), this type of molecular interaction is large enough to produce appreciable frequency and intensity changes in the vibrational spectra of the examined compounds. In fact, the disturbances are so significant that IR and Raman spectroscopy provide the most informative source of criteria for the presence of hydrogen bonds [221-223].

Hydrogen bonding involves the interaction between a proton donating group (R_1-X-H) and a proton acceptor (Y-R_2) and may be described schematically by

$$R_1 - X - H \cdots Y - R_2$$

The origin of this interaction has been described in electrostatic and quantum-mechanical terms [224, 225]. The formation of a hydrogen bond is generally favored by highly electronegative X and Y atoms with relatively small atomic radii (for example, O, N, F) and requires Y atoms with lone-pair electrons [226-228].

The quantum-mechanical aspects of the hydrogen bond have been discussed by some authors in terms of various potential curves for the proton involved in the hydrogen bond [227, 230-232]. Both, experimental and theoretical aspects of hydrogen bonding are comprehensively covered in numerous books and review articles [221-223, 228, 230, 235-244].

The displacements involved in the various modes of the X, H, and Y atoms participating in the hydrogen bond and the corresponding wavenumber ranges are shown in Table 4-2.

As a consequence of the hydrogen bonding forces the ν(XH)- and ν(YR$_2$)-stretching frequencies will be lowered, whereas the deformation frequencies associated with the motions of the H and Y atoms perpendicular to their respective bonds will be increased.

The energy of the hydrogen bond is directly reflected by the ν(XH$\cdots$Y)-stretching and δ(XH$\cdots$Y)-deformation vibrations. Unfortunately, these vibrations are of extremely low frequency and relatively few reliable data are available. Low-frequency Raman and FIR spec-

TABLE 4-2   Vibrational Modes of the Hydrogen Bond

Displacements involved	Nomenclature	Wavenumber range
$R_1 - \overset{\leftarrow}{X} - \vec{H} \cdots Y - R_2$	ν(XH)-stretching vibration	3600-2500 cm^{-1}
$R_1 - X - \overset{\uparrow}{\underset{\downarrow}{H}} \cdots Y - R_2$	δ(XH)-in-plane deformation vibration	1650-1000 cm^{-1}
$R_1 - \overset{-}{X} - \overset{+}{H} \cdots \overset{-}{Y} - R_2$	γ_w(XH)-out-of-plane deformation vibration	900-300 cm^{-1}
$R_1 - \overset{\leftarrow}{X} - \overset{\leftarrow}{H} \cdots \vec{Y} - R_2$	ν(XH$\cdots$Y)-stretching vibration of hydrogen bond	250-30 cm^{-1}
$R_1 - X - H \cdots \overset{\uparrow}{\underset{\downarrow}{Y}} - R_2$	δ(XH$\cdots$Y)-deformation vibration of hydrogen bond	<100 cm^{-1}

troscopic investigations of hydrogen bonding (primarily in low-molecular-weight compounds) have been summarized in Refs. 222, 237, and
245 to 250.

Most of the investigations so far reported deal with the observed
frequency shift and intensity increase of the ν(XH)-stretching vibration upon hydrogen bonding. Especially the frequency shift $\Delta\nu$(XH) has
been correlated with various chemical and physical properties of the
hydrogen bond {e.g., enthalpy of formation, bond distance $R_{X...Y}$,
ν(XH) half bandwidth [221-223, 251-259]}. Some authors have obtained
linear relations between these parameters and the frequency shift of
the ν(YR$_2$)-stretching vibrations [221-223, 260, 261]. The significant
IR intensity enhancement which accompanies the formation of the hydrogen bond can be interpreted in terms of the augmentation of the X-H
and electron donor dipole moment as a consequence of the charge re-
distribution produced by the new bond [242, 262].

Dependent on the geometry of the molecule intra- or intermolecular
hydrogen bonds may be encountered. Intramolecular hydrogen bonds can
be distinguished from the intermolecular type, since the latter variety is concentration dependent while the former is not. Commonly the
type of hydrogen bonding has been found to be characteristic of the
stereochemistry or conformation of the investigated compound [263].

Thus, the helix conformation of polypeptides and proteins is sta-
bilized by intramolecular hydrogen bonds, while intermolecular hydrogen bonding occurs in the extended structure [264, 265].

Hydrogen bonding is particularly important in polymers and copolymers containing the following molecular groups: amide (polypeptides, proteins, polyamides), urethane (polyurethanes), hydroxyl
(cellulose, polyvinyl alcohol), carboxyl (polyacrylic acid). Hydrogen
bonds of the type CH$\cdots$Cl have been suggested of importance for the
structure of polyvinyl chloride [30, 266, 267].

The ability of a CH group to act as a proton donor in hydrogen
bonding apparently depends on the carbon hybridization [C(sp)-H >
C(sp^2)-H > C(sp^3)-H] and increases with the number of adjacent elec-
tron-withdrawing groups [268].

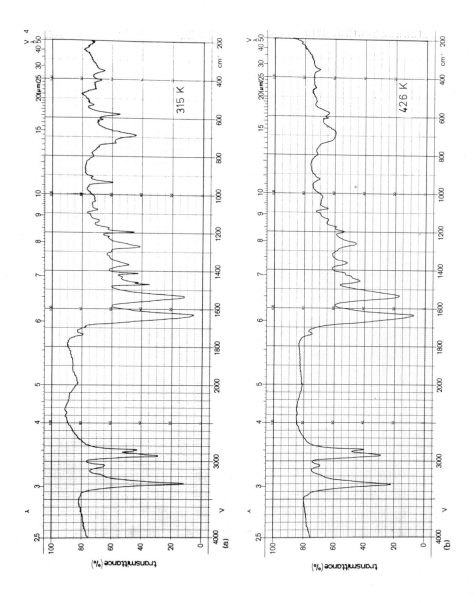

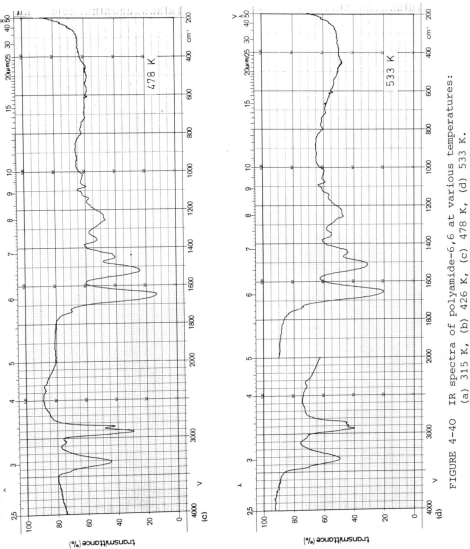

FIGURE 4-40   IR spectra of polyamide-6,6 at various temperatures:
(a) 315 K, (b) 426 K, (c) 478 K, (d) 533 K.

IR spectroscopy has proved an excellent tool to study the hydrogen bonding behavior in polyamides and polyurethanes. Thus, it has
been shown that at room temperature over 99% of the NH protons in
both even and odd members of the aliphatic polyamide series are hydrogen bonded [198, 200]. Valuable information regarding the temperature dependence of hydrogen bonding may be derived from IR studies at
elevated temperatures [209, 213, 217, 218, 229]. The IR spectra of a
polyamide-6,6 film (Fig. 4-40) recorded in the room temperature to
melting range interval (with the aid of the variable temperature
cell of Fig. 3-30) exhibit some characteristic temperature-dependent
features. Although no detailed information can be derived from the
order-sensitive absorption bands in the 800 to 1200 cm^{-1} wavenumber
region owing to the reduction in intensity and diffuse nature of
these bands far below the melting point, significant changes can be
observed for the intensity, wavenumber position, and shape of the
ν(NH) absorption band as a consequence of changes in the hydrogen
bonding state. While the intense absorption band at 3300 cm^{-1} can be
assigned to the ν(NH)-stretching vibration of the NH groups associated through hydrogen bonds [ν_{ass}(NH)], the small shoulder at about
3450 cm^{-1} is characteristic of the ν(NH) vibration of the non-hydrogen bonded NH groups [ν_{free}(NH)]. With increasing temperature, the
following spectral changes are observed:

1. The intensity of the ν_{free}(NH) band increases at the cost of the
   ν_{ass}(NH) band.
2. The peak maximum of the ν_{ass}(NH) band is shifted toward larger
   wavenumbers.
3. The half width of the ν_{ass}(NH) band increases considerably with
   increasing temperature.

The intensity decrease and increase of the ν_{ass}(NH) and ν_{free}(NH)
absorptions, respectively, are indicative of the shift in equilibrium concentration of the hydrogen bonded and non-hydrogen bonded
NH groups. Unfortunately, these absorption bands overlap strongly
and cannot be resolved in the components. Bessler and Bier [209]
have applied a base-line method for the estimation of the percentage

of non-hydrogen bonded NH groups by adopting a value of 1.6 for the absorptivity ratio a_{ass}/a_{free}. For the investigated polyamides (poly-amide-6, polyamide-6,10, polyamide-10,10, polyamide-11) they derived approximate values between 10 and 20% at 503 K. A total ν(NH) absorbance area procedure for the quantitative assessment of hydrogen bonding in polyamides has been recently proposed by Schroeder and Cooper [218]. For various polyamide homo- and copolymers they obtained somewhat higher values for the fraction of nonbonded NH groups (for example, 20% at 393 K for polyamide-6,10). The enthalpies of dissociation of hydrogen bonds reported in this paper range from 34 to 50 kJ/mol.

The wavenumber shift and increase in bandwidth of the ν_{ass}(NH) band at higher temperatures are the result of a general weakening of the hydrogen bonds and a concomitant broader distribution of their energies. In previous investigations on polyamide-6, Frigge and Dechant have shown [213] that the spectral characteristics mentioned under 1, 2, and 3 undergo drastic changes in the melting range of the polymer. T_g and T_m transitions have also been observed in plots of the total ν(NH) absorbance area of various polyurethanes against temperature [212, 269].

Similar temperature-dependent spectral changes can be observed for the ν(C=O) absorption band of polyamides and polyurethanes (see Fig. 4-40). The frequency shift in the vibration of bonded and free carbonyl groups however, is less pronounced than that of NH groups. This is to be expected since the acceptor atom is certainly less displaced than the hydrogen atom of the donor group [217].

The intensity and wavenumber changes of the ν(NH) absorption band occurring in the IR spectra of aliphatic polyamides and polyurethanes at elevated temperatures could not be observed with poly-(p-phenylene terephthalamide). In Fig. 4-41 the IR spectra of a poly-(p-phenylene terephthalamide) film recorded in a high-temperature cell at 313, 423, and 573 K, respectively, are shown. The lack of similar, significant spectral changes up to temperatures as high as 573 K can be interpreted as a consequence of the stability of the involved

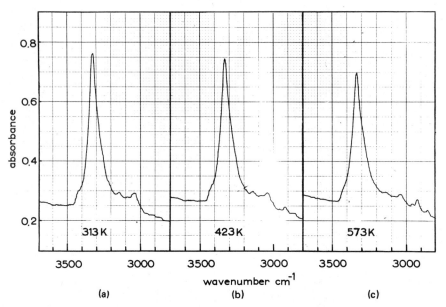

FIGURE 4-41   IR spectra of poly(p-phenylene terephthalamide) in the
ν(NH)-stretching vibration region at various temperatures: (a) 313 K, (b) 423 K, (c) 573 K.

hydrogen bonds and is consistent with the exceptional thermal and
mechanical properties of this aromatic polyamide.

The bandwidth (up to 500 cm^{-1} and more) and complex nature of the
ν(XH)-stretching vibration has been treated theoretically as a consequence of proton tunneling between different potential wells [230,
270-273] or coupling with other vibrations of the molecule [274-277].
Sometimes it may simply be attributed to a superposition of the ν(XH)
absorption bands belonging to various hydrogen bonds. This has been
demonstrated for cellulose where the primarily crystalline ν(OH) band
remaining upon deuteration is split into several maxima which could
be assigned to inter- and intramolecular hydrogen bonds from their
dichroic behavior in polarization spectra (Fig. 4-69).

Hydrogen bonding plays a very important role in the structure
of biological macromolecules such as polypeptides, proteins, and polynucleotides. Changes in the conformation of these polymers (e.g., denaturation, α-β transformation) are invariably accompanied by rupture
and reorganization of the vital hydrogen bonds [264, 265, 278-280].

Most IR and Raman spectroscopic investigations of hydrogen bonding in biopolymers deal with the ν(NH) (about 3300 cm^{-1}) and ν(C=O) (1600-1700 cm^{-1}) vibrations [127, 281-284]. Thus, the high ν(NH)-stretching frequency (3325-3330 cm^{-1}) in the IR spectra of collagen and collagen-like polypeptides has been interpreted by the degree of supercoiling in these polymers which may sterically prevent optimal hydrogen bond angles. The shift of this absorption toward 3300 cm^{-1} upon denaturation confirms that the high ν(NH) frequency is characteristic of some level of collagen structure which is destroyed by heating [285].

In spectroscopic studies of polynucleotides the ν(C=O) absorption of the uracil base was found extremely sensitive to interbase hydrogen bonds [284, 286]. Peticolas and Small [284] have characterized phase changes of the helical system of a polyriboadenylic/uridylic acid by the temperature dependence of Raman bands in this region. One strong band was observed at low temperatures for the double helix (1681 cm^{-1}), thought to be due to one of the ν(C=O)-stretching vibrations of the hydrogen-bonded uracil. Above the transition temperature (332 K) the band at 1681 cm^{-1} is replaced by two bands at 1698 and 1660 cm^{-1} which were assigned to the two ν(C=O)-stretching vibrations of the uracil moiety. The plot of the 1660 cm^{-1} band intensity with increasing temperature has been interpreted as a clear indication of the loosening and eventual break up of the interbase hydrogen bonding in the double helix.

Various authors [287-290] have shown that the intensity ratio of the 827/852 cm^{-1} doublet in the Raman spectra of polypeptides containing the tyrosine residue is determined by the nature of hydrogen bonding of the phenolic hydroxyl, which determines the extent of the Fermi resonance between the two components [289]. Lord and co-workers have studied the association of certain adenine and uracil crystals by low-frequency IR and Raman spectroscopy [291, 292]. FIR spectra of polyribocytidylic and polyriboinosinic acid have been obtained [293], but owing to the diffuse nature of the bands, no detailed assignments to low-frequency vibrational modes (e.g., stretching of base pair hydrogen bonds) could be made.

4.3.4 Orientation

The increasing interest in the elucidation of orientation phenomena
in polymeric materials stems from both theoretical and practical
points of view. Thus, the measurement of orientation not only provides a basis for establishing molecular mechanisms of deformation
in polymers but also contributes to a better understanding of the
interrelationship between technological properties and polymer structure. Among other techniques (e.g., x-ray diffraction, birefringence,
NMR, polarized fluorescence, sonic techniques) IR and laser Raman
spectroscopy are useful tools for the investigation of anisotropy in
polymers. Owing to the different selection rules of the physical processes involved in IR and Raman spectroscopy, the theory of IR dichroism and Raman polarization measurements will be treated separately. Nevertheless, a uniform coordinate system and angle nomenclature
was introduced for the description of the geometrical relations involved in the different spectroscopic techniques.

4.3.4.1 Infrared Dichroism

When an oriented polymer is investigated with linearly polarized
radiation the absorbance of a single group in the polymer chain is
proportional to the square of the scalar product of its transition
moment vector $M = \partial\vec{\mu}/\partial r$ and the electric vector E of the incident
polarized radiation [see also Eqs. (2.19) and (2.20)]:

$$a \propto \left(\frac{\partial\vec{\mu}}{\partial r}E\right)^2 = (ME)^2 \cos^2\gamma \tag{4.26}$$

where $|\partial\vec{\mu}/\partial r| = M$, $|E| = E$, and γ is the angle between the transition
moment and the electric vector. Thus, maximum absorption takes place
when the electric vector is parallel to the transition moment of the
vibrating group, but no light will be absorbed when its electric
vector is perpendicular to the transition moment. The actually observed absorbance A is equal to the sum of absorbance contributions
from all structural units:

$$A \propto \int_n (ME)^2 \, dn \tag{4.27}$$

where the integral refers to the summation of all molecules. If these
atomic groups and their associated transition moments are randomly
oriented within the polymer, the measured absorbance A is independent
of the polarization direction of the incoming light. For anisotropic
distribution of the transition moments, the observed net absorbance
varies with the direction of the electric vector E of the polarized
radiation. The effect of anisotropy on a particular absorption band
in the IR spectrum of a polymer is characterized by the dichroic
ratio R

$$R = \frac{A_{\parallel}}{A_{\perp}} \qquad\qquad (4.28)$$

where $A_{\parallel}$ and $A_{\perp}$ are the integrated absorbances measured with radiation polarized parallel and perpendicular to a direction of refer-
ence (e.g., draw direction), respectively.[†] Under the assumption of
equivalent band shapes the integrated absorbances may be replaced by
the absorbances measured at the peak maxima. The determination of
dichroic ratios is of significant importance for three aspects:

1. Complete assignment of vibrations to specific symmetry races (in
   combination with Raman depolarization measurements)

2. Elucidation of the molecular geometry by detemination of transition moment directions of particular groups relative to each
   other or to the molecular chain axis

3. Quantitative relation of the dichroic ratio to certain orientation functions characteristic of the average orientation of structural units within the polymer.

It must be emphasized, however, that the exact direction of the
vibrational transition moment relative to the direction of the chemical bond is not always known. Owing to the relatively small mass of
the hydrogen atom, the transition moment directions of certain
stretching vibrations such as C-H or N-H will be localized along the
chemical bond. The participation of the hydrogen atom in a hydrogen
bond however, may result in distortion of the transition moment di-

---

[†]The experimental background of polarization measurements is treated
in Sec. 3.1.1.1.

rection. Because of the large force constants involved, transition
moments of stretching modes belonging to multiple bonded groups will
also coincide closely with the bond direction as long as no significant delocalization of the bonding electrons occurs [for example,
ν(C=O) in oxidized polyethylene or ν(C≡N) in polyacrylonitrile]. Despite coupling of the two C-H bond vibrations in the CH_2 group the
transition moment direction for the stretching and deformation modes
of this group are also fairly well defined (Fig. 4-42). Strong mechanical and electrical coupling of the C=O and C-N stretching vibrations in the planar amide group -CONH- cause appreciable deviation
(approximately 15°) of the ν(C=O) transition moment from the corresponding bond direction [294-297]. Similar discrepancies have been reported for the vibrations involving the -COO- group in polyesters
[297, 298] and for the ν(CCl) vibration in polyvinyl chloride [192].

In general, the crystalline and amorphous phase of a semicrystalline polymer can be characterized separately by evaluating the dichroic effects of absorption bands which are peculiar to these regions.

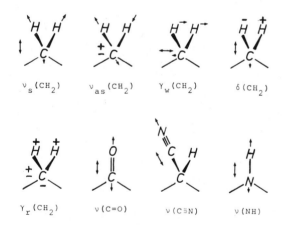

$\nu_s(CH_2)$ $\nu_{as}(CH_2)$ $\gamma_w(CH_2)$ $\delta(CH_2)$

$\gamma_r(CH_2)$ $\nu(C=O)$ $\nu(C≡N)$ $\nu(NH)$

FIGURE 4-42   Transition moment directions of some vibrational modes
              (the symbols + and - denote the movement of atoms per-
              pendicular to the plane of the paper).

The vibrational transition moments of symmetrical molecules are parallel or perpendicular to their symmetry elements, and therefore
dichroic measurements on crystalline absorption peaks can be employed
to study the orientation of certain crystallographic axes of the conformationally regular structure within the crystalline regions. When
more than one polymer chain runs through the unit cell, the splitting
of conformational regularity bands in two components which may be polar-

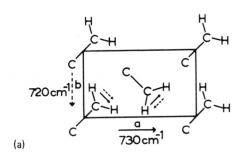

(a)

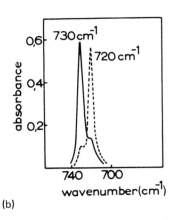

(b)

FIGURE 4-43   (a) Transition moment directions of the in-phase and
out-of-phase CH_2-rocking vibration in polyethylene (see
text). (b) IR spectrum of a single crystal of monoclinic
n-$C_{36}H_{74}$ (see also Ref. 299). The solid line shows radiation with electric vector polarized along a-axis; the
dashed line shows radiation with electric vector polarized along b-axis.

ized along different crystal axes can be observed as a consequence
of intermolecular forces (Fig. 4-43) [299]. Although the chain conformation is no longer uniform in the amorphous regions, it has been
shown [300] that certain absorptions may be assigned to specific sequences of local conformations within these regions. However, the
dichroic effects observed are generally small and naturally depend
on the conformation sequence represented by the absorption band under
examination [301].

A very useful technique for separating the crystalline and amorphous contributions to the absorption band of a particular group is
deuterium exchange (see Sec. 4.3.5). Neglecting relaxation phenomena
during the deuteration process, the dichroism of the residual X-H
and frequency-shifted X-D-stretching vibration, for example, will
approximately reflect the orientation of the crystalline and amorphous domains, respectively (Sec. 4.3.5, Fig. 4-85). The principal
advantage of such an experiment is that it offers the possibility
of studying the preferential alignment of the transition moments of
identical vibrational modes in different states of order.

4.3.4.2 Relation between Orientation and Infrared Dichroism

For the derivation of general relations between the dichroic ratio
and orientation parameters let us introduce the following three coordinate systems (Fig. 4-44): (x,y,z) refers to the structural unit
or the molecule, (u,v,w) to the sample frame, and (X,Y,Z) to the
laboratory. This nomenclature for the coordinate systems is used
throughout the book. For conventional infrared dichroic measurements
of film samples the stretching (w), transverse (u), and thickness (v)
directions coincide with the laboratory coordinate axes Z, X, and Y,
respectively. Let us specify the direction of the transition moment
associated with the vibrational mode of a structural unit with reference to the coordinate system of the sample frame (u,v,w) by the
polar and azimuthal angles γ and δ, respectively (Fig. 4-45). The
orientation of the sample may then be described by a distribution
function $a(\gamma,\delta)$ characterizing the density of transition moments

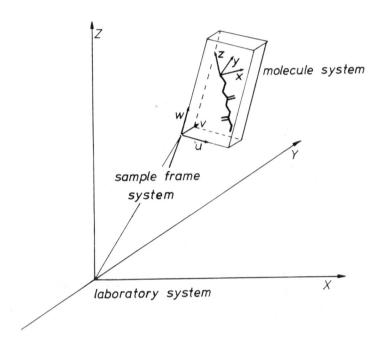

FIGURE 4-44   Definition of the coordinate systems: (x,y,z) molecule
system, (u,v,w) sample frame system, and (X,Y,Z) laboratory system.

pointing in the segment γ, $\gamma + d\gamma$; δ, $\delta + d\delta$. For an isotropic sample
this distribution function is a constant a_o which can be derived from
the normalization condition

$$\int_{\gamma=0}^{\pi} \int_{\delta=0}^{2\pi} a(\gamma,\delta) \, \sin\gamma \, d\delta \, d\gamma = 1 \qquad\qquad (4.29)$$

as $a_o = 1/4\pi$.

With the distribution function $a(\gamma,\delta)$, the absorbance $A_{uvw}(\alpha,\beta)$
can be expressed in terms of the coordinate system (u,v,w) of the
sample frame as follows:

$$A_{uvw}(\alpha,\beta) = K \int_{\gamma=0}^{\pi} \int_{\delta=0}^{2\pi} a(\gamma,\delta) \, \sin\gamma \, (ME)^2 \, d\delta \, d\gamma \qquad (4.30)$$

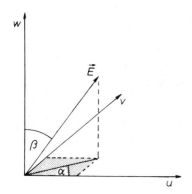 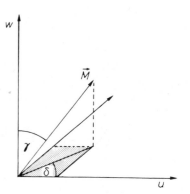

FIGURE 4-45   Characterization of E and M by the angles α, β and γ, δ, respectively.

where α and β are the angles between the projection of E in the uv-plane and the u-axis and E and the w-axis, respectively (Fig. 4-45). Let us assume E as unit vector. With $M = |\mathbf{M}|$ we can represent the vectors E and M by

$$E = (\sin\beta \cos\alpha, \sin\beta \sin\alpha, \cos\beta) \tag{4.31}$$

$$M = M(\sin\gamma \cos\delta, \sin\gamma \sin\delta, \cos\gamma) = (M_u, M_v, M_w) \tag{4.32}$$

Then the square of the scalar product ME is given by

$$
\begin{aligned}
\left(\frac{\partial\vec{\mu}}{\partial r}E\right)^2 &= M^2[\sin^2\gamma \cos^2\delta \sin^2\beta \cos^2\alpha + \sin^2\gamma \sin^2\delta \sin^2\beta \sin^2\alpha \\
&\quad + \cos^2\gamma \cos^2\beta + 2(\sin^2\gamma \sin\delta \cos\delta \sin^2\beta \sin\gamma \cos\alpha \\
&\quad + \sin\gamma \cos\gamma \cos\delta \sin\beta \cos\beta \cos\alpha \\
&\quad + \sin\gamma \cos\gamma \sin\delta \sin\beta \cos\beta \sin\alpha)] \\
&= M_u^2\sin^2\beta \cos^2\alpha + M_v^2\sin^2\beta \sin^2\alpha + M_w^2\cos^2\beta \\
&\quad + 2(M_uM_v\sin^2\beta \sin\alpha \cos\alpha + M_uM_w\sin\beta \cos\beta \cos\alpha \\
&\quad + M_vM_w\sin\beta \cos\beta \sin\alpha)
\end{aligned}
\tag{4.33}
$$

Introduction of Eq.(4.33) in Eq.(4.30) results in

$$A_{uvw}(\alpha,\beta) = K \int_{\gamma=0}^{\pi} \int_{\delta=0}^{2\pi} a(\gamma,\delta) \, \sin\gamma \, [M_u^2 \sin^2\beta \, \cos^2\alpha + M_v^2 \sin^2\beta \, \sin^2\alpha$$

$$+ M_w^2 \cos^2\beta + 2(M_u M_v \sin^2\beta \, \sin\alpha \, \cos\alpha + M_u M_w \sin\beta \, \cos\beta \, \cos\alpha$$

$$+ M_v M_w \sin\beta \, \cos\beta \, \sin\alpha)] \, d\delta \, d\gamma \qquad (4.34)$$

For light polarized along the u, v, or w-axis the corresponding absorbances A_u, A_v, A_w of the transition moments M_u, M_v, M_w can be derived from Eq.(4.34) as

$$A_u = A_{uvw}(0^\circ,90^\circ) = K \int_{\gamma=0}^{\pi} \int_{\delta=0}^{2\pi} a(\gamma,\delta) \, \sin\gamma \, M_u^2 \, d\delta \, d\gamma \qquad (4.35a)$$

$$A_v = A_{uvw}(90^\circ,90^\circ) = K \int_{\gamma=0}^{\pi} \int_{\delta=0}^{2\pi} a(\gamma,\delta) \, \sin\gamma \, M_v^2 \, d\delta \, d\gamma \qquad (4.35b)$$

$$A_w = A_{uvw}(\alpha,0^\circ) = K \int_{\gamma=0}^{\pi} \int_{\delta=0}^{2\pi} a(\gamma,\delta) \, \sin\gamma \, M_w^2 \, d\delta \, d\gamma \qquad (4.35c)$$

Then Eq.(4.34) reads

$$A_{uvw}(\alpha,\beta) = A_u \sin^2\beta \, \cos^2\alpha + A_v \sin^2\beta \, \sin^2\alpha + A_w \cos^2\beta$$

$$+ A_{uv} \sin^2\beta \, \sin\alpha \, \cos\alpha + A_{uw} \sin\beta \, \cos\beta \, \cos\alpha$$

$$+ A_{vw} \sin\beta \, \cos\beta \, \sin\alpha \qquad (4.36)$$

where the terms A_{uv}, A_{uw}, and A_{vw} have the form

$$A_{ij} = 2K \int_{\gamma=0}^{\pi} \int_{\delta=0}^{2\pi} a(\gamma,\delta) \, \sin\gamma \, M_i M_j \, d\delta \, d\gamma \qquad (4.37)$$

The A_{ij} vanish if the function $a(\gamma,\delta)$ is even for the singularities of $\sin\gamma \, M_i M_j$:

$$A_{uv} = 0: \quad \sin^3\gamma \, \sin\delta \, \cos\delta = 0 \quad \text{then} \quad \gamma = 0, \ \delta = 0, \ \frac{\pi}{2}, \ \pi, \ \frac{3\pi}{2}$$

$$A_{uw} = 0: \quad \sin^2\gamma \, \cos\gamma \, \cos\delta = 0 \quad \text{then} \quad \gamma = 0, \ \frac{\pi}{2}, \ \delta = \frac{\pi}{2}, \ \frac{3\pi}{2} \qquad (4.38)$$

$$A_{vw} = 0: \quad \sin^2\gamma \, \cos\gamma \, \sin\delta = 0 \quad \text{then} \quad \gamma = 0, \ \frac{\pi}{2}, \ \delta = 0, \ \pi$$

According to Fig. 4-45, it can be derived from the above values that the distribution function $a(\gamma, \delta)$ has to be symmetric to at least two of the planes uv, uw, and vw.

Since in an absorption process a transition moment is equivalent to its opposite one, the distribution function always contains a center of inversion, thereby creating additional symmetry elements [29]. Practically, most of the transition moment distribution functions induced by deformation processes fulfil the above symmetry requirements and the cross terms A_{uv}, A_{uw}, and A_{vw} disappear. Equation (4.36) is then the equation for an ellipsoid whose axis lengths are given by $(1/A_u)^{1/2}$, $(1/A_v)^{1/2}$, and $(1/A_w)^{1/2}$. $[1/A_{uvw}(\alpha, \beta)]^{1/2}$ is then the distance from the center O to a point P at the surface of the ellipsoid, where $\overline{OP}$ is parallel to the electric vector of the incident polarized radiation. Thus, the dichroic behavior of an anisotropic sample is completely defined by these three parameters. However, several different orientation distributions can give rise to the same intensity ellipsoid, and the form of the distribution function generally cannot be unambiguously determined from the dichroic measurements.

For every absorption band a structural absorbance A_o can be defined [301-304]:

$$A_o = \frac{(A_u + A_v + A_w)}{3} \qquad (4.39)$$

representing the absorbance of the investigated absorption band exclusive of contributions due to orientation of the polymer. Orientation parameters $A_u/3A_o$, $A_v/3A_o$, and $A_w/3A_o$ can be constructed to determine the fraction of molecules oriented in the three mutually perpendicular directions of the sample. For uniaxially oriented samples $(A_u = A_v \neq A_w)$ the structural factor reads

$$A_o = \frac{(A_\parallel + 2A_\perp)}{3} \qquad (4.40)$$

Thus, uniaxial orientation with cylindrical symmetry about the draw direction can be characterized by the classical two-dimensional dichroic ratio R, but for the complete characterization of the inten-

sity ellipsoid in more complex orientation types (e.g., biaxial, where
the two axes of the crystallites are oriented in preferential directions), three-dimensional measurements are generally required. In the
case of an oriented polymer film the absorbance components in the w
(stretching), u (transverse), and v (thickness) directions have to be
determined for the absorption band under investigation. The determination of A_u and A_w with polarized light incident perpendicular to the
uw-plane is trivial, but A_v has to be determined on microtome sections
along the w- or u-axis [305] or with the film tilted about the X- or
Z-axis. The absorbance component in the v direction can then be obtained from Eq.(4.36) via

$$A_{XYZ}(\alpha, 90^{\circ}) = A_u \cos^2\alpha + A_v \sin^2\alpha$$

$$A_v = \frac{A_{XYZ}(\alpha, 90^{\circ}) - A_u \cos^2\alpha}{\sin^2\alpha} \qquad (4.41)$$

where $A_{XYZ}(\alpha, 90^{\circ})$ is the absorbance measured with light polarized
perpendicular to the Z direction of the film sample tilted about the
Z direction and α is the true angle the beam makes with the v direction in the sample interior. Owing to the increase in path length
upon tilting α may be determined from separate absorbance measurements
with polarized light in a conventional and a tilted (for the angle α')
sample film geometry [303]. The experimental angle of tilt α' and the
true angle α the beam makes with the v-axis are related by the index
of refraction through Snell's law:

$$n = \frac{\sin\alpha'}{\sin\alpha} \qquad (4.42)$$

The detailed experimental procedure and the sources of error (reflection losses, refractive index) encountered in these measurements are
described in a number of papers [302-304, 306]. Alternatively, the
three absorbance components of anisotropic samples may be determined
with the aid of the attenuated total reflection (ATR) technique [307,
308].

In the practical application of the general analysis, mathematical

relations between the parameters of distribution function models and
the dichroic ratio are derived and correlated with the experimentally
observed values. Although this approach is restricted to relatively
simple models, it has been extensively applied to establish the basic
concepts of orientation mechanisms in a great variety of polymeric
systems. In what follows the results of such calculations for some
selected distribution models will be discussed.

For the simplest model of perfect uniaxial order it is assumed
that the polymer chains are all oriented parallel to the draw directions, and that the transition moments associated with the vibrations
of the absorbing groups lie in a cone with a semiangle ψ and the draw
direction as axis. For perfect axial orientation ψ thus coincides
with γ. From Eqs. (4.35a) and (4.35c) it follows that

$$A_\perp = A_{uvw}(0°,90°) = K \sin^2\psi \int_{\gamma=0}^{\pi} \int_{\delta=0}^{2\pi} a(\delta) \sin\gamma \cos^2\delta \, d\delta \, d\gamma$$

$$= \frac{1}{2}K \sin^2\psi \tag{4.43}$$

and

$$A_\parallel = A_{uvw}(\alpha,0°) = K \cos^2\psi \int_{\gamma=0}^{\pi} \int_{\delta=0}^{2\pi} a(\delta) \sin\gamma \, d\delta \, d\gamma = K \cos^2\psi \tag{4.44}$$

where the constant factors $\sin^2\psi$ and $\cos^2\psi$ of M_u and M_w, respectively
[Eq. (4.32)], are extracted from the integral. The dichroic ratio is
then expressed by [294]

$$R_0 = 2 \cot^2\psi \tag{4.45}$$

As ψ varies from 0 to $\pi/2$, R_0 varies from ∞ to 0, and no dichroism
($R_0 = 1$) will be observed for $\psi = 54°44'$.

In practice the orientation of the molecular chains is never perfect and the real situation can be described by introducing a factor
f, defined by pretending a fraction f of the polymer to be perfectly
uniaxially oriented, while the remaining fraction (1 - f) is randomly
distributed. The dichroic ratio R is then given by [309]

$$R = \frac{f \cos^2\psi + (1/3)(1 - f)}{(1/2) f \sin^2\psi + (1/3)(1 - f)} \qquad (4.46)$$

where ψ is the angle between the transition moment of the absorbing group and the chain axis of the polymer molecule. Equation (4.46) may be put into a different form:

$$f = \frac{(R - 1)(R_O + 2)}{(R_O - 1)(R + 2)} \qquad (4.47)$$

Here R is the measured dichroic ratio of the investigated absorption band and R_O is defined as in Eq. (4.45).

One method of estimating the value of f is to make assumptions on the transition moment angle ψ of a certain group frequency. The value of f, calculated with the aid of the experimentally measured dichroic ratio R, may then be applied to determine the transition moment angle of other vibrations [310]. Fraser has discussed a method which permits the determination of a minimum value f_m for the fraction of perfectly oriented molecules [311].

As an alternative to Eq. (4.46), the effect of imperfect orientation may be represented by supposing all the molecular chains to be displaced by the same angle θ from parallelism with the draw direction. The transition moments of the absorbing groups then lie in a cone with semiangle ψ whose axis (the polymer chain) itself lies in a cone with semiangle θ and the draw direction as axis (Fig. 4-46). The expression for R then becomes

$$R = \frac{2 \cot^2\psi \cos^2\theta + \sin^2\theta}{\cot^2\psi \sin^2\theta + (1 + \cos^2\theta)/2} \qquad (4.48)$$

Hence, if the direction of the transition moment with respect to the chain axis is known, the average orientation of the chain segments can be determined from the measured dichroic ratio. A further analysis of these expressions reveals that the fraction of oriented chains [Eqs. (4.46) and (4.47)] can also be expressed in terms of the more generally applicable orientation function f [312-315], which is not

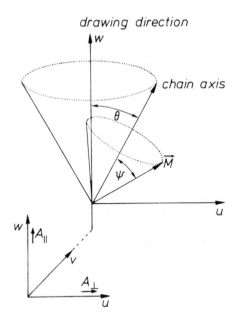

FIGURE 4-46   Distribution of transition moments in oriented polymers
with respect to the drawing direction (see text).

restricted to a certain model or orientation distribution:

$$f = \frac{(3 \langle \cos^2\theta \rangle - 1)}{2} \tag{4.49}$$

Here θ is the average angle between a reference axis (e.g., the chain
axis) and the principal deformation direction of the sample and

$$\langle \cos^2\theta \rangle = \frac{\int_{\theta=0}^{\pi} F(\theta) \cos^2\theta \sin\theta \, d\theta}{\int_{\theta=0}^{\pi} F(\theta) \sin\theta \, d\theta} \tag{4.49a}$$

with $F(\theta)$ being the probability distribution function for the reference axis. For parallel alignment f becomes unity, for perpendicular
alignment -1/2, and for random orientation with $\langle \cos^2\theta \rangle = 1/3$ f be-

comes O. The primary advantage of this deduction is that the orientation functions determined by independent methods (e.g., x-ray diffraction, birefringence, sonic modulus) may be coupled with the results
derived from IR dichroic measurements of absorption bands which are
representative of the appropriate phase [315]. For example, the dichroic ratio of a crystallinity band in a uniaxially oriented polymer
can be correlated with the orientation function for the crystalline
regions determined quantitatively from azimuthal wide-angle x-ray
diffraction measurements by Wilchinsky's method [316, 317] to calculate the transition moment angle of the corresponding vibration. Provided the polymer under investigation can be treated as a two-phase
system in a further step, the orientation functions of the amorphous
(f_{am}) and crystalline (f_c) regions have been expressed in terms of
an average orientation function f_{av} weighted by the amount of each
phase present [318]:

$$f_{av} = \eta f_c + (1 - \eta) f_{am} \qquad (4.50)$$

where η and $1 - \eta$ are the fractions of the crystalline and amorphous
materials, respectively. However, in view of the rather complex nature of orientation phenomena in polymeric systems, the analysis of
anisotropy in terms of this relationship can only be treated as a
rough approximation of the real situation.

Some uniaxially oriented polymeric materials may be defined by a
model containing a fraction f in which all chains make an angle θ
with the direction of stretch, while the remaining fraction $(1 - f)$
is unoriented. The dichroic ratio of an absorption band which results
from absorption in both the crystalline and amorphous regions is then
given by [319]

$$R = \frac{f \cos^2\psi + (2 \sin^2\theta)/(2 - 3 \sin^2\theta) + (1/3)(1 - f)}{(1/2) f \sin^2\psi + (2 \sin^2\theta)/(2 - 3 \sin^2\theta) + (1/3)(1 - f)} \qquad (4.51)$$

where θ is the inclination of the transition moment to the molecular
axis. For the quantitative application of Eq. (4.51), an approximate
value of the angle θ can be determined by x-ray diffraction measure-

ments [319-321]. For the specification of biaxial orientation in
polymer films Stein [322] introduced a set of six orientation functions:

$$f_{\rho_1} = (3 <\cos^2\rho_1> - 1)/2$$

$$f_{\rho_2} = (3 <\cos^2\rho_2> - 1)/2$$

$$f_{\rho_3} = (3 <\cos^2\rho_3> - 1)/2$$

$$(4.52)$$

$$f_{\sigma_1} = 2 <\cos^2\sigma_1> - 1$$

$$f_{\sigma_2} = 2 <\cos^2\sigma_2> - 1$$

$$f_{\sigma_3} = 2 <\cos^2\sigma_3> - 1$$

which define the state of orientation of the crystallites of chain
segments with respect to a unique direction in the film (f_{ρ_1}, f_{ρ_2},

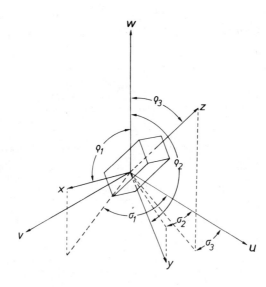

FIGURE 4-47   Coordinate system applied to the specification of the
              state of orientation of a crystallite in a biaxially
              oriented polymer (see text).

f_{ρ_3}) and the plane of the film (f_{σ_1}, f_{σ_2}, f_{σ_3}), respectively (Fig. 4-47).
Further accounts on the interpretation of IR dichroism measurements
are available in the literature [29, 323-326]. Ward et al. presented
a theoretical treatment for IR absorption by an oriented polymer,
dealing quantitatively with reflection corrections and the internal
field problem for the case of uniaxial symmetry and for the situation
where the electric field vector is incident along a principal axis
[327].

4.3.4.3 Raman Polarization

The Raman scattered radiation has directional properties governed
by the symmetry of the corresponding molecular vibration and the
orientation of the molecule relative to the propagation and electric
vector direction of the exciting beam and the viewing direction of
the spectrometric system [328, 329].

The intensity of a Stokes-shifted Raman line of frequency ν_{lm}
polarized in a direction j, excited by a laser beam polarized in a
direction i, is given by [330]

$$I_{lm}(j) = \frac{2\pi^2(\nu_o - \nu_{lm})^4 hN}{\mu c^4 \nu_{lm}\left[1 - e^{-h\nu_{lm}/kT}\right]}(\alpha'_{ij})^2_{lm} I_o(i) \tag{4.53}$$

where μ is the reduced mass, N is the number of scattering centers
per unit volume, $I_o(i)$ is the i-polarized laser intensity, ν_o is the
frequency of the excitation line, l and m are the initial and final
levels, respectively, of the vibrational transition, and α'_{ij} is a
component of the differential polarizability tensor which characterizes the change in polarizability during the vibration.

Despite random orientation of the molecule with respect to the
direction of laser-beam polarization, the intensity of a Raman line
for substances in the gaseous and liquid phase depends on the polarization of the incident beam and the symmetry of the vibrational mode.
For the geometries of Raman experiments as illustrated in Fig. 3-12,
the ratio of the X-polarized ($I_\perp$) to the Z-polarized ($I_\parallel$) Raman in-

tensity at any frequency is, by definition, the depolarization ρ:

$$\rho = \frac{I_\perp}{I_\parallel} = \frac{I(X(ZX)Y)}{I(X(ZZ)Y)} \tag{4.54}$$

The symbols outside the parentheses are the directions of the incident (left) and scattered radiation (right). The symbols inside denote the polarization directions of the incident (left) and scattered light (right). For plane-polarized incident radiation a band is called *depolarized* for $\rho = 0.75$ and *polarized* for $\rho < 0.75$ [331].

The complex nature of the Raman effect of oriented solids will become evident from inspection of the sample geometry of uniaxial orientation (Fig. 4-48). For the situation illustrated in Fig. 4-48 where β_1 and β_2 define the directions of polarization of the incoming and scattered radiation with respect to the fiber axis orientation, the intensity of the scattered radiation depends on products of terms involving $\cos^2\psi$, $\cos^2\theta$, $\cos^2\beta_1$, and $\cos^2\beta_2$. Thus, the scattered intensity depends upon the experimentally chosen values of β_1 and β_2, the angle ψ, and orientation functions $\cos^2\theta$ and $\cos^4\theta$. In contrast to IR spectroscopy, $\cos^4\theta$ is now determined in the Raman experiment which enables a better definition of the orientation distribution function. The correct interpretation of the results, however, is limited to good optical quality specimens which do not show bire-

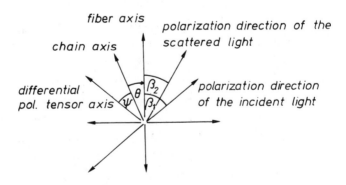

FIGURE 4-48   Definition of angles in the Raman polarization experiment for uniaxially oriented samples.

fringence and scramble the polarization of the light appreciably as
a consequence of scattering.

Apart from the aspect of characterizing the degree of orientation
in polymers Koenig [332] and Hendra [333] have demonstrated the value
of Raman and IR polarization studies for qualitative differentiation
of possible conformations in monosubstituted vinyl polymers. Since
the factor group for all conformationally regular isotactic helical
polymers with n chemical units and m turns per crystallographic unit
cell is $C_{2m\pi/n}$ and is isomorphous to the point group C_n, the Raman-
active vibrations belong to the A, E_1, and E_2 species and the IR-active vibrations to the A and E_1 species. According to the transformation behavior of the derived tensor components $\alpha'_{ij}{}^{\dagger}$ and the translation vector upon symmetry operations, the A mode is polarized and
and the E_1 and E_2 modes are depolarized in the Raman experiment. The
infrared bands with parallel dichroism can be assigned to the A species and those with perpendicular dichroism to the E_1 species. A more
general treatment which is applicable to partially oriented samples
was worked out by Snyder [334], Fanconi et al. [335], and Boerio and
Bailey [336]. The basic concept is to express the Raman scattering
activities which originally refer to a molecular coordinate system
(x,y,z) in terms of the laboratory coordinate system X, Y, Z. One
axis of the molecule is assumed to be unique by virtue of its being
an axis of rotation or its being perpendicular to a plane of reflection. This unique axis can be aligned either parallel (e.g., polyethylene, polypropene, polybutene, polyamides-x) or perpendicular
(e.g., polyalkylene terephthalates, polyamides-x,y, polyesters-x,y).
Let us assume the molecular z-axis to be parallel to the stretching
(fiber) axis. Then, the x-and y-axes are randomly oriented in a plane
perpendicular to the fiber axis. Therefore, after the transformation
of the derived tensor components α_{ij}(xyz) into the corresponding
quantities α_{ij}(XYZ) averaging over all possible orientations of the
x- and y-axes has to be carried out.

The relation between the components α_{ij}(xyz) and α_{ij}(XYZ) is
given by the orthogonal transformation

---

†To avoid confusion with the symbolism of transposed matrices in what
follows only α will be used instead of α'.

$$\alpha(XYZ) = \Phi' \alpha(xyz) \Phi \qquad (4.55)$$

where α is the tensor with the components α_{ij}. The intensity of the scattered radiation is proportional to the term $[\alpha_{ij}(XYZ)]^2$. The transformation matrix Φ depends on the geometrical arrangement of the sample.

Figure 4-49 shows the commonly used geometry for these depolarization measurements. Let ϕ be the angle between the laboratory Z-axis and the molecular y-axis. Then the corresponding transformation matrices for the following three sample arrangements read:

$$\Phi(I) = \begin{pmatrix} -\cos\phi & \sin\phi & 0 \\ 0 & 0 & 1 \\ \sin\phi & \cos\phi & 0 \end{pmatrix} \qquad (4.56a)$$

$$\Phi(II) = \begin{pmatrix} 0 & 0 & 1 \\ -\cos\phi & \sin\phi & 0 \\ \sin\phi & \cos\phi & 0 \end{pmatrix} \qquad (4.56b)$$

$$\Phi(III) = \begin{pmatrix} -\cos\phi & \sin\phi & 0 \\ \sin\phi & \cos\phi & 0 \\ 0 & 0 & 1 \end{pmatrix} \qquad (4.56c)$$

By inserting these matrices into Eq. (4.55) the observable quantities

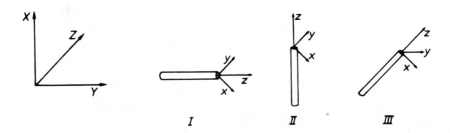

FIGURE 4-49   Geometry and coordinate systems in Raman polarization measurements of uniaxially oriented samples.

α_{ij}(XYZ) can be expressed as linear combination of the α_{ij}(xyz) for a given angle ϕ. However, since the x- and y-axes are randomly oriented, we have to average over all the possible orientations:

$$\alpha(XYZ)^2 = \frac{1}{2\pi} \int_{\phi=0}^{2\pi} [\Phi' \, \alpha(xyz) \, \Phi]^2 \, d\phi \tag{4.57}$$

Since

$$\frac{1}{2\pi} \int_{\phi=0}^{2\pi} \cos^2\phi \, \sin^2\phi \, d\phi = \frac{1}{8} \tag{4.58a}$$

$$\frac{1}{2\pi} \int_{\phi=0}^{2\pi} \cos^4\phi \, d\phi = \frac{1}{2\pi} \int_{\phi=0}^{2\pi} \sin^4\phi \, d\phi = \frac{3}{8} \tag{4.58b}$$

$$\frac{1}{2\pi} \int_{\phi=0}^{2\pi} \sin^2\phi \, d\phi = \frac{1}{2\pi} \int_{\phi=0}^{2\pi} \cos^2\phi \, d\phi = \frac{1}{2} \tag{4.58c}$$

$$\frac{1}{2\pi} \int_{\phi=0}^{2\pi} \sin\phi \, \cos\phi \, d\phi = \frac{1}{2\pi} \int_{\phi=0}^{2\pi} \sin^3\phi \, \cos\phi \, d\phi$$

$$= \frac{1}{2\pi} \int_{\phi=0}^{2\pi} \cos^3\phi \, \sin\phi \, d\phi = 0 \tag{4.58d}$$

The term $\alpha(XYZ)^2$ can be represented for the three discussed cases by

$$\begin{pmatrix} \alpha_1 & \alpha_3 & \alpha_2 \\ \alpha_3 & \alpha_4 & \alpha_3 \\ \alpha_2 & \alpha_3 & \alpha_1 \end{pmatrix}_I, \quad \begin{pmatrix} \alpha_4 & \alpha_3 & \alpha_3 \\ \alpha_3 & \alpha_1 & \alpha_2 \\ \alpha_3 & \alpha_2 & \alpha_1 \end{pmatrix}_{II}, \quad \begin{pmatrix} \alpha_1 & \alpha_2 & \alpha_3 \\ \alpha_2 & \alpha_1 & \alpha_3 \\ \alpha_3 & \alpha_3 & \alpha_4 \end{pmatrix}_{III} \tag{4.59}$$

The four different terms α_i are given by

$$\alpha_1 = \frac{1}{8}\left[2(\alpha_{xx} + \alpha_{yy})^2 + (\alpha_{xx} - \alpha_{yy})^2 + 4\alpha_{xy}^2\right]$$

$$\alpha_2 = \frac{1}{8}\left[(\alpha_{xx} - \alpha_{yy})^2 + 4\alpha_{xy}^2\right]$$

$$\alpha_3 = \frac{1}{2}\alpha_{yz}^2 + \alpha_{zx}^2 \tag{4.60}$$

$$\alpha_4 = \alpha_{zz}^2$$

Since every element of the matrices of Eq. (4.59) belongs to a distinct position of polarizer and analyzer we can observe the quantities α_i for the shown geometrical arrangement under the following conditions:

	I	II	III
x(zz)y	α_1	α_1	α_4
x(zx)z	α_2	α_3	α_3
x(zx)y	α_2	α_3	α_3
x(yx)z	α_3	α_3	α_2
x(yx)y	α_3	α_3	α_2
x(zy)z	α_3	α_2	α_3
x(yz)y	α_3	α_2	α_2
x(yy)z	α_4	α_1	α_1

Generally, it will not be possible to determine the aboslute values of the α_i. However, from the observed changes in intensity between the spectra recorded with different sample arrangement and different directions of analyzer and polarizer, we can derive the symmetry of the vibration. This becomes feasible by inspection of the character table belonging to the molecule under consideration. Since the polarizability tensor components $\alpha_{ij}(xyz)$, which are observable in every symmetry species, are listed in the character tables, comparison of Eq. (4.60) and the matrices of Eq. (4.59) on the one side and the $\alpha_{ij}(xyz)$ on the other side enables us to find out those symmetry species observable under a given geometry. Tables with all the Raman scattering activities for the case that the unique axis of the molecule is parallel to the molecular z-axis are published in Ref. 334.

Although several reports on the examination of Raman scattering in anisotropic polymeric materials are available [328, 333, 337-339], so far only Bower [340] has given a thorough account of the theoretical background required to derive useful information about orientation distributions.

4.3.4.4   Infrared Dichroism and Raman Depolarization Studies of

   Selected Polymers

As an introduction to the experimental results of dichroic and depolarization measurements, the IR and Raman spectra of a drawn poly-
amide-6 film and fiber, respectively, will be discussed. In the IR
spectra [Fig. 4-85(a)] the most striking dichroic effects are observed with the $\nu(NH)$ (3300 cm^{-1}), $\nu(CH_2)$ (2800-3000 cm^{-1}), amide I
(1640 cm^{-1}), amide II (1550 cm^{-1}), and amide III (1260 and 1201 cm^{-1})
absorption bands. The σ-dichroism of the $\nu(NH)$, $\nu(CH_2)$, and amide I
bands unambiguously reflects the preferential alignment of the corresponding transition moments perpendicular to the direction of
stretch. The somewhat higher σ-dichroism of the $\nu(NH)$ band (R = 0.67)
in comparison to the amide I band (which is assigned to the $\nu(C=O)$-
stretching vibration) (R = 0.80) has been interpreted in terms of
a slight displacement of the $\nu(C=O)$ transition moment from the bond
direction [294-297]. The π-dichroism of the amide II and amide III
bands can be primarily attributed to the contribution of the $\nu(C-N)$-
stretching vibration of these coupled modes [341]. The IR dichroism
data are in agreement with the results derived from the Raman intensities of a polyamide-6 fiber measured with the electric vector of
the incident radiation polarized parallel and perpendicular to the
fiber axis, respectively (Fig. 4-50). Here, the preferential alignment of the polymer backbone in the fiber axis direction can be derived from the polarization intensities of the $\nu(C-N)$- (1081 cm^{-1})
and $\nu(C-C)$- (1126 cm^{-1}) stretching vibrations [165, 342].

   As a further illustration of IR and Raman polarization measurements and their interpretation the IR and Raman spectra (Fig. 4-51
and 4-52) of uniaxially drawn polyethylene oxide films will be treated
in some detail. Upon crystallization polyethylene oxide forms a 7:2
helix. Therefore, when intermolecular coupling effects are neglected
the symmetry properties can be described by the factor group $D_{4\pi/7}$
which is isomorphous to the point group D_7. The symmetry species, number of normal vibrations, and selection rules are listed in Table 4-3.

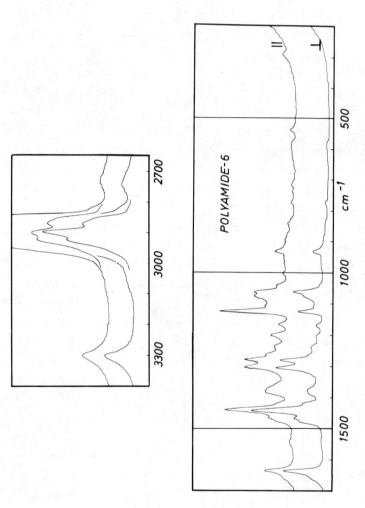

FIGURE 4-50   Raman spectra of polyamide-6 filament (295 K, draw ratio λ = 4).
Symbols: ∥, electric vector of the laser radiation polarized
parallel to drawing direction; ⊥, electric vector of the laser
radiation polarized perpendicular to drawing direction.

TABLE 4-3   Species, Number of Normal Vibrations, and Selection Rules
for Polyethylene Oxide (Point Group D_7)

Species	Number of normal vibrations	Selection rules IR	Raman
A_1	10	$f^\dagger$	$a^\dagger$ (zz, xx + yy)
A_2	9	a (z, R_z)	f
E_1	20	a (x, y, R_x, R_y)	a (xz, yz)
E_2	21	f	a (xy, xx - yy)

† f = forbidden, a = allowed

The subscripts x, y, and z refer to the molecular coordinates. The
simplest method to differentiate between vibrations perpendicular
or parallel to the draw direction and to assign the observed bands
to the corresponding species is outlined in what follows. According
to Table 4-3 the A_2 modes cause more intense IR bands when using parallel polarization, while the E_1 modes are more intense in the spectra
measured with the electric vector perpendicular to the draw direction
(Fig. 4-51). In the Raman experiment the following geometry for the
incident and scattered radiation can be used: X(ZZ)Y, X(ZX)Y, X(YZ)Y,
and X(YX)Y. Taking into account that polyethylene oxide exhibits uniaxial orientation on stretching, we have to average over the angle ϕ
between the X-axis of the space-fixed coordinate system and the x-
axis of the molecular coordinate system. Then, for ideal uniaxial
orientation, the observable intensities can be derived from Eqs. (4.59)
and (4.60) as follows:

$$I(X(ZZ)Y) \propto \alpha_{zz}^2 \quad \rightarrow A_1$$

$$I(X(ZX)Y) \propto \frac{1}{8}(\alpha_{xz} + \alpha_{zy})^2 \quad \rightarrow E_1$$

$$I(X(YZ)Y) \propto \frac{1}{8}(\alpha_{xz} + \alpha_{zy})^2 \quad \rightarrow E_1$$

$$I(X(YX)Y) \propto \frac{1}{8}(\alpha_{yy} + \alpha_{xx})^2 \quad \rightarrow E_2$$

for case I   (4.61)

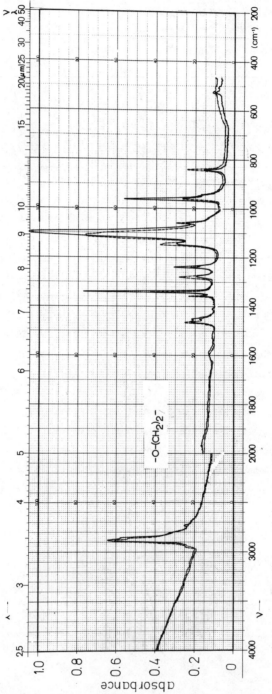

FIGURE 4-51   IR spectra of polyethylene oxide (320 K, draw ratio λ = 4). The solid line shows the electric vector of the polarized radiation parallel to the draw direction; the dashed line shows the electric vector of the polarized radiation perpendicular to the draw direction.

and

$$I(X(YZ)Y) \propto \frac{1}{8}(\alpha_{xz} + \alpha_{zy})^2 \quad \rightarrow E_1$$

$$I(X(YX)Y) \propto \frac{1}{8}(\alpha_{xz} + \alpha_{zy})^2 \quad \rightarrow E_1$$

$$I(X(ZZ)Y) \propto \frac{1}{8}(\alpha_{yy} - \alpha_{xx})^2 + \frac{1}{2}\alpha_{xy}^2 \quad \rightarrow E_2$$

$$I(X(ZX)Y) \propto \frac{1}{8}(\alpha_{yy} - \alpha_{xx})^2 + \frac{1}{2}\alpha_{xy}^2 \quad \rightarrow E_2$$

$$\left. \right\} \text{ for case III} \quad (4.62)$$

As indicated in these equations, it should be possible to observe all the species separately in a perfectly oriented sample when using a polarizer and an analyzer. However, without analyzer (which is the simplest way to obtain depolarization data) we observe:

$$I_\| = I(X(ZZ)Y) + I(X(ZX)Y) \propto \alpha_{zz}^2 - \frac{1}{8}(\alpha_{xz} + \alpha_{zy})^2 \rightarrow A_1 + E_1$$

$$(4.63)$$

$$I_\perp = I(X(YZ)Y) + I(X(YX)Y) \propto \frac{1}{8}(\alpha_{xz} + \alpha_{zy})^2 + \frac{1}{8}(\alpha_{yy} - \alpha_{xx})^2$$

$$+ \frac{1}{2}\alpha_{xy}^2 \rightarrow E_1 + E_2$$

Figure 4-52 shows the Raman spectra of a polyethylene oxide fiber which was scanned with different geometrical conditions. By comparing the corresponding spectra, the A_1, E_1, and E_2 modes can be determined. This assignment can be verified by comparing the IR (Fig. 4-51) and Raman spectra (Fig. 4-52), since according to Table 4-3, the E_1 modes are Raman- and IR-active, while the A_1, A_2, and E_2 modes show either Raman or IR activity. The assignment of the individual absorption bands is listed in Table 4-4 [343-347].

The structural changes in polyacrylonitrile upon uniaxial orientation of film samples have been studied in some detail by IR spectroscopy with linearly polarized radiation and wide-angle x-ray diffraction [348]. In Fig. 4-53 the IR polarization spectra of the $\nu_{as}(CH_2)$ (2940 cm^{-1}), $\nu(C{\equiv}N)$ (2241 cm^{-1}), and $\delta(CH_2)$ (1452 cm^{-1}) absorption bands are shown alongside the corresponding wide-angle x-ray diagrams for polyacrylonitrile films of various draw ratios ($\lambda = 1$,

TABLE 4-4　Assignment of Calculated and Observed Wavenumbers for Polyethylene Oxide

Species	Calculated wavenumber (cm^{-1})	Observed wavenumber (cm^{-1}) Raman	IR	Assignment
A_1	2940	2905		CH_2 asymmetric stretching
	2833	2833		CH_2 asymmetric stretching
	1479	1489		CH_2 bending
	1423	1448		CH_2 wagging
	1252	1240		CH_2 twisting
	1137	1129		C-C stretching
	1073	1077		COC stretching and CH_2 rocking
	866	863		CH_2 rocking and CH_2 stretching
A_2	2943		2885	CH_2 asymmetric stretching
	2883		2855	CH_2 symmetric stretching
	1470		1462	CH_2 bending
	1344		1342	CH_2 wagging
	1264		1240	CH_2 twisting
	1087		1095	C-O stretching
	964		965	CH_2 rocking
	533		530	CCO deformation and CH_2 rocking
E_1	2943	2942	2940	CH_2 asymmetric stretching
	2883	2887	2886	CH_2 symmetric stretching
	2873	2874	2875	CH_2 symmetric stretching
	1476	1473	1474	CH_2 bending
	1353	1365	1362	CH_2 wagging
	1286	1283	1276	CH_2 twisting
	1234	1234	1232	CH_2 twisting
	1142	1145	1148	COC stretching and CH_2 rocking
	1112	1115	1115	COC symmetric stretching
	1060	1067	1060	COC stretching and CH_2 rocking
	941	945	947	CH_2 rocking and COC stretching
	847	847	945	CH_2 rocking and COC stretching
	524	532	531	OCC deformation
E_2		1489		CH_2 bending
		1400		CH_2 wagging
		590		OCC deformation

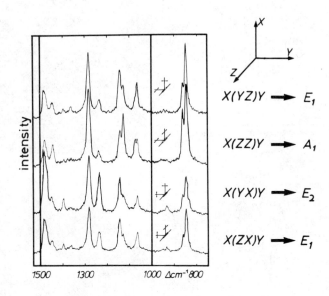

FIGURE 4-52   Raman spectra of polyethylene oxide obtained with various polarization geometries (295 K, draw ratio λ = 14).

2, and 3). The extent of orientation induced by the drawing operation
at 403 K is reflected in the increasing σ-dichroism of the absorption
bands under consideration as well as in the contraction of the intense
equatorial (100) reflex in the x-ray diagrams. In Fig. 4-54 the dichroic ratios of the respective absorption bands are plotted in dependence of the draw ratio. The differences in the dichroic ratios
of the CH_2-stretching and -bending vibrations whose transition moments
ought to be both perpendicular to the direction of the extended chain
can be explained in terms of the interaction of the CH_2-stretching
and -wagging vibrations. This interaction which has been thoroughly
discussed by Zbinden [29] tends to reduce the σ-dichroism of the
stretching vibration. The better perpendicular alignment of the $\nu(C\equiv N)$
in comparison to the $\delta(CH_2)$ transition moments has been attributed
to intermolecular dipole association of the nitrile groups.

From three-dimensional measurements on tilted films the orienta-
tation parameters of the $\nu(C\equiv N)$ and $\delta(CH_2)$ absorption bands which
represent the orientation of the side groups and the polymer chains,

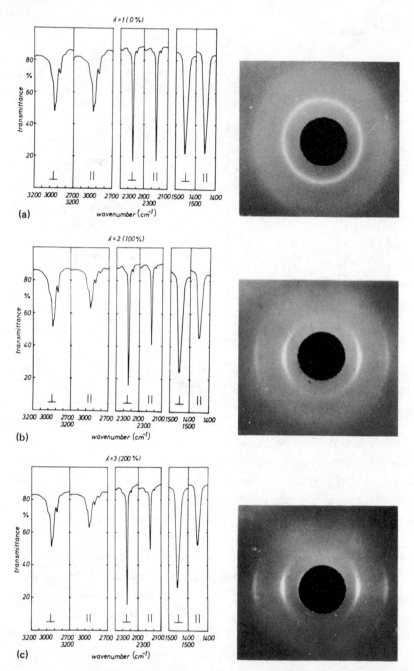

FIGURE 4-53 IR polarization spectra of the ν_{as}(CH$_2$) (2940 cm^{-1}), ν(C≡N) (2241 cm^{-1}), and δ(CH$_2$) (1452 cm^{-1}) absorption bands and wide-angle x-ray patterns of variously drawn polyacrylonitrile films: (a) draw ratio $\lambda = 1$, (b) draw ratio $\lambda = 2$, (c) draw ratio $\lambda = 3$. Symbols: ⊥, electric vector of polarized IR radiation perpendicular to drawing direction; ‖, electric vector of polarized IR radiation parallel to drawing direction.

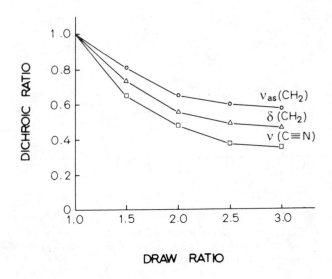

FIGURE 4-54   Dichroic ratios $R = A_{\parallel}/A_{\perp}$ of the $\nu_{as}(CH_2)$, $\nu(C\equiv N)$, and
$\delta(CH_2)$ absorption bands of polyacrylonitrile in relation
to draw ratio.

respectively, have been determined and plotted versus draw ratio
(Fig. 4-55). The small deviations from the value of statistical
orientation (0.33) indicate relatively low orientation of the atomic
groups under consideration. At higher draw ratios the measured orientation is close to the uniaxial type $(A_u = A_v \neq A_w)$ with somewhat
preferential alignment of the $\nu(C\equiv N)$ and $\delta(CH_2)$ transition moments
perpendicular to the film plane $(A_v > A_u)$. Assuming an angle of 90°
between the transition moment of the $\nu(C\equiv N)$ vibration and the polymer
chain axis the IR dichroism data of the 200% drawn sample yield [with
the aid of Eq. (4.48)] an average orientation angle of the polymer
chains with the stretching direction of about 33°. The discrepancy
between this value and the much lower value of 15° derived from x-ray
data of the same sample suggests that the polymer chains occur in a
kinked rather than in an extended chain conformation [349]. Other
authors have ascribed the lower degree of orientation calculated
from IR dichroism to the superposition of the absorptions of ordered

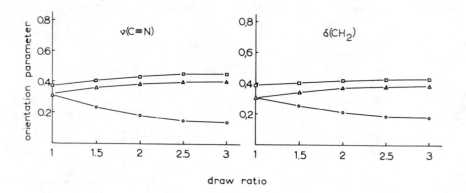

FIGURE 4-55   Orientation parameters $A_u/3A_O$ (Δ), $A_v/3A_O$ (o), and $A_w/3A_O$
($\square$) of the $\nu(C\equiv N)$ and $\delta(CH_2)$ absorption bands of variously
drawn polyacrylonitrile films.

and disordered regions [350]. However, to account for the measured
IR and x-ray data and the relatively low degree of crystallinity
(about 30%) reported for polyacrylonitrile [351, 352] model calculations for a two-phase system with Eq. (4.51) indicate that some preferential alignment of polymer chains in the drawing direction must
also occur in the less ordered regions [348].

Full advantage can be taken of IR dichroic measurements when the
relative orientation of different atomic groups, for example, in a
copolymer, has to be studied. In Fig. 4-56 the IR polarization spectrum of a 150% drawn acrylonitrile methylacrylate (94:6) copolymer
film containing about 5% dimethylformamide (DMF) solvent residues
is shown. The dichroic ratios of the polymer specific $\nu(C\equiv N)$ (2241 cm^{-1}),
$\delta(CH_2)$ (1450 cm^{-1}), and $\nu(C=O)$ (1730 cm^{-1}) as well as the solvent specific $\nu(C=O)$ (1667 cm^{-1}) absorption bands have been plotted in dependence of the draw ratio in Fig. 4-57. Adopting a procedure proposed by
Ruscher and Schmolke [350], the relative orientation of the acrylonitrile and methylacrylate segments can be compared. Thus, for equivalent orientation behavior the orientation factor ratio [calculated
with the aid of Eq. (4.47)] of representative absorption bands [$\nu(C\equiv N)$

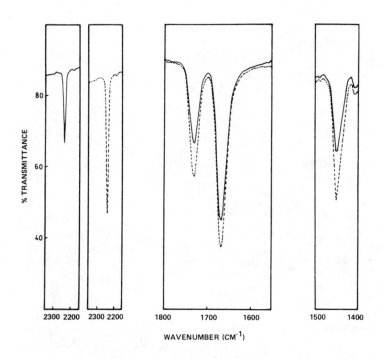

WAVENUMBER (CM⁻¹)

FIGURE 4-56   IR polarization spectra of the $\nu(C{\equiv}N)$ (2240 cm^{-1}),
$\nu(C{=}O)_{polymer}$ (1730 cm^{-1}), $\nu(C{=}O)_{DMF}$ (1667 cm^{-1}), and
$\delta(CH_2)$ (1450 cm^{-1}) absorption bands of a drawn (λ = 2.5)
acrylonitrile-methylacrylate (94:6) copolymer film.
The solid line shows the polarization direction parallel
to the direction of elongation; the dashed line shows
the polarization direction perpendicular to the direction of elongation.

for the acrylonitrile segments, $\nu(C{=}O)$ for the methylacrylate seg-
ments] must be independent of sample elongation provided that no
change in the transition moment inclination toward the chain axis
takes place during the deformation. This effect has been verified
for the copolymer segments under examination (Table 4-5). Additionally, from Figs. 4-56 and 4-57 conclusions may be drawn with respect
to the interaction between polymer and solvent molecules [353]. In
the process of spinning fibers or casting films from DMF solutions
of polyacrylonitrile or its copolymers solvent residues are ten-

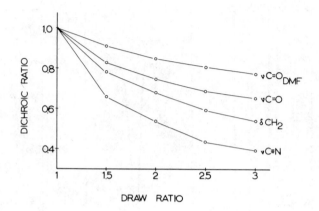

FIGURE 4-57   Dichroic ratios of the $\nu(C{\equiv}N)$, $\delta(CH_2)$, $\nu(C{=}O)$, and
$\nu(C{=}O)_{DMF}$ absorption bands of variously drawn acrylonitrile-methylacrylate (94:6) copolymer films in relation to draw ratio.

aciously retained by the polymer [354]. In Fig. 4-56 the $\nu(C{=}O)$ absorption band of the DMF residues is shifted for approximately 7 cm^{-1}
toward lower wavenumbers (1667 cm^{-1}) in comparison to the spectrum
of the pure solvent. Furthermore, the dichroic ratios of the polymer
and solvent specific absorption bands exhibit an analogous trend in
dependence of sample elongation (Fig. 4-57). These phenomena have
been attributed to an interaction of the DMF cabonyl groups with the
nitrile groups of polymer segments via dipole association [353, 355-
357]

Valuable structural information can be derived from the IR and
Raman spectra of oriented specimens of polyethylene terephthalate.
Schmidt [302], for example, has shown for a series of uniaxially
stretched polyethylene terephthalate films that the *gauche-trans*
transformation of the aliphatic segments, which has been observed
as a consequence of heat treatment (see Sec. 4.3.1), is also induced
by mechanical drawing. By comparison of one-way drawn films and heat-
set, nonoriented films of various densities he could prove that the
extra amount of *trans* content as measured by the 975 cm^{-1} absorption

TABLE 4-5   Orientation Factor (f) Ratios for the $\nu(C\equiv N)/\nu(C=O)$ Vibrations in Relation to Draw Ratio λ for Acrylonitrile-Methyl-Acrylate 94:6 Copolymer

Draw Ratio λ	$c \dfrac{f_{\nu(C\equiv N)}}{f_{\nu(C=O)}}$ polymer [†]
1.5	2.03
2.0	1.99
2.5	1.99
3.0	1.89

[†] The term c is a constant factor [350].

band in the oriented specimen must be localized in the amorphous regions.

IR polarization spectra of polyethylene terephthalate (Fig. 4-58) have been the subject of numerous investigations [358-364]. Especially the dichroic effects of absorption bands which are characteristic of certain structural units (aliphatic or aromatic), rotational isomers (*trans* or *gauche*), and/or different states of order (amorphous or crystalline) have been applied to study the deformation mechanism of polyethylene terephthalate in dependence of the drawing conditions (temperature, speed, draw ratio). Thus, in a recent paper Urbanczyk [363] characterized the relative orientation of aliphatic and aromatic segments as well as the amorphous and crystalline phases in terms of orientation functions derived from IR dichroism studies of the 875, 1343, 1473, 1580, and 1368 cm^{-1} absorption bands and x-ray measurements. Ward et al. [364] combined the IR dichroism results from the 975, 896, 875, and 795 cm^{-1} absorption bands with measurements of the refractive indexes to provide quantitative information regarding both molecular orientation and conformational changes. The analysis of these IR data has been based on theoretical results which take into account reflectivity corrections and the internal field effect [327].

Although the assignment to symmetry species determined for the IR-active modes in polyethylene terephthalate by vibrational analysis

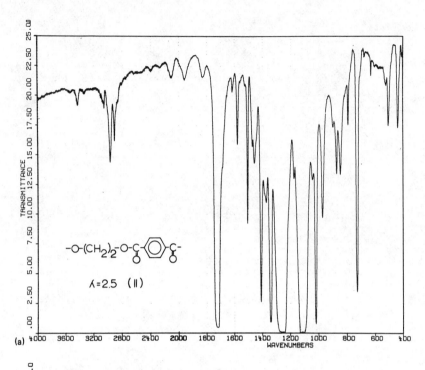

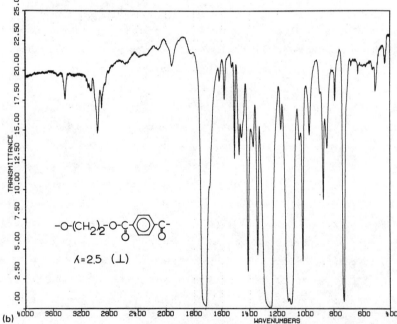

FIGURE 4-58   IR spectra of a drawn polyethylene terephthalate film (draw
ratio λ = 2.5) :   (a) electric vector of polarized radiation
parallel to drawing direction.   (b) electric vector of
polarized radiation perpendicular to drawing direction.

are in good agreement with measurements of infrared dichroism for oriented samples [152, 156, 365] a few Raman assignments remain doubtful. To assign all the Raman lines to symmetry species it is necessary to study the polarization behavior of the Raman lines (Fig. 4-59).

In polymers with uniaxial orientation usually the principal symmetry axes of the molecules are aligned along the fiber axis while the x- and y-axes are randomly oriented in a plane perpendicular to the fiber axis, and the Raman scattering activities can be derived according to Eq. (4.60) [338]. However, in polyethylene terephthalate the axis of rotation is approximately perpendicular to the fiber axis. Therefore, to assign the bands of the corresponding Raman-acitve species, it has to be taken into account that the observed Raman intensities are related to space-fixed coordinates while the polarizability tensor components (which are summarized in the character tables) refer to the coordinates (x,y,z) of the molecule [366].

The transformation of the coordinates can be described by the

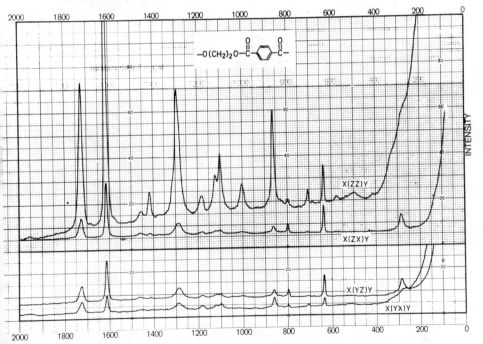

FIGURE 4-59   Raman polarization spectra of polyethylene terephthalate filament (draw ratio λ = 5) (see text for nomenclature of the polarization geometry).

matrix Φ. Using the polarization geometry of Figs. 3-12 and 4-59 we observe light scattered in the direction Y and polarized either along X (α_{XY} and α_{XZ}) or along Z (α_{ZZ} and α_{ZY}).

According to Fig. 3-12 it is also possible to measure α_{YZ} and α_{XZ} or α_{YY} and α_{XY} when the light is scattered in the Z direction. Since the fiber axis is placed along Z the space-fixed coordinate system (laboratory system) (X,Y,Z) and the molecular coordinate system (x, y,z) are related via

$$
\begin{pmatrix} X \\ Y \\ Z \end{pmatrix} = \begin{pmatrix} \cos\phi & -\sin\phi & 0 \\ \sin\phi & \cos\phi & 0 \\ 0 & 0 & 1 \end{pmatrix} \begin{pmatrix} x \\ y \\ z \end{pmatrix} \tag{4.64}
$$

where ϕ denotes the angle between X and x. From Eq. (4.55) we find for the required tensor components in terms of the coordinates (X,Y,Z) by averaging over ϕ

$$
\alpha_{ij}^2 (XYZ) = (\sum_{kl} \phi_{ik}\phi_{jl}\alpha_{kl})^2 \tag{4.65}
$$

From these equations we can derive the components to be

$$
\left.\begin{aligned}
\alpha_{XX}^2 &= \tfrac{3}{8}\alpha_{xx}^2 + \tfrac{3}{8}\alpha_{yy}^2 + \tfrac{1}{4}\alpha_{xx}\alpha_{yy} + \tfrac{1}{2}\alpha_{xy}^2 \\
\alpha_{YY}^2 &= \tfrac{3}{8}\alpha_{xx}^2 + \tfrac{3}{8}\alpha_{yy}^2 + \tfrac{1}{4}\alpha_{xx}\alpha_{yy} + \tfrac{1}{2}\alpha_{xy}^2 \\
\alpha_{ZZ}^2 &= \alpha_{zz}^2
\end{aligned}\right\} A_g
$$

$$ \tag{4.66} $$

$$
\left.\begin{aligned}
\alpha_{ZX}^2 &= \alpha_{XZ}^2 = \alpha_{ZY}^2 = \alpha_{YZ}^2 = \tfrac{1}{2}(\alpha_{zy} + \alpha_{zx})^2 \\
\alpha_{XY}^2 &= \alpha_{YX}^2 = \tfrac{1}{8}(\alpha_{yy} - \alpha_{xx})^2 + \tfrac{1}{2}\alpha_{xy}^2
\end{aligned}\right\} A_g, B_g
$$

According to the character table of the point group C_{2h} to which polyethylene terephthalate belongs and with reference to Fig. 3-12 the A_g modes can be observed under X(ZZ)Y geometry and the A_g and B_g modes can be observed for the X(YZ)Y, X(YX)Y, and X(ZX)Y geometries. The B_g modes should be more intense for the X(YX)Y and X(ZX)Y geometries in comparison to the A_g modes than for the X(YZ)Y geometry. From the intensity of the band near 800 cm^{-1}, it can be derived that

this band has to be assigned to a B_g mode (Fig. 4-59). For a highly
drawn fiber this band vanishes completely when using X(ZZ)Y geometry
[336]. More detailed studies on the molecular orientation in uniaxially drawn samples of polyethylene terephthalate are given in Refs. 367
to 369.

Apart from estimating the degree of orientation in polymers, polarization measurements are a powerful tool for the differentiation
of various chain conformations [370]. In proteins, for example, the
molecular chains may occur either as helix or in the extended form.
These chain conformations were confirmed by the observations that
the ν(NH) and ν(C=O) vibrations show parallel dichroism in the helix
conformation and perpendicular dichroism in the extended form [371].

The interesting phenomenon of IR dichroism inversion on drawing
has been observed with some absorption bands in the spectra of poly-
amide-6,6, polyethylene terephthalate [372], polyethylene [373], polyvinyl chloride [374], and polyurethanes [229]. This anomaly may be
visualized as the successive change from the preferential orientation
of crystallites to the orientation of molecular chains.

Experiments on the influence of annealing temperature on IR dichroism have been reported [375, 376] and the observed changes explained in terms of the glass transition of the polymer under investigation.

IR dichroism and Raman polarization measurements have been employed in structural and orientation studies on a wide range of polymers, and for further details the reader is referred to summaries of
references in books and review articles [29, 30, 32, 192, 326, 328,
377-380].

The characterization of orientation by vibrational spectroscopy
is not necessarily restricted to solid, anisotropic materials. Thus,
accounts are also available on the measurement of macromolecular
alignment by IR dichroism in polymer solutions which had been exposed to shearing forces [381] and electric [382, 383] and magnetic
fields [384].

4.3.4.5 Spectroscopic Studies of Deformation, Stress Relaxation,
         Fracture, and Fatigue in Polymers

In recent years there has been steadily increasing interest in the
effect of mechanical stress on the vibrational spectra of polymers
and in the measurement of spectroscopic parameters which may provide
some insight into the deformation and fracture mechanism of polymers.
Thus, IR and Raman spectroscopic investigations of polymers under
mechanical load, during elongation, stress relaxation, creep, and
after fracture have been reported [161, 385-412].

    In these studies information concerning the stressed or fractured
polymer is primarily derived from examination of the shift in peak
position, change in shape, polarization properties, and the occurrence of new absorption bands. Owing to the long scan duration of
dispersive instruments however, the IR and Raman spectroscopic characterization of molecular changes in deformation processes was so far
restricted to stepwise procedures. Any spectroscopic data then refer
to the sample at the relaxed stress levels of the individual elongation steps which commonly differ significantly from the corresponding unrelaxed stress levels of a continuously measured stress-strain
curve. Hence, strictly speaking, the structural conclusions derived
on a microscopic scale from spectroscopic measurements in stepwise
elongation procedures should not be correlated with the macroscopic
deformation properties reflected by the stress-strain diagrams. Alternatively, dynamic IR (DIR) [393] and Raman investigations can be
performed with conventional instrumentation by monitoring the intensity of selected absorption bands at constant wavenumber. This technique however, is not generally applicable (apart from the fact that
it requires multiple experiments for various absorption bands) because any frequency shift of a stress-sensitive absorption band (see
below) would severely bias the results of such investigations. This
situation has been alleviated by the introduction of rapid-scanning
FTIR spectrometers. With such systems intensity changes and dichroic
properties can be monitored in short-time intervals over the entire
mid-infrared region during elongation, stress relaxation, and creep

and even extremely small changes in the spectra can be accentuated by mathematical procedures.

In a series of papers various authors using IR spectroscopy have demonstrated both theoretically and experimentally that the frequency and shape of skeletal vibrations are stress-dependent [385, 390, 393-396, 401-403]. They typically observed

1. Symmetrically shaped bands deforming asymmetrically
2. Small frequency shifts $\Delta \nu$ to longer wavelength which are a function of the applied stress σ:

$$\Delta \nu = \alpha \sigma \tag{4.67}$$

where α is a proportionality constant.

These phenomena are qualitatively illustrated in Fig. 4-60 with

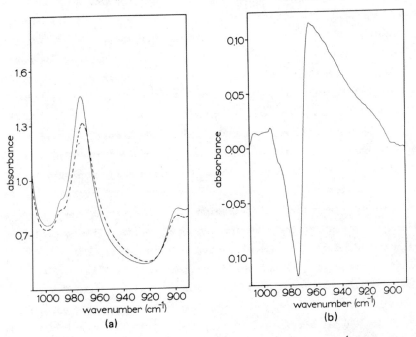

(a)

(b)

FIGURE 4-60   Effect of mechanical stress on the 973 cm^{-1} $\nu(O-CH_2)$ absorption band of polyethylene terephthalate: (a) the solid line indicates an unstressed polymer, the dashed line the stressed (230 MN/m^2) polymer. (b) difference spectrum: stressed polymer − unstressed polymer.

reference to the 973 cm^{-1} absorption band of the $\nu(O-CH_2)$ skeletal vibration of polyethylene terephthalate. Figure 4-60(a) shows the shape of this absorption band in the unstressed state and under load (230 MN/m^2). The subtle differences observable in the spectra can be enhanced by the absorbance subtraction technique. In the difference spectrum [Fig. 4-60(b)] the shift of the peak maximum toward longer wavelengths and the low-frequency tailing are reflected by a pronounced asymmetric dispersion-shaped profile.

Such results have been interpreted in terms of a nonuniform distribution of the external load resulting in a nonsymmetrical displacement of the individual atomic absorption frequencies about the maximum of the band [393, 396, 401, 403]. While the long-wavelength tail was attributed to strongly overloaded bonds of disordered regions the band peak has been assigned to the absorption of bonds which are primarily situated in the crystalline regions and which are stressed very similar to the external stress [402-404]. External load distribution functions have been derived for various polymers, and it has been shown that this distribution not only varies from one polymer to another but also depends on the conformational regularity of the polymer under examination [385].

In a theoretical treatment Wool [391, 393] has shown that two basic mechanisms can influence the molecular orientation as determined by IR dichroism measurements. The rigid body mechanism involves rotation and/or translation of entire chains or chain segments and affects the spatial distribution of chains with respect to some frame of reference. The relationship among various orientation distributions induced by this mechanism and the experimentally determined dichroic ratio R has been discussed in detail in Sec. 4.3.4.2. The mechanism of extensional orientation arises by homogeneous deformation of linear chain segments leading to changes of the transition moment directions of IR vibrations with respect to the chain axis. Hence, although the chain axis orientation actually remains unaltered variations of IR polarization intensities may occur.

Though less frequently, Raman spectroscopy has also been applied to study stress-sensitive phenomena in polymers. In fact, the con-

venience of using bulk samples with almost no restriction on sample
preparation gives it some advantage over the IR technique. Thus, the
frequency shifts, band intensity changes, and depolarization ratios
of Raman-active vibrations of a series of polymers (polypropene,
polycarbonate, polystyrene, polyamide-6,6, and polydiacetylenes) have
been investigated as a function of strain [388, 389, 398, 406] and
it could be shown that the skeletal modes are not the only modes
which are stress-sensitive [389]. On the other hand, the fact that
no measurable frequency shift of bands was observed in the Raman
spectra of stressed poly(p-phenylene terephthalamide) fibers [386]
indicates that no or minimal stretching or bending of chemical bonds
in this polymer takes place as a function of stress.

The characterization of structural changes during elongation and
recovery of various polymers by simultaneous FTIR and stress-strain
measurements has been recently reported [395]. For this purpose a
stretching machine has been constructed (see Fig. 3-29) which allows
a polymer film to be uniaxially drawn with a specified elongation
rate while mounted in the sampling beam of the FTIR spectrometer.
In polarization measurements the polarization direction of the radiation is alternately adjusted parallel and perpendicular to the drawing direction and specific values of the dichroic ratio of any absorption band at small strain intervals relative to the total elongation may then be obtained by relating the mean absorbance value
of two subsequent parallel polarization spectra to the absorbance
value of the corresponding perpendicular polarization spectrum and
vice versa. As will be outlined in the following sections, this FTIR
technique generally provides a more detailed understanding of the
microscopic deformation behavior of the material under investigation
by correlating the spectroscopic data with the various deformation
stages reflected by the stress-strain diagram.

In Fig. 4-61 the FTIR polarization spectra recorded during uniaxial deformation of an isotactic polypropene film are shown separately for the parallel and perpendicular polarization directions as
functions of strain. With increasing elongation several absorption
bands exhibit significant parallel (π) and perpendicular (σ) dichroism.

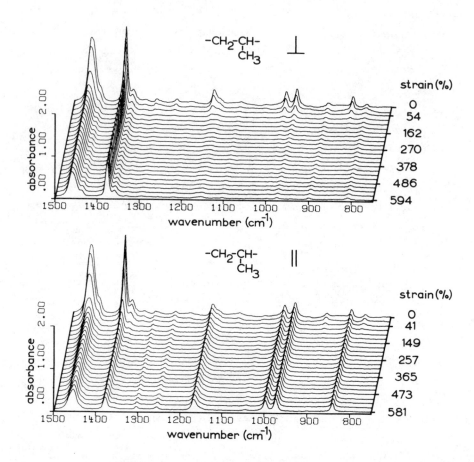

FIGURE 4-61   FTIR spectra recorded at 13.5% strain intervals during
the uniaxial deformation of isotactic polypropene with
radiation polarized alternately parallel and perpendicular to the direction of elongation.

For the quantitative characterization of chain alignment during the
deformation procedure the 1378, 999, and 975 cm^{-1} absorption bands
have been selected.

In the IR spectrum of crystalline isotactic polypropene absorption
bands can be assigned to A or E mode vibrations whose transition moment directions are oriented parallel and perpendicular to the 3:1
helix axis, respectively. The σ-dichroic absorption band at 1378 cm^{-1}

has been predominantly assigned to the $\delta_s(CH_3)$ E mode [192]. The potential energy distributions of the π-dichroic 999 and 975 cm^{-1} A mode absorptions have been calculated by several authors [475, 476]. Thus, the 999 cm^{-1} band belongs to strongly coupled $\gamma_r(CH_3)$, $\nu(C-CH_3)$, $\delta(CH)$ and $\gamma_t(CH_2)$ vibrations while the 975 cm^{-1} absorption involves strongly coupled $\gamma_r(CH_3)$ and $\nu(C-C)$ backbone vibrations. In a recent FTIR absorbance subtraction study of isotactic polypropene Painter et al. [144] have isolated IR bands which are characteristic of the regular 3:1-helical and irregular conformations in the ordered and amorphous phases, respectively, from the spectrum of the semicrystalline polymer. Hence, the 1378 and 975 cm^{-1} absorptions have been shown to contain contributions of both phases, while the 999 cm^{-1} band is predominantly characteristic of conformationally regular chains in the ordered domains.

The stress-strain diagram corresponding to the polarization spectra of Fig. 4-61 is shown in Fig. 4-62. The formation of a neck which propagates through the specimen is indicated by the yield point at

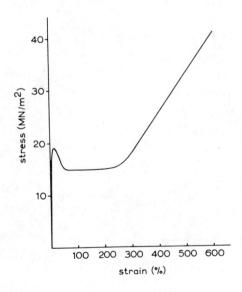

FIGURE 4-62   Stress-strain diagram of isotactic polypropene.

about 20% strain and the subsequent plateau region up to 250% strain.
Beyond 250% strain a linear increase of stress with strain is observed
in the so-called strain-hardening region [315].

For a correlation of the mechanical deformation behavior with the
spectroscopic results the dichroic ratios of the 1378 cm^{-1} and 999
and 975 cm^{-1} absorption bands (determined from the integrated absorbances) have been plotted as functions of strain in Figs. 4-63
and 4-64, respectively. Common to all profiles, large σ-dichroism
(Fig. 4-63) and π-dichroism (Fig. 4-64) effects are observed in the
region from 50 to about 150% strain. These drastic changes which are
indicative of a preferential perpendicular alignment of the CH_3 side
groups and a parallel alignment of the polymer helix axes with reference to the direction of stretch correspond to the propagation of
the neck past the sampling area in the spectrometer. They may shift
slightly on the strain scale in different experiments dependent upon
the position of the initial formation of the neck relative to the

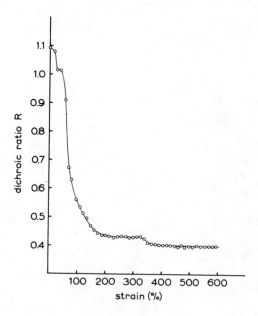

FIGURE 4-63   Dichroic ratio R of the $\delta_s(CH_3)$ 1378 cm^{-1} absorption
band of isotactic polypropene in relation to strain.

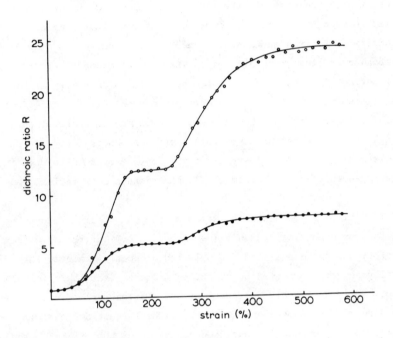

FIGURE 4-64   Dichroic ratios R of the 975 cm^{-1} (●) and 999 cm^{-1} (○) absorption bands of polypropene as functions of strain.

sampling area. The dichroism measured in this strain range represents the average orientation of the inhomogeneous sample area exposed to the IR beam. Once the shoulder-neck region has completely moved past the sampling area, no significant changes in orientation can be detected spectroscopically until the onset of the strain-hardening region. Different polarization phenomena for the absorption bands under examination are observed beyond 250% strain where the specimen has been reduced to uniform cross section. While the σ-dichroic ratio of the 1378 cm^{-1} band tails off as a function of strain the 975 and 999 cm^{-1} bands reflect further π-dichroism increases. An interpretation of these effects has to take into account the assignment of the absorption bands. Thus, the large dichroic changes of all absorption bands during formation of the neck indicate that both, the amorphous and crystalline domains are oriented with respect to the direction of

stretch. With the onset of the strain-hardening region however, a deformation mechanism begins to operate which contributes mainly to the
orientation of conformationally regular chains and favors polymer
backbone over side group alignment. These spectroscopic results may
be interpreted in terms of slippage processes which take place in the
crystalline phase and lead to a more pronounced orientation effect
on the polymer backbone than the side groups due to intermolecular
interactions. The comparatively small changes observed for the dichroic ratios beyond 500% strain up to sample failure show that in
this advanced stage of deformation further chain orientation tends
to zero.

In Table 4-6 the dichroic ratios R, the orientation functions f,
and the corresponding angles θ (see Sec. 4.3.4.2) of the discussed
absorption bands are listed for selected elongation values. The initially negative orientation functions indicate that the original
sample shows a sligthly preferential orientation perpendicular to
the direction of subsequent elongation. The largest decrease in
angle θ during elongation has been observed for the 999 cm^{-1} band.
Thus, a value of 16° was derived for the 500% strained sample. From
the azimuthal intensity distribution of the equatorial wide-angle
x-ray reflex of a 500% drawn sample at relaxed stress level a value
of about 8° was obtained for the average angle of disorientation of
the polymer chains in the crystalline regions [319, 321]. These experimental results strongly indicate that portions of conformationally regular sequences also occur in less oriented amorphous domains
[418]. The polarization data of the 1378 and 975 cm^{-1} absorption
bands which contain contributions both of the amorphous and crystalline phase yield significantly larger angles θ then the 999 cm^{-1}
band and confirm that the polymer chains in the amorphous regions
are on the average less regularly aligned in the direction of stretch
than the polymer chains in the crystalline domains. The orientation
effects of the 1378 cm^{-1} band are further reduced by the superposition of this E mode vibration with the less intense π-dichroic A
mode vibration [192].

TABLE 4-6 Orientation Parameters of Isotactic Polypropene Derived from FTIR Polarization Spectra Recorded during Uniaxial Deformation

Strain (%)	Dichroic ratio			Orientation function			$\theta(°)$		
	R_{1378}	R_{999}	R_{975}	f_{1378}	f_{999}	f_{975}	θ_{1378}	θ_{999}	θ_{975}
0	1.09	0.85	0.82	-0.06	-0.05	-0.06	57	57	57
50	0.98	1.4	1.3	0.01	0.12	0.09	54	50	51
100	0.56	5.9	3.4	0.34	0.62	0.44	41	30	38
150	0.47	11.8	4.9	0.43	0.78	0.56	38	22	33
200	0.44	12.5	5.1	0.46	0.79	0.58	37	22	32
250	0.44	13.1	5.3	0.46	0.80	0.59	37	21	31
300	0.43	18.0	6.4	0.47	0.85	0.64	36	18	29
400	0.41	22.1	7.4	0.49	0.87	0.68	35	17	28
500	0.40	23.2	7.7	0.50	0.88	0.69	35	16	27
575	0.40	23.6	7.9	0.50	0.88	0.70	35	16	27

FTIR spectroscopy with simultaneous stress-strain monitoring has
also been applied to the study of segmental orientation in polyester
urethane films during uniaxial deformation. In fact, segmented polyurethanes are particularly suited to IR investigation because they
contain functional groups such as N-H, C=O, and CH_2 which can be assigned to characteristic absorption bands and may be expected to re-
side in specific domain locations [217, 229].

Linear polyurethanes are usually prepared by condensation of a
diisocyanate with a low-molecular-weight chain-extender glycol and
a high-molecular-weight (1000-2000) macroglycol. A typical polyester
urethane might be synthesized from diphenyl methane-4,4'-diisocyanate,
butane diol, and a dihydroxy-terminated adipic acid/butane diol/
ethylene glycol polyester (molecular weight about 2000). The soft
segments of this type of polyurethane basically consist of the reaction products of the diisocyanate component and the macroglycol,
whereas the hard segments contain largely aromatic and butane diol
moieties linked together by urethane groups. Hence, most of the
N-H and urethane-carbonyl groups will be located in the hard-segment
domains, while the majority of the CH_2 and ester-carbonyl groups re-
side in the soft-segment-phase. The IR spectra of three typical re-
presentatives of the above-mentioned polyester urethane with different diisocyanate:polyester molar ratios are shown in Fig. 4-65.
On the basis of established frequency correlations for the functional
groups of polyurethanes [217] the extent of soft- and hard-segment
orientation can be monitored by means of the polarization properties
of the $\nu(CH_2)$ (2960 cm^{-1}), $\nu(C=O)_{ester}$ (1735 cm^{-1}) and $\nu(NH)$ (3330 cm^{-1}),
$\nu(C=O)_{urethane}$ (1705 cm^{-1}), and $\delta(NH)+\nu(C-N)$ (1530 cm^{-1}) absorption
bands, respectively. The analysis of the $\nu(C=O)$ absorption region
however, is complicated by the overlap of the hydrogen-bonded and
non-bonded ester- and urethane-absorptions [217] and only approximate
values for the dichroic ratios may be derived with the aid of a band-
separation technique. To illustrate the typical dichroic effects the
FTIR polarization spectra taken during a loading-unloading cycle of
a polyester urethane based on the above-mentioned components have been
plotted in three-dimensional representations in Figs. 4-66 and 4-67.

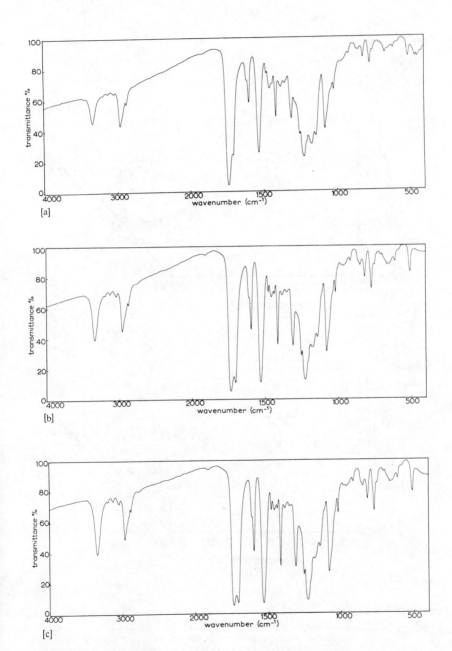

FIGURE 4-65   IR spectra of polyester urethane films based on various
              diisocyanate:polyester molar ratios: (a) 3.4, (b) 6.6,
              (c) 8.7 (see text).

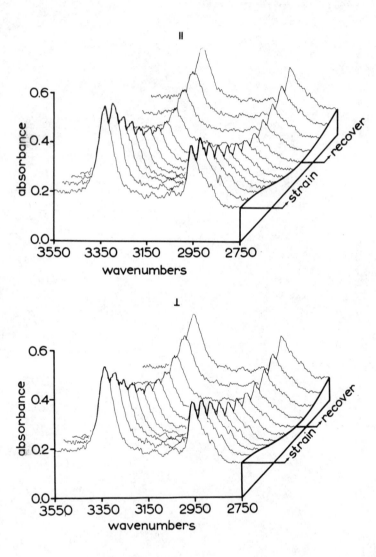

FIGURE 4-66   FTIR spectra of a polyester urethane film in the ν(NH)-
and ν(CH₂)-stretching vibration region recorded during
uniaxial elongation to 220% strain and subsequent recovery to zero stress with radiation polarized alternately parallel (‖) and perpendicular (⊥) to the direction of elongation.

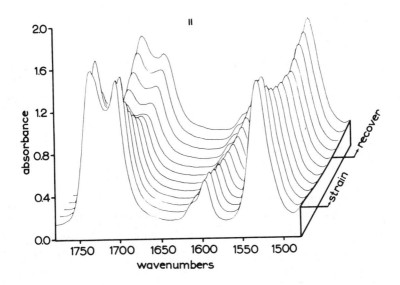

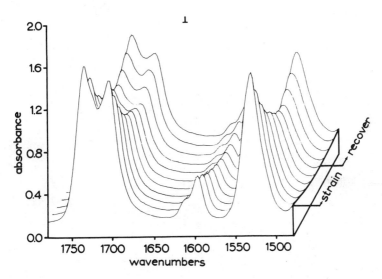

FIGURE 4-67   FTIR spectra of a polyester urethane film in the $\nu(C=O)$-
and $\delta(NH) + \nu(CN)$-vibration region recorded during uniaxial elongation to 220% strain and subsequent recovery
to zero stress with radiation polarized alternately
parallel (∥) and perpendicular (⊥) to the stretching
direction.

Evaluation of such spectra in terms of the σ-dichroism of the ν(NH),
ν(CH$_2$), and ν(C=O) absorptions and the π-dichroism of the δ(NH) + ν(CN)
absorption has demonstrated that the hard segments generally exhibit
a better chain alignment during the deformation process. In contrast
to the soft segments the orientation of the hard segments is also
more effectively retained upon recovery to zero stress. Under identical experimental conditions of the mechanical treatment the maximum
degree and the retention of segmental orientation in cyclic loading-
unloading procedures increases with the hard-segment content of the
polymer under investigation.

The close relation between the composition and the mechanical
properties of these polymers is also reflected in the stress-strain
diagrams of elongation-recovery cycles (Fig. 4-68). Hence, with increasing hard-segment content a distinct increase in elastic modulus
and in stress hysteresis (ratio of area bounded by a strain cycle to
the total area underneath the elongation curve) was observed.

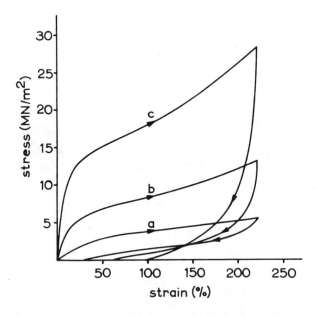

FIGURE 4-68   Stress-strain curves of the polyester urethane films
             of Fig. 4-65 for loading-unloading cycles.

From the preliminary investigations so far available [395] it is
considered that the data derived from simultaneous FTIR spectroscopic
and mechanical measurements will certainly contribute to a more pro-
found understanding of the relationship between the structural and
technological properties of this class of polymers.

Several authors have shown that under conditions of stress the
orthorhombic modification of crystalline polyethylene is partially
transformed to a monoclinic structure [147, 471]. The progress and
extent of this morphological change in dependence of strain may be
conveniently studied by FTIR spectroscopic investigation of the CH_2-
rocking mode. Unlike the 720:730 cm^{-1} band doublet of the orthorhombic
modification the monoclinic structure is characterized by a single
band near 717 cm^{-1}. Figure 4-69(a) shows the FTIR spectra taken during uniaxial elongation and stress relaxation of a high-density poly-

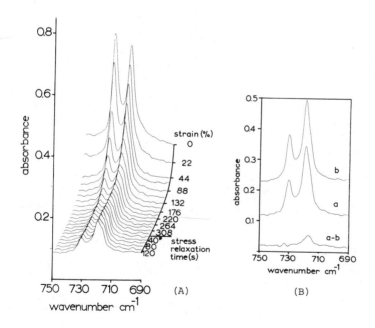

FIGURE 4-69   FTIR spectra recorded during elongation and stress relaxation of a high-density polyethylene film (A) and absorbance subtraction of successively recorded spectra (B)
(a: 88% strain, b: 66% strain, a-b: difference spectrum).

ethylene film. The appearance of the 717-cm^{-1} shoulder during elongation can be readily accentuated by absorbance subtraction [Fig. 4-69
(b)]. A detailed analysis of the difference spectra revealed that the
structural transformation takes place almost entirely before 100%
elongation is reached and no significant molecular changes could be
detected during stress relaxation up to 120 s.

In some polymers a reversible transformation from one conformation to another takes place under tension. The reversible change observed for the crystalline structure of polybutylene terephthalate
(PBT) fibers and films [160, 161, 395, 398, 400, 407-409] shall serve
as an example here. The crystal structure of the relaxed modification
of PBT has been studied by x-ray diffraction, and the deviation of
the fiber identity period (1.16 nm) from the length of the fully extended repeating unit (1.32 nm) has been primarily attributed to a
crumpled *gauche-trans-gauche* conformation of the aliphatic chain
segments [160, 407-409]. Further x-ray investigations have established that under stress the crystal structure changes from the relaxed modification to a strained form in which the polymer chains

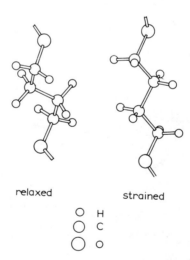

relaxed   strained

○ H
○ C
◯ O

FIGURE 4-70   Conformation of the aliphatic segments of PBT in the
relaxed and strained crystal modification. [Reproduced
from I. H. Hall and M. G. Pass, Polymer 17:807 (1976)
by permission of the publishers, IPC Business Press Ltd.]

are essentially fully extended with an all-*trans* sequence of the
aliphatic segments (Fig. 4-70). The corresponding increase in the
fiber identity period (1.30 nm) can be derived from the shift of the
($\bar{1}$04) and ($\bar{1}$06) meridional reflexes of the wide-angle x-ray diagrams
(Fig. 4-71) [160, 398, 407, 409]. The phenomenon of crystal structure
transformation under stress in PBT has also been studied by IR spectroscopy [161, 395, 400, 409]. In Fig. 4-72 (a) and (b) the IR spectra of a relaxed and a strained PBT film sample are shown. The absorption bands which have been previously demonstrated by the FTIR
absorbance subtraction technique (Fig. 4-28) to be characteristic of
the relaxed crystal modification (1460, 1450, 1390, 1324, 1210, 1030,
917, 810, and 750 cm^{-1}) decrease in intensity with increasing strain
of the sample. The new absorptions observable at 1485, 1470, 1393,
960, and 845 cm^{-1} have been assigned to CH_2-bending, -wagging, and
-rocking vibrations, respectively, of the extended all-*trans* methylene
segments in the strained specimen [161]. In a recent FTIR spectroscopic investigation [400] these spectral changes have been monitored in
small strain intervals during elongation and recovery procedures of
PBT film samples. A typical stress-strain diagram for a loading-unloading-loading cycle of a PBT film is shown in Fig. 4-73. The polymer film under investigation has been extended at constant strain
rate (20%/min) to 15% strain (trace 1) returned to the original po-

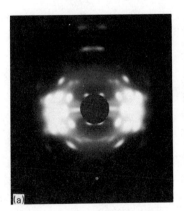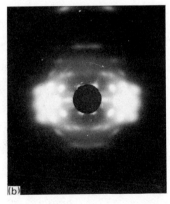

FIGURE 4-71   Wide-angle x-ray diffraction patterns of PBT fibers:
(a) unstressed, (b) stressed.

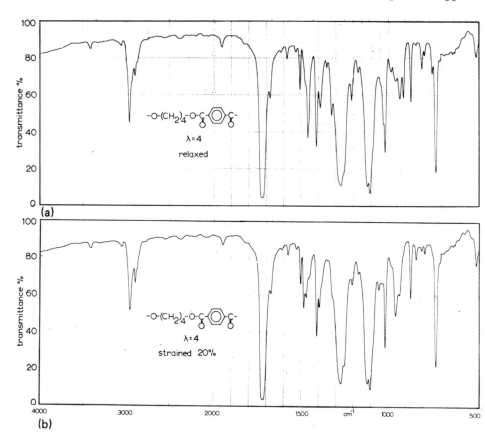

FIGURE 4-72   IR spectra of oriented (draw ratio λ = 4) and annealed
(15 hr at 483 K) PBT film: (a) relaxed, (b) strained 20%.

sition (trace 2) and finally reextended until failure occurred (trace 3).

The stress-strain curve of PBT can be roughly separated into
three portions. An initial steep rise of stress with strain is followed by a comparatively flat portion between 5 and 15% strain, and
beyond 15% strain the stress again rises steeply until fracture occurs [400, 407]. Apart from a small permanent deformation (about 2%),
the induced strains are recoverable as shown by the unloading curve
(trace 2) of Fig. 4-73.

The FTIR spectra taken during the loading-unloading-loading cycle
are shown in Fig. 4-74. They clearly reflect the significant intensity changes of the above-mentioned absorption bands in relation to

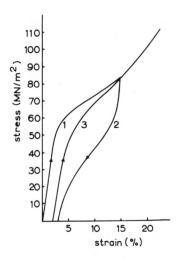

FIGURE 4-73   Typical stress-strain diagram for (1) loading, (2) unloading, and (3) loading cycle of a PBT film.

strain. A more informative representation of the reversible transi-

tion is obtained by plotting the intensity of absorption bands which

are characteristic of the stressed and relaxed segments, respectively,

as a function of strain. For this purpose the absorbance variations

of the $\delta(CH_2)$ absorption bands at 1485 and 1460 cm^{-1} have been chosen

in Fig. 4-75. The absorbances plotted in Fig. 4-75 have been corrected

for changes in sample thickness with the aid of the aromatic $\nu(C-C)$

absorption at 1510 cm^{-1} as reference band.

Unlike detailed x-ray diffraction measurements [160, 398, 407]

IR spectroscopic changes characteristic of the conformational tran-

sition can already be observed at 2% strain and extend beyond 15%

strain up to sample failure (Fig. 4-75). A reasonable explanation for

this discrepancy may be presented in terms of the microscopic behav-

ior of the material. While the occurrence of isolated stressed poly-

mer units at low strains in the crystal lattice of the relaxed modi-

fication is directly reflected in the vibrational spectrum the ap-

pearance of the corresponding near-meridional wide-angle x-ray re-

flexes requires the spatial coherence of the scattering centers.

Owing to this requirement, the ($\bar{1}$04) and ($\bar{1}$06) reflexes of the stres-

sed modification are first observed at distinctly higher strains than

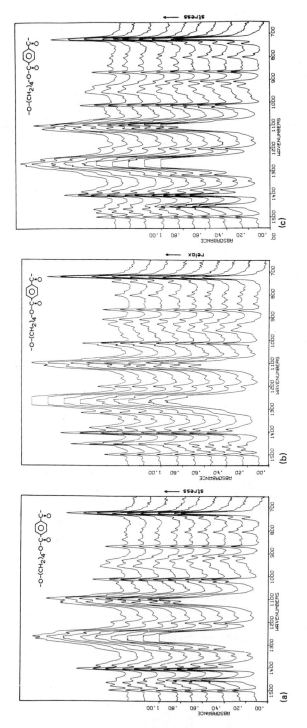

FIGURE 4-74 FTIR spectra recorded during (a) loading, (b) unloading, and (c) loading cycle of a PBT film.

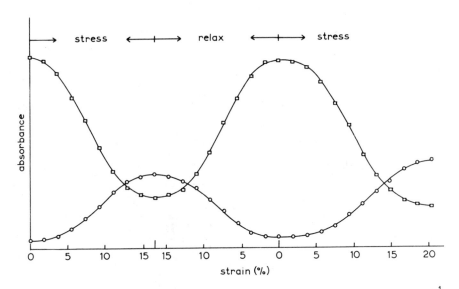

FIGURE 4-75   Absorbances of the $\delta(CH_2)$ absorption bands at 1460 cm^{-1}
(□) (relaxed form) and 1485 cm^{-1} (o) (stressed form) in
relation to strain.

the IR spectroscopic changes. In analogy, unstressed segments are
still detected in the IR spectrum beyond 15% strain, although the
($\bar{1}$04) and ($\bar{1}$06) reflexes of the relaxed crystal form have disappeared
in the x-ray diagram [400, 407]. The percentage of residual unstressed modification in different samples as derived from the decrease
of the 1460 cm^{-1} absorption band at strains of 20% varied between
10 and 20%. A reasonable explanation for this observation may be a
nonuniform stress distribution in the samples as a consequence of
nonuniformities in their thickness and differences in the perfection
of their crystal structure.

As an interesting detail of these investigations an isosbestic
point characteristic of the equilibrium established between the relaxed and strained units at different stress levels could be detected
in the region of the $\delta(CH_2)$ absorptions (Fig. 4-77).

Intensity changes accompanying this reversible conformational
transformation have also been observed in the Raman spectra of PBT

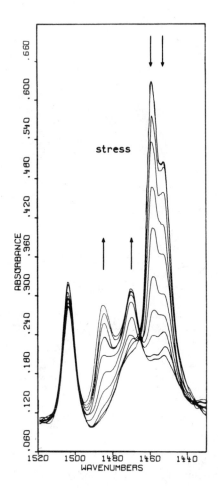

FIGURE 4-76   Isosbestic point observed in the $\delta(CH_2)$ absorption re-
              gion of FTIR spectra recorded during a loading procedure
              of a PBT film.

fibers [161, 398, 407] (see also Fig. 4-38). Vibrational spectroscopy
thus provides a reliable tool to monitor the extent of such structural
changes alongside x-ray diffraction.

The elucidation of the molecular mechanism responsible for the
stress relaxation in semicrystalline polymers is a subject of considerable interest [414, 419, 420] and major advances in this field
of research can certainly be achieved by the application of vibra-

tional, especially FTIR, spectroscopy. Generally, the structural consequences of stress relaxation can be studied spectroscopically by
focusing intensity changes of absorption bands which are sensitive
to stress and orientation as a function of time after application of
the stress. In dynamic IR studies of isotactic polypropene [390, 393]
stress relaxation has been found to be divisible into a fast and a
slow decay region. In the fast decay region the aligned chains de-
velop an overstress and the nonaligned chains are quickly relieved
of their initial overstress. In the slow decay region the aligned
chains gradually loose their overstress and tend to disorient or dis-
tort their helical conformation. During the initial period of fast
stress relaxation the average molecular orientation increased, went
through a maximum, and then decreased gradually in the slow stress
relaxation region.

In an effort to obtain more detailed information on the molecular
mechanism responsible for stress relaxation FTIR spectroscopy and
mechanical measurements have been applied to characterize the time
dependence of the conformational transition of PBT during stress
relaxation. Figure 4-77(a) shows the stress-strain/time diagram of
a PBT film sample which has been initially extended to 12.5% strain
and then held at constant strain for four minutes. The detailed intensity changes monitored during elongation and stress relaxation
at the peak maximum of the 1460 and 1485 cm^{-1} absorption bands which
are characteristic of the unstressed and elongated segments in the
crystal phase, respectively, are shown in Fig. 4-77(b). Analogous
profiles to the 1485 cm^{-1} absorption intensity have also been observed for the 960 and 845 cm^{-1} absorption bands which are representative of the elongated segments in the amorphous and crystalline
regions [161, 395, 409]. Although the spectroscopic changes occurring
during elongation as a function of stress have been shown to be
largest in the region from 5 to 15% strain [395, 400, 407] and despite a pronounced decrease of stress in the fast decay region [Fig.
4-77(a)], no significant intensity changes of the conformation sensitive absorption bands under examination have been detected by FTIR
spectroscopy while the sample is held at constant strain of 12.5%

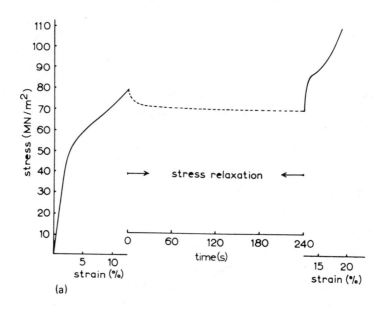

(a)

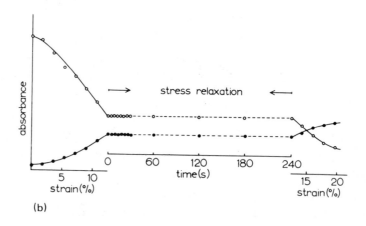

(b)

FIGURE 4-77   Elongation and stress relaxation of a PBT film: (a)
stress-strain/time diagram, (b) absorbance variations
of the 1460-cm^{-1} (o) and 1485-cm^{-1} (●) absorption bands
monitored by FTIR spectroscopy during elongation and
stress relaxation.

[Fig. 4-77(b)]. These results clearly indicate that the proportion of the crumpled and elongated segments in PBT is not appreciably effected by stress relaxation.

As the most probable mechanism which does not necessarily require a change in the proportion of relaxed and elongated structural units stress relaxation can occur by slippage processes of intercrystalline segments. Depending on the shear strength of the crystallites this mechanism may also involve the crystal phase [413]. According to the FTIR spectroscopic and mechanical results the process is characterized by an overall conservation of the elastically stored deformation and a more uniform distribution of stress over the chain segments, respectively. The last-mentioned consequence becomes evident in the stress-strain/time diagram upon reloading of the sample after a certain period of stress relaxation when the stress increases rapidly toward a level which is significantly higher than before stress relaxation. These findings also support the view that chain rupture does not play an important role in the actual stress relaxation mechanism because fractured segments would certainly not increase the mechanical strength.

Several electron spin resonance (ESR) investigations have established that free radicals are formed as a result of interatomic bond scission under mechanical stress and during fracture of polymers [410-412, 416, 421]. However, the sensitivity of existing spectrometers and the reduction of radical population by abstraction, combination, and termination reactions, especially in the presence of air or at elevated temperatures, may seriously limit the application of ESR spectroscopy. Alternatively, IR spectroscopy has proved of considerable value in detecting the products of the scavenging reactions by difference spectroscopy of the stressed or fractured sample against the unstressed specimen. Thus, Zhurkov et al. [405, 410] and Koenig et al. [74] have estimated the concentration of ruptured bonds in mechanically loaded or fractured polyethylene and polypropene by measuring the intensity of characteristic absorption bands. It was shown that under tension and after fracture the absorption bands at 910, 965, 1378, and 1735 cm^{-1} increased in intensity. These absorp-

tion bands can be assigned, respectively, to γ_w(CH) vinyl, γ_w(CH)
vinylene, δ_s(CH$_3$) methyl, and ν(C=O) aldehyde vibrations. In a more
recent paper the relation between viscometry, ESR and IR estimates
of bond rupture and their relevance to mechanical properties of several polymers have been considered by Crist et al. [397].

Polymeric engineering materials subjected to repeated cyclic or
random mechanical stresses exhibit a deterioration of material properties termed fatigue [411]. Through simultaneous application of
mechanical spectroscopy, electron microscopy, x-ray diffraction, and
FTIR spectroscopy Sikka [392] has been able to characterize molecular
rearrangements occurring in polystyrene films as a consequence of
cyclic elongations. Minute changes in the FTIR spectra caused by the
fatigue process could be accentuated by subtracting the spectrum of
the unfatigued sample from the spectrum of the fatigued specimen.
Thus, changes in the absorption frequency of various vibrational
modes of the phenyl side groups have been detected. Furthermore IR
bands associated with bending and stretching vibrations of the CH$_2$
groups and bending modes of the CH group have reflected frequency
shifts and band distortion. These results suggest that fatigue has
modified the intra- and intermolecular interactions of the polymer
chains.

4.3.5 Isotope Exchange

For studies of isotope exchange in chemical compounds, those spectroscopic techniques which reflect specific properties of the nucleus,
such as spin, mass, etc., will prove valuable. Owing to the mass dependence of the vibrational frequency, the observation of frequency
shifts upon isotopic substitution is to be expected in IR and Raman
spectroscopy. Since the force constants remain essentially unchanged,
the frequency shifts can be predicted in terms of the altered atomic
masses alone. More often however, from the actual changes in the corresponding spectra the nature of the vibrations and the atoms involved will be deduced. Deuterium and its compounds are readily available and with few exceptions [422-424] have been used for the inves-

tigations of isotope effects in the vibrational spectra of polymers
[379, 425-427]. Complete or partial replacement of hydrogen atoms
by deuterium will assist in the definite assignment of hydrogen vibrations and their coupling interactions with other vibrational modes.
In cases where intra- or intermolecular interactions occur between
hydrogen vibrations and other modes, thus giving rise to absorption
bands of complex origin, such interactions will usually disappear
upon selective deuterium exchange [341, 428].

The extent of the frequency shift has been discussed in the form
of more or less approximate isotopic frequency rules. The simplest
approximation considers the hydrogen atoms of mass m attached to an
atom of much larger mass M. With the equation for the vibrational
frequency of the harmonic oscillator, it can be shown that the upper
limit for the frequency ratio of a localized X-H and the corresponding X-D vibration will be

$$\frac{\nu_{X-H}}{\nu_{X-D}} = \left(\frac{\mu'}{\mu}\right)^{1/2} \leq 2^{1/2} \qquad (4.68)$$

with $\mu = mM/(m + M)$ and $\mu' = 2mM/(2m + M)$. If there were no coupling
between vibrational modes of the same symmetry species, this frequency ratio would be either 1 for pure nonhydrogen or about $2^{1/2}$
for pure hydrogen vibrations. Any value in between indicates that
more or less coupling takes place.

Krimm has derived an approximation rule for individual frequencies
which applies quite well to the case of hydrogen-deuterium substitution [429]:

$$\frac{\nu_k}{\nu_k'} = \left[1 - \frac{\sum_i \Delta T_i}{\rho T}\right]^{-1/2} \qquad (4.69)$$

where ν_k and ν_k' are the zero-order frequencies of the k-th vibration
before and after isotope substitution, respectively. T is the total
kinetic energy of the vibration, $\sum_i \Delta T_i$ is the change in kinetic energy
upon isotope exchange, and $\rho = m_i'/m_i$ is the ratio of the isotopic to

the normal mass. The results of Krimm's rule applied to some molecular groups of relevance to polymer spectroscopy have been summarized
in Ref. 192.

The only exact rule, the Redlich-Teller product rule [430], unfortunately holds only for the product of all frequencies belonging
to a certain symmetry species:

$$\prod_i \frac{\nu_i}{\nu_{i'}} = \left[\left(\frac{m'}{m}\right)^n \left(\frac{M}{M'}\right)^t \left(\frac{I_x}{I'_x}\right)^{\delta_x} \left(\frac{I_y}{I'_y}\right)^{\delta_y} \left(\frac{I_z}{I'_z}\right)^{\delta_z} \right]^{1/2} \tag{4.70}$$

Here ν_i are the frequencies or wavenumbers; m, the atomic masses; M,
the masses of the molecules; I_x, I_y, and I_z, the moments of inertia
about the x-, y-, and z-axes. The primed symbols belong to the isotope substituted molecule. n is the number of vibrations including
the zero modes, which the substituted atoms contribute to the symmetry class under consideration; t is the number of translations;
δ_x, δ_y, and δ_z are 1 or 0, depending on whether the rotation about
the corresponding axis belongs to this symmetry class or not. However,
the application of this equation is limited to small, highly symmetrical molecules whose moments of inertia are known. Because of the
usually low symmetry of polymer molecules, each symmetry species
contains many frequencies and the rule is not very helpful.

Polymers deuterated at carbon atoms are almost exclusively synthesized by polymerization of the deuterated monomers. References on
the synthesis and IR and Raman spectra of polymers with CD groups
have been summarized in Table 4-7.

The IR and Raman spectra of the simplest, though very illustrative example, polyethylene and its deuterated analogue, polydeutero-
ethylene, are reproduced in Fig. 4-78. The values for the observed
wavenumber ratios of the $\nu_{as}(CH_2)/\nu_{as}(CD_2)$ (1.34), $\nu_s(CH_2)/\nu_s(CD_2)$
(1.37), $\delta(CH_2)/\delta(CD_2)$ (1.35), and $\gamma_r(CH_2)/\gamma_r(CD_2)$ (1.39) vibrations
in the IR spectra compare well with those calculated by the applica-
tion of Krimm's approximation rule.

The vibrational spectra of polymers where specific hydrogen atoms
have been replaced by deuterium have often supported an unequivocal

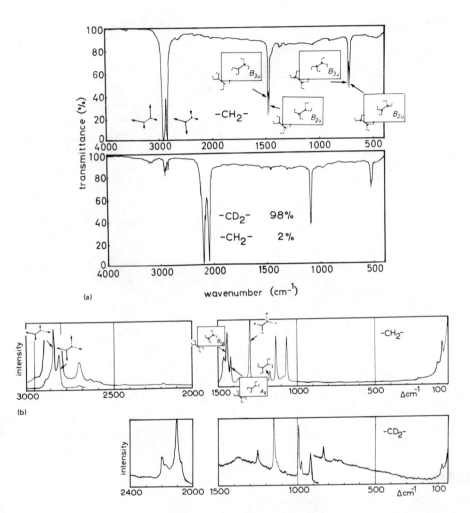

FIGURE 4-78   (a) IR and (b) Raman spectra of polyethylene and poly-
deuteroethylene and some vibrational assignments.

assignment of absorption bands. Polyacrylonitrile (PAN) and its selectively deuterated derivatives have been studied by Yamadera et
al. [431]. The wavenumber positions of the C-H- and C-D-stretching
vibration bands of PAN, PANαd$_1$, PANβd$_2$, and PANd$_3$ have been summarized in Table 4-8. The values of the calculated wavenumber ratios
indicate the isolated nature of the vibrations under consideration.

TABLE 4-7   References to the Synthesis and Vibrational Spectra of
            Polymers with CD Groups

Polymer	Refs.
Polyethylene	[469-472]
Polypropene	[473-483]
Polyalkylethylenes	[182, 183, 484, 485]
Polystyrene	[432-434, 486-488]
Polyvinylcyclohexane	[480]
Polybutadiene	[489-491]
Polyacrylonitrile	[431, 492-494]
Polyvinyl chloride	[495-498]
Polyvinylidene chloride	[499]
Polyvinyl alcohol	[456]
Polyvinyl acetate	[456]
Polymethylmethacrylate	[500-503]
Aliphatic polyethers	[504-509]
Polyethylene sulfide	[510]
Polyethylene terephthalate	[151, 154, 155, 365, 435]
Polybutylene terephthalate	[512]
Aliphatic polyamides	[511, 513]
Polypeptides	[514]
Cellulose	[515, 516]

TABLE 4-8   Wavenumbers (cm^{-1}) for the C-H- and C-D-Stretching Vibra-
            tions in PAN, PANαd$_1$, PANβd$_2$, and PANd$_3$

PAN	2950 $\nu_a(CH_2)$	2870 $\nu_s(CH_2)$	2930 $\nu(CH)$
PANαd$_1$	2950 $\nu_a(CH_2)$	2870 $\nu_s(CH_2)$	2170 $\nu(CD)$
PANβd$_2$	2180 $\nu_a(CD_2)$	2130 $\nu_s(CD_2)$	2930 $\nu(CH)$
PANd$_3$	2180 $\nu_a(CD_2)$	2130 $\nu_s(CD_2)$	2170 $\nu(CD)$
Wavenumber ratio $\dfrac{\nu(CH)}{\nu(CD)}$	1.35	1.35	1.35

In the interpretation of the vibrational spectra of polystyrene
and aromatic polyesters, the substitution of the aliphatic and/or
aromatic protons by deuterium has contributed to a more definite assignment of the aliphatic- and aromatic-segment vibrations [152, 155,
156, 365, 432-435]. Selective deuteration of the aromatic protons in
polybutylene terephthalate, for example, leaves most of the previously discussed conformation sensitive IR bands of the aliphatic seg-

ments unchanged whereas the absorption bands of vibrations involving
ring atoms or strongly coupling with ring motions undergoe distinct
frequency shifts (Fig. 4-79).

An interesting alternative to the synthetic route of deuterated
polymers is direct isotope substitution with liquid or gaseous D_2O.
However, this technique is limited to polymers containing loosely
bonded protons in OH-, NH- or SH- groups because hydrogen atoms attached to carbon will not be exchanged. Nevertheless, a great number
of polymers (cellulose, polyvinyl alcohol, polyamides, polyurethanes)
can be studied by this technique [32, 165, 379, 436-440]. Its applicability to the characterization of molecular order in polymers is
based on the fact that hydroxyl and amino groups in chain segments

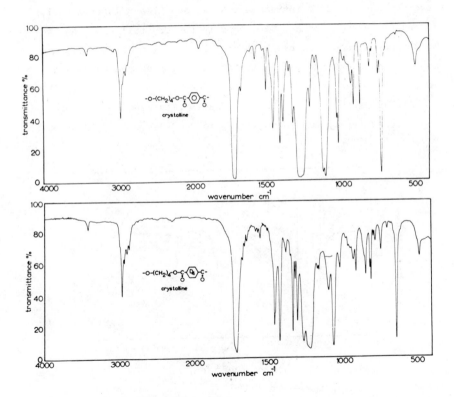

FIGURE 4-79   IR spectra of polybutylene terephthalate and polybutylene
terephthalate-D_4.

of crystalline regions are less exposed to substitution than are
those in amorphous domains. Thus, the rate and extent of the exchange
reaction will strongly depend on the mechanical and thermal history
of the polymer under investigation.

The structure of cellulose has been extensively studied by this
technique [438, 441-445]. In Fig. 4-80 the deuteration progress of
100% drawn regenerated cellulosic fibers with D_2O vapor in the earlier
mentioned ATR sample holder (Fig. 3-25) is followed in the IR spectra
of the OH-, CH-, and OD-stretching vibration region (4000-2000 cm^{-1}).
With progressing deuteration the broad $\nu(OH)$ absorption band at about
3500 cm^{-1} decreases, while the $\nu(OD)$ band at 2500 cm^{-1} increases in
intensity. In addition, the $\nu(OH)$ band splits into four bands which
have been assigned to intramolecular (1 and 2) and intermolecular
(3 and 4) hydrogen-bonded OH groups in the crystallites [444, 445].
This view is supported by the parallel dichroism of the absorption
bands 1 and 2 and the perpendicular dichroism of the bands 3 and 4
in the ATR polarization spectra (Fig. 4-81).

Higher rates and extents in accessibility have been reported for
the deuteration by immersion in liquid D_2O [446, 447]. Adopting this

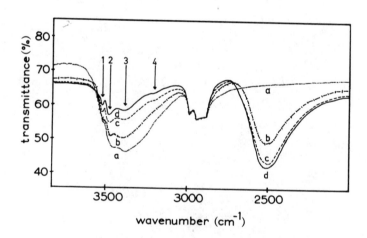

FIGURE 4-80   IR spectra of a 100% drawn cellulosic fiber: (a) un-
              deuterated, (b) deuterated 24 hr, (c) deuterated 48 hr,
              (d) deuterated 60 hr.

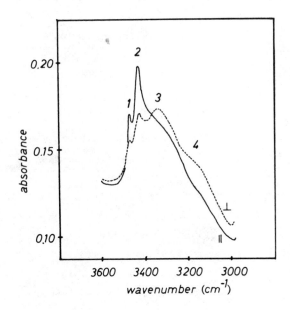

FIGURE 4-81   IR polarization spectra of the ν(OH) absorption band
of a 100% drawn cellulosic fiber. The solid line shows
parallel polarization, the dashed line perpendicular
polarization.

procedure, the deuterated polymer sample usually cannot be completely
rehydrogenated by H_2O [441, 443, 444]. The resolution of the residual
ν(OD) absorption band into separate bands upon rehydrogenation is
evidence for the penetration of D_2O molecules into the crystalline
regions of the original polymer sample.

Given the applicability of Beer's law and assuming that the rea-
son for a proton not to exchange is due to it being inaccessible to
D_2O the percentage of accessible regions Z at any stage of deuteratio
can be determined from the relation [447-449]:

$$Z = \frac{A_{\nu(XD)}}{kA_{\nu(XH)} + A_{\nu(XD)}} \; 100 \; (\%) \tag{4.71a}$$

where $A_{\nu(XH)}$ and $A_{\nu(XD)}$ are the integrated absorbances of the ν(XH)
and ν(XD) absorption bands under examination and the constant k is

the ratio of the corresponding absorptivities $a_{\nu(XH)}/a_{\nu(XD)}$. Alternatively, Z can be calculated from the expression

$$Z = \left[1 - \frac{A_{\nu(XH)}}{A_{\nu(XH)_i}} \right] 100 \ (\%) \tag{4.71b}$$

Here the subscript i refers to the integrated absorbance of the $\nu(XH)$ absorption band of the undeuterated sample. For quantitative evaluation of accessibility data in a first approximation, the difference in absorptivities of the $\nu(XH)$ absorption bands of the amorphous and crystalline regions and of the $\nu(XH)$ and the corresponding $\nu(XD)$ absorption bands [449] can be neglected. Significant differences in the perfection of molecular order have been detected by this technique in various cellulosic samples [438, 450, 451].

In studies of the orientation phenomena in drawn polyvinyl alcohol films [452] the deuteration technique has been used to separate the $\nu(CH)$ absorption bands from the $\nu(OH)$ absorption background in the undeuterated polymer (Fig. 4-82). Upon deuteration the dichroism of the $\nu(CH)$-stretching vibrations can be evaluated without interference from the neighboring $\nu(OH)$ band which has considerably decreased in intensity – from the integrated intensity of the residual $\nu(OH)$ absorption band an accessibility of about 90% was derived with Eq. (4.71b) [453]. In marked contrast to the relatively large σ-dichroism of the $\nu(CH)$ and $\nu(CH_2)$ absorption bands, no significant dichroism is observable for the $\nu(OH)$ and $\nu(OD)$ absorption bands [454, 455]. Krimm [454] has shown that these experimental results are essentially in agreement with the hydrogen bonding geometry of the polyvinyl alcohol structure proposed by Bunn [456].

In Fig. 4-83(a) and (b) the IR spectra of undeuterated and partially deuterated (with D_2O vapor at 323 K for 120 hr in the cell shown in Fig. 3-23) polyamide-6 films [(a) cast from HCOOH solution and annealed at 433 K for 3 hr and (b) prepared by hot pressing (30 N/mm^2) at 513 K and subsequent quenching in liquid nitrogen] are shown. The most obvious spectral changes upon deuteration directly reflect those absorption bands which belong to vibrations involving the amide-hydrogen atom. Thus, the H-D isotope exchange re-

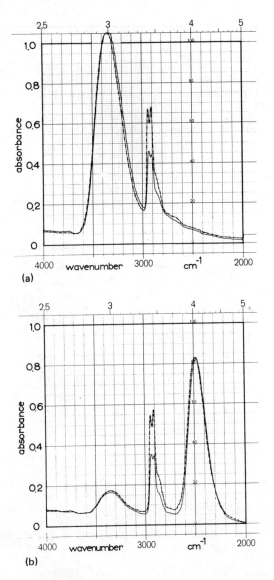

FIGURE 4-82   IR spectra of drawn (200%) polyvinyl alcohol film:
(a) undeuterated, (b) deuterated. The solid line indicates the electric vector parallel to the drawing direction; the dashed line indicates the electric vector
perpendicular to the drawing direction.

sults in partial replacement of the $\nu(NH)$ band at 3300 cm^{-1} and absorption bands at 3200 and 3060 cm^{-1} by a band doublet of the $\nu(ND)$-
stretching vibration at 2465 and 2410 cm^{-1} which has been assigned

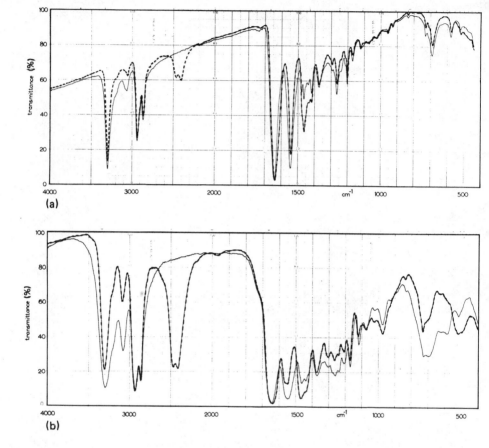

FIGURE 4-83   IR spectra of polyamide-6 films (a) cast from HCOOH and
              annealed at 433 K for 3 hr and (b) prepared by hot pres-
              sing (30 N/mm^2) at 513 K and subsequent quenching in
              liquid nitrogen: Solid line, undeuterated; dashed line,
              upon deuteration with D_2O vapor at 333 K for 120 hr.

to Fermi resonance [341] (it has also been attempted to interprete
the doublet in terms of the state of order and different hydrogen
bonding of the sample [457, 458]. While the amide I band at 1635 cm^{-1},
essentially attributed to the ν(C=O)-stretching vibration, is almost
unaffected by the deuteration (apart from a small wavenumber shift
to 1630 cm^{-1}), the amide II (1540 cm^{-1}) and amide III (1260 cm^{-1})
bands have been interpreted as coupled modes in terms of resonance

between the symmetric ν_s(OCN)-stretching and the δ(NH)-deformation vibrations. Upon substitution of H by D, their resonance is destroyed and these bands are partially replaced by the unperturbed ν(CN) mode at 1465 cm^{-1} and the δ(ND) band at about 970 cm^{-1}. Pronounced effects can also be observed for the amide V (690 cm^{-1}) and amide VI (580 cm^{-1}) bands [437].

In Fig. 4-84 the progress in deuteration [evaluated as percentage of accessible regions from the integrated absorbance of the ν(NH) absorption at 3300 cm^{-1}] for the two polyamide-6 samples of Fig. 4-83 is shown. The relatively large accessibility of more than 60% in the quenched specimen is in contrast to the much lower amount of accessible regions (30-35%) of the sample cast from formic acid solution and annealed at 433 K for 3 hr. These differences can unambiguously be attributed to the higher content of amorphous phase in the quenched sample which is also reflected in the densities of the specimens under examination (quenched sample, 1.136 g/cm^3; sample cast from formic acid solution, 1.161 g/cm^3) [459].

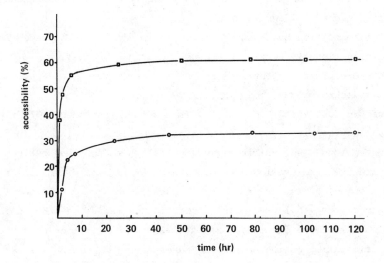

FIGURE 4-84   Progress in H-D exchange (expressed as accessibility)
             in relation to time. (o) sample (a) of Fig. 4-83, ($\square$)
             sample (b) of Fig. 4-83.

Similar investigations have been applied to aromatic polyamides
and the very low extent of isotope substitution (10-15%) in poly(p-
phenylene terephthalamide) fibers, for example, was found in agreement
with the model of a densely packed, fibrillar structure consisting
of rigid, almost parallel, regularly hydrogen-bonded polymer chains
[460].

Valuable information about the orientation of polymer chains in
regions of different molecular order has been obtained by a combination of deuterium exchange with IR dichroism measurements on drawn
polyamide-6 films [461]. In Fig. 4-85 the IR polarization spectra of
a 250% elongated film (a) before and (b) after deuteration by the
above-mentioned technique are shown. While the dichroism of the ν(NH)
absorption band in the undeuterated sample is characteristic of the
average ν(NH) transition moment orientation of the total polymer the
dichroic properties of this absorption band remaining upon deuteration
can then be assumed to represent the orientation of the crystalline
phase only. This conclusion is also supported by the considerable reduction of the bandwidth of the ν(NH) absorption as a consequence of
isotope exchange (Figs. 4-83 and 4-85). In contrast to the ν(CH$_2$) and
ν(C=O) absorption bands the dichroic ratio of the ν(NH) absorption
band significantly decreases upon deuteration (e.g., from R = 0.67
to R = 0.60 in the spectra of Fig. 4-85). From numerical application
of the dichroism and accessibility values to the polymer models involved in Eqs. (4.48) and (4.51), it has been concluded [461] that
in analogy to the crystalline domains the polymer chains of the amorphous regions also exhibit a preferential, although less perfect,
alignment in the drawing direction. Similar effects can be observed
for the π-dichroism of the amide II band at 1540 cm^{-1}. The complex
nature of the ν(ND) band doublet complicates the use of the observable
σ-dichroism for a detailed quantitative characterization of the average orientation of the amorphous phase.

The diffusive transport of low-molecular-weight compounds in
polymeric solids is of considerable importance for many technical
applications such as separation processes by polymer membranes, dye
diffusion in fibrous materials, etc.. Despite numerous practical and

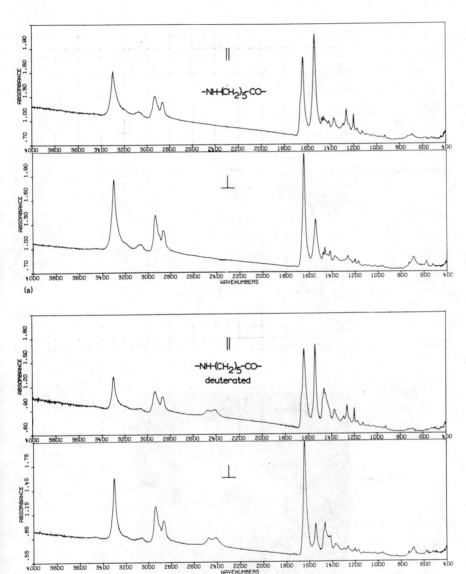

FIGURE 4-85   IR polarization spectra of drawn (250%) and annealed
               (3 hr at 433 K) polyamide-6 film: (a) undeuterated, (b)
               deuterated with D_2O vapor at 333 K for 120 hr. Symbols:
               ‖, electric vector parallel to the drawing direction;
               ⊥, electric vector perpendicular to the drawing direction.

theoretical studies [437, 462-467] in an effort to gain a more complete understanding of the processes involved and their dependence on
the chemical and physical properties of the polymer and the permeant
molecule further systematic investigations appear essential for a
confident interpretation of the results so far available. In this
respect the isotope exchange technique offers the possibility to
study a wide range of polymer-sorbate systems. As an illustration

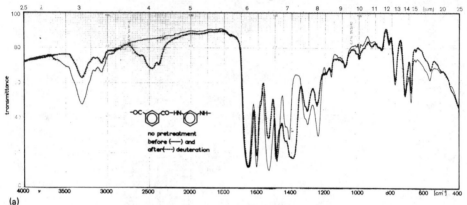

(a)

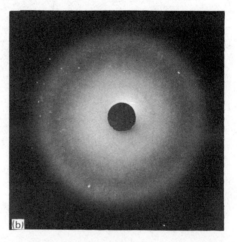

FIGURE 4-86   (a) IR spectra of poly(m-phenylene isophthalamide) film
               (cast from dimethylacetamide solution). Solid line, un-
               deuterated polymer; dashed line, polymer deuterated with
               D_2O vapor at 333 K for 130 hr. (b) Wide-angle x-ray dif-
               fraction diagram of poly(m-phenylene isophthalamide)
               film used for the deuteration experiments.

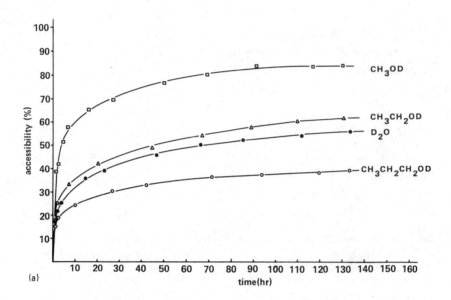

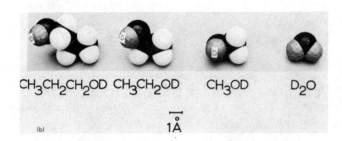

FIGURE 4-87   Isotope exchange of poly(*m*-phenylene isophthalamide)
films with deuteration agents of different molecular
dimensions: (a) progress of H-D exchange (expressed as
accessibility) for different deuteration agents in dependence of time, (b) Stuart models of the various
deuteration agents.

the accessibilities [monitored spectroscopically by the ν(NH) absorb-
ance] of primarily amorphous poly(*m*-phenylene isophthalamide) film
(Fig. 4-86) for deuterating agents of different molecular dimensions
[Fig. 4-87(b)] are shown in Fig. 4-87(a) in dependence of time. An
inverse relation between the size of the permeant and the extent of

isotope exchange may be derived for the homologous deuterated alco-
hols. In the case of D_2O, however, it becomes obvious that not only
is the diffusive transport governed by the molecular dimensions but
other factors such as the polarity of the penetrating molecules and
the polymer segments play an important role [468].

4.4 LOW-FREQUENCY VIBRATIONS

The separation between the mid-infrared (MIR) and far-infrared (FIR)
region is primarily a consequence of instrumental factors. The experimental difficulties of dispersive IR spectroscopy in the FIR compared to the MIR wavenumber region can be summarized as follows [517]:

1. Low radiant power of available continuous sources
2. High power of unwanted radiation at short wavelengths
3. Low absorption coefficients for molecular vibrations
4. Strong emission effects
5. Powerful and ubiquitous absorption due to pure vibrational tran-
    sitions of water vapor
6. Lack of good optical materials

Although some of these difficulties are now overcome by using FTIR
spectrometers most spectroscopic papers so far published deal with
the MIR region only. This also applies to low-frequency Raman spectroscopy, although no comparable experimental difficulties occur (see
Chap. 3). In what follows the region below 600 cm^{-1} (or $\Delta 600\ cm^{-1}$ in
Raman spectroscopy) will be treated as FIR or low-frequency region,
but it should be emphasized that there exists no strict rule for such
a borderline.

   In the low-frequency region the following types of vibrations can
be observed:

1. Stretching vibrations of heavy atoms.
2. Angle deformation (bending) vibrations of heavy atoms. The bending
    vibrations of the polymer skeleton belong to this group (for ex-
    ample, C-C-C, C-O-C, C-N-C, C-S-C). Furthermore, the one-dimen-
    sional accordionlike lattice vibrations [longitudinal acoustical

mode (LAM)] of long conformationally regular monomers (e.g.,
n-alkanes) and polymers must be classified here. These vibratons
cause low-frequency bands in the Raman spectra and will be discussed in more detail in Sec. 4.4.2.

3. Torsional vibrations. These motions correspond to hindered rotations of molecular parts around a certain bond (for example,
   C-C, C-N, C-O). In a first approximation neither bond lengths
   nor bond angles are changed.

4. Vibrations of hydrogen bonds. Owing to the comparatively small
   force constants involved, the vibrations of hydrogen bonds give
   rise to absorptions in the low-frequency region.

5. Lattice vibrations (intermolecular vibrations between adjacent
   chains within a crystallographic unit cell). Here, the single
   chains vibrate against each other. Since the intermolecular forces
   are of an order of magnitude smaller than the intramolecular
   forces and the masses of the (in a first approximation) rigid
   chains are large, the resulting vibrational frequencies are low.

6. Defect-induced absorptions. These absorptions can be explained by
   interaction of the radiation with dipoles originating from local
   defects within the crystal.

In the following sections particular vibrations are discussed in more
detail.

4.4.1 Stretching, Bending, and Torsional Vibrations below 600 cm^{-1}

For the MIR region an empirical interpretation of the corresponding
IR and Raman spectra is found in most papers, and theoretical dis-
cussions based on a normal-coordinate analysis are less frequently
encountered. However, the opposite is observed for FIR spectroscopy;
the majority of assignments result from normal-coordinate analysis,
whereas in only a few papers are empirical interpretations discussed.

The low-frequency region is of great interest in the study of
both the vibrational behavior and the state of order of polymers,
because many normal vibrations occur in this region and their frequencies often reflect the intermolecular interactions of the polymer

under investigation. Some frequently encountered vibrations and their
wavenumbers are listed in Table 4-9. Table 4-10 gives a summary of
publications concerned with the assignment of stretching, bending
and torsional vibrations in the FIR and low-frequency Raman spectra
of selected polymers.

4.4.2 Longitudinal Acoustical Modes

Many polymers with long methylene sequences show in the low-frequency
range of their Raman spectra bands whose positions depend on the
length of the conformationally regular sequences of the chains.
Earlier, it was reported [561-565] that Raman spectra of solid
n-alkanes contain a series of bands between 20 and 150 cm^{-1} whose
frequency is inversely proportional to the chain length. These bands
were assigned to the longitudinal acoustical modes. The results can
be described by the classical vibrating-spring model which leads to

TABLE 4-9 Approximate Frequency Ranges for Some Commonly Encountered
FIR Vibrations[†]

Frequency (cm^{-1})	Type of vibration
730-630	C=O in-plane deformation
590-530	C=O out-of-plane deformation
600-520	C-C-C deformation
520	CH_2-CO-O deformation
450-430	O-CH_2-CH_2 deformation
370-295	CH_2-CO-O-CH_2 deformation
235-145	CO-O torsion
150-70	C-C torsion
540-490	Ring-breathing coupled with C-X stretching[††]
240-140	Ring deformation
500-440	C-X out-of-plane deformation of benzene benzene ring containing polymers[††]
250-200	CF_2 rocking
320-280	CCl_2 rocking
300-270	CF_2 wagging
370-340	CCl_2 wagging

[†]It has to be taken into account that strong coupling generally occurs in the FIR region.

[††]X: NO_2, OH, NH_2, CH_3, F, CD_3, CN, CHO.

TABLE 4-10   References to FIR and Low-Frequency Raman Spectra of
Polymers

Polymer	Refs.
Polyethylene	[530, 533, 579-589]
Polypropene	[524, 527, 533, 540]
Polyalkylethylenes	[523, 557]
Natural rubber	[519]
Poly-p-xylylene	[525]
Polyethers	[505, 509, 533, 536, 539, 543, 550, 551, 558]
Polystyrene	[528, 559]
Polythioether	[510]
Polyvinyl chloride	[427, 532, 533]
Polyvinylidene chloride	[532]
Polyvinylidene fluoride	[520]
Polychlorotrifluoroethylene	[535]
Polytetrafluoroethylene	[522, 533, 541, 542, 545-549, 553-555]
Polyacrylonitrile	[533, 534]
Polyvinylformate	[518]
Polymethylmethacrylate	[526, 528, 529, 556]
Polyester-x	[544]
Polyester-x,y	[531]
Polyethylene terephthalate	[533, 537, 538, 556]
Polyamide-x	[521, 544, 552]
Polyamide-x,y	[521]
Proteins and polypeptides	[281, 282, 517, 556, 560, 643]

the frequency relation:

$$\nu = \frac{m}{2L} \sqrt{\frac{E}{\rho}} \qquad (4.72)$$

where m is the order of the vibration, E is Young's modulus, ρ is
the density, and L is the length of the extended methylene sequence.
The LAM appear solely in the Raman spectrum and then only in its odd
orders. With the aid of the experimental data a value for Young's
modulus can be evaluated and compared with the results derived from
other methods. Thus, taking into account weak interlamellar forces,
Strobl and Eckel [566] have determined the limiting elastic modulus
of crystalline polyethylene from Raman spectroscopic investigation
of the LAM frequencies of a series of n-alkanes.

Since many polymers crystallize from the melt in spherulites composed of lamellar structural units involving polymer chains folded on the lamellar surfaces, the LAM with nodes in the fold zone are observable in these polymers, too [567-570]. It was shown [567] that the LAM of polyethylene has a frequency which is inversely proportional to the length of the extended methylene sequence within a lamellar unit and can be used to estimate the lamellar thickness. The results are in reasonable agreement with the long spacing obtained from small-angle x-ray scattering [568, 571]. Peterlin and coworkers [568] have pointed out by a model calculation that the vibrational antinode assumed to lie at the fold-core interface may, in fact, lie within the fold zone. Thus, the length estimated from the measurement of the acoustic mode may be significantly greater than the planar zigzag polymethylene sequences. However, the following theoretical considerations and experimental evidence suggest that the antinode lies close to the fold-core interface, but the vibrational force constant may not be exactly the same as for alkanes:

1. The frequency of the LAM strongly depends on the sample pretreatment. Thus, for example, the bands of the first order vibration of polyethylene occur at 16, 11.8, 9, and 7 cm^{-1} for samples as supplied, cooled from the melt at $10°/min$, annealed at 373 K, and annealed at 401 K, respectively. However, the bands are not temperature-sensitive for the same pretreatment. Since at low temperatures the modulus of elasticity and the density of the fold phase should be different from that at higher temperatures, a shift in frequency would be expected upon cooling if this phase contributed significantly to the vibration.

2. No change in frequency occurs upon etching the fold zone with fuming nitric acid [568, 572].

3. Chains tilted toward the lamellar surface do not cause any frequency shift [573].

4. Calculations showed [574] that CH_2 units in the *gauche* conformation will confine the vibration, whereas methyl side groups will little influence the frequency.

The molecular vibrational characteristics of a polymer such as

polyethylene contain features characteristic of all the molecular
species present in the sample, modified by any environmental con-
straints placed on individual molecular sequences. Thus, since molten

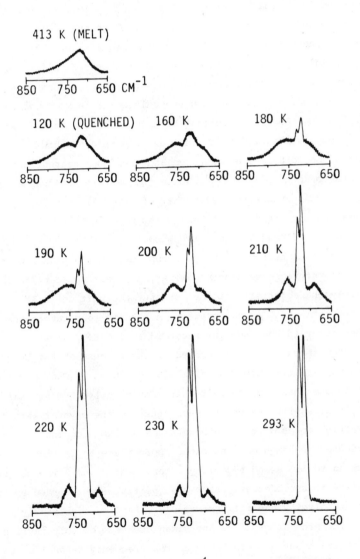

FIGURE 4-88   IR spectra near 700 cm^{-1} of polyethylene melt and
quenched melt at temperatures between 120 and 290 K.

polyethylene is not composed of all-*trans* molecular sequences but
rather of a complex mixture of rotameric isomers, the vibrational
spectrum, as studied in infrared absorption or Raman scattering, is
diffuse and lacks sharp features. Crystalline polyethylene has a
small number of well-known vibrations whose appearance in the spectra
is largely understood [189]. Further, the requirement that the unit
cell contains two chains results in a characteristic splitting of the
vibrational bands. This phenomenon (observed for motions occurring
in phase and out of phase within the unit cell) is known as *correlation splitting*. Thus, vibrational techniques enable us to discern the
onset of crystallization in a specimen. Figure 4-88 shows the IR
spectra of polyethylene as quenched and then when heated to various
temperatures up to 295 K [575]. While the last recorded spectrum can
be said to be characteristically that of crystalline polyethylene,
the first is more akin to that of molten polyethylene.

A sharp band near 720 cm^{-1} is characteristic of the B_{2u} class
CH_2-rocking motion of a methylene group in either a planar or non-
planar polymethylene sequence. However, its partner near 730 cm^{-1} is
definitely characteristic of the orthorhombic unit cell. It is quite
clear that this band is not present in the low temperature spectrum
but does occur in the spectrum recorded at 180 K. Further, the sharp
band system appears to be generated at the expense of the broad diffuse characteristic extending from 680 to 820 cm^{-1}. The presence of
more than one chain in the unit cell enables polyethylene chains to
move as bodies with respect to each other in the crystal lattice.
Such lattice modes give well-known bands at low frequencies. At 123 K
the quenched specimen gave no bands appropriate to lattice modes but
on warming to 183 K and recooling a band near 72 cm^{-1} was observed.
To confirm these observations, Raman spectra were recorded on small
pieces of polyethylene film prepared and handled identically to those
used in infrared absorption. The Raman spectrum of polyethylene contains at least seven sharp bands in the frequency range 1500 to 1000
cm^{-1} and well-understood origin [189]. These bands are reduced to
three diffuse features in molten polyethylene [Fig. 4-89(b)]. The
quenched melt at about 120 K gives a spectrum similar to the melt,

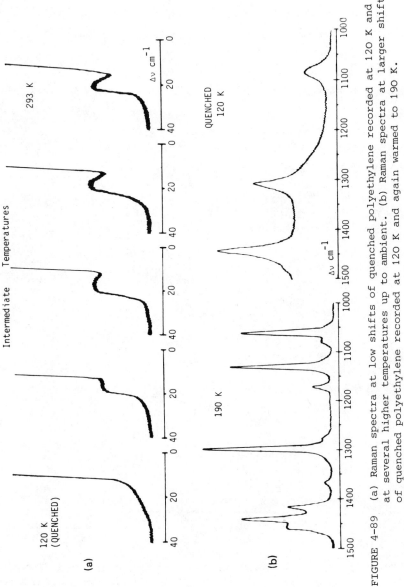

FIGURE 4-89   (a) Raman spectra at low shifts of quenched polyethylene recorded at 120 K and at several higher temperatures up to ambient. (b) Raman spectra at larger shifts of quenched polyethylene recorded at 120 K and again warmed to 190 K.

while on warming to 190 K the spectrum becomes that of crystalline
material [575]. At very low shifts from the exciting laser frequency,
bands are found in the Raman spectrum of polyethylene due to acoustic
motions of the extended methylene species within the lamellae [562,
571]. In Fig. 4-89(a) the Raman spectra of quenched polyethylene
warmed within a range of temperatures between 120 and 290 K are
shown. It appears that the acoustic vibration characteristic of lamellar structure is not present initially on quenching but becomes more
prominent with warming. The lamellar thickness derived from the wavenumber shift of the acoustic mode would appear to center on 15 nm,
but this may change during the warming process.

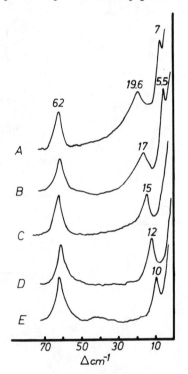

FIGURE 4-90   Low-frequency Raman spectra of polyethylene sulfide
              (A) melt, rapidly quenched, (B) to (E) specimen A an-
              nealed for 2 hr at 393 K, 423 K, 443 K, and 463 K, re-
              spectively. [Reprinted with permission from P. J. Hendra
              and H. A. Majid, J. Mat. Sci. 10:1871 (1975). Copyright
              by Chapman & Hall Ltd. 1975.]

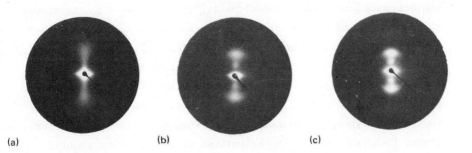

(a)   (b)   (c)

FIGURE 4-91   Small-angle x-ray diffraction pattern of polyethylene
sulfide fibers (a) as spun, (b) annealed at 433 K for
60 min, (c) annealed at 453 K for 60 min. (Courtesy of
Prof. H. D. Noether, Celanese Research Company, N.J.)

Raman spectroscopy has also been used to establish the presence
of lamellae in crystalline polyethylene sulfide and their thickening
upon annealing as a function of temperature [570]. In Fig. 4-90 the
low-frequency Raman spectra of polyethylene sulfide quenched from the
melt (A) and annealed for 2 hr at 393, 423, 443, and 463 K, respectively (B-E), are shown. While the band at 62 cm^{-1} has been tentative-
ly assigned to a lattice mode [510], the sensitivity of the absorptions between 20 and 10 cm^{-1} to thermal history have been interpreted
in terms of third-order longitudinal acoustic vibrations [570]. Although the relationship between the frequency and the lamellar thickness is not known, it can be concluded from the frequency shift that
the annealing operation leads to a gradual increase in lamellar thickness until at 463 K the original thickness has almost doubled. Evidence for the occurrence of lamellar morphology and lamellar thickening upon thermal treatment has also been obtained from small-angle
x-ray diffraction on polyethylene sulfide fibers [576]. In as-spun
fibers the center of the fanlike meridional diffraction pattern
[Fig. 4-91(a)] lies at about 7.5 nm. Upon annealing this pattern breaks
up into two meridional spots with spacings of about 10 and 20 nm,
respectively [Fig. 4-91(b) and (c)]. Thus, the appearance of the
interior strong diffraction spot suggests an approximate doubling of

the original long period. Additionally, the difference in the degree
of orientation of the two small-angle x-ray spots supports the assumption that the lamellar thickening preferentially takes place in
domains where the chain orientation is parallel to the fiber axis.

So far most papers deal with the discussion of the observed results. Recently Hsu et al. [569, 577] observed the LAM in helical
isotactic polypropene. Although the wavenumber is inversely proportional to L [Eq. (4.72)] the calculated frequency is much higher
than is expected from the fold period obtained from small-angle
x-ray scattering. Since this discrepancy cannot be accounted for by
noncrystalline chain ends alone, theoretical studies of the vibrational properties of a generalized elastic rod model were carried
out [577]. An equation was derived which governs the longitudinal
acoustical vibrations of a composite crystalline-to-amorphous elastic
rod with perturbing forces at the ends. It is shown that this model
can explain the behavior of the observed Raman-active low-frequency
bands associated with this mode in polyethylene and isotactic polypropene. Normal coordinate calculations [578] show that a single conformational defect of an otherwise all-*trans* n-alkane molecule dis-
rupts the LAM associated with the all-trans molecule. Similar calculations for a chain in the conformation produced by smoothly
twisting a planar zigzag through 180° about the chain axis show that
the decrease in the LAM observed in some polyethylene samples cannot be accounted for by smoothly twisted chains, but by defects which
involve large localized deviations from the all-*trans* conformation.
Snyder [592] has described a procedure whereby not only the average
straight-chain length within lamellae of crystalline polyethylene
may be calculated from the LAM-1 frequency, but also the straight-
chain distribution from the shape of this band.

4.4.3 Vibrations of Hydrogen Bonds

Generally, hydrogen bonds originate from an interaction of a proton
donating group and a proton acceptor (Sec. 4.3.3). However, while
the absorptions of $\nu(XH)$ vibrations involved in hydrogen bonds are

usually isolated and seldomly overlapped by other absorptions in the
MIR, an assignment of the corresponding vibrations of the hydrogen
bond itself in the FIR region is often complicated because of coupling
and the occurrence of other absorptions at comparable wavenumbers.
For a more detailed treatment of the vibrations associated with hydrogen bonds the reader is referred to Sec. 4.3.3.

4.4.4 Lattice Vibrations

Lattice modes in polymers are caused by vibrations of the whole chains
against each other. The most frequently studied vibrations of this type
are those of polymers with long methylene sequences. The FIR spectra
of polyethylene, for example, show two bands at about 73 and 108 cm^{-1}.
The first one is observable at room temperature and can occur in other
polymers or monomers with a polyethylene-like arrangement of the CH_2-
groups within the unit cell [523, 579, 580]. The 108 cm^{-1} band shows
weak intensity and occurs only at low temperatures [581, 582]. With
decreasing temperature the wavenumber of the low-frequency band shifts
to higher values [583, 584] and has been classified as a crystallinity
band [585]. The wavenumber of this band also depends on the chain
packing as shown in Refs. 579 and 586. The band which was not observed in the case of methylene sequences crystallizing in a tri-
clinic unit cell [586] has been assigned to a translatory lattice mode
[587]. Calculations taking into account hydrogen-hydrogen interactions
between adjacent chains confirmed this assumption [588, 589] and to-
day the following assignment is well proved: the 73 cm^{-1} band is
caused by a translatory vibration parallel to the b-axis of the unit
cell (Fig. 4.43) with a transition moment parallel to the a-axis and
belongs to the B_{3u} representation. The corresponding vibration parallel to the a-axis with a transition moment parallel to the b-axis
(B_{2u} mode) causes the band at 108 cm^{-1} (calculated frequency: 113 cm^{-1}).
While the older calculations considered hydrogen-hydrogen interactions,
only, recent recalculations [590] use semiempirical potential functions
[589, 591] considering in addition the C-H as well as the C-C interactions.

4.4.5 Defect-Induced Absorptions

In the FIR region a high background absorption can often be observed.
The spectra show a definite structure even outside the optically active modes. These additional absorptions were interpreted as defect
modes. This means that the lattice modes become active as soon as the
translatory symmetry of the lattice breaks down because of the presence of defects. These defects cause new lattice vibrations located
within the defects and do not depend on the phonons of the ideal
crystal, or they couple with the phonons of a dispersion branch and
cause the optical activity of the whole branch. This explains the
high background absorption of all amorphous polar structures in the
FIR. With increasing number of defects or for the transition from the
crystalline to the amorphous state, the selection rules k = 0 are no
longer valid and all vibrations accompanied by a change in the dipole
moment become IR-active. An assignment of the extra peaks due to folds,
kinks, and other conformational defects by a classical normal coordinate analysis is not practicable because of the irregularity of the
defects. For detailed accounts of the experimental results obtained
from various polymers and the theoretical methods available for the
interpretation of these absorption bands the reader is referred to
the special literature [539, 541-543, 555, 586, 593-600].

4.5 NEAR-INFRARED SPECTROSCOPY

4.5.1 Introduction

The near-infrared (NIR) region of the spectrum covers the interval
from about 0.7 μm (14285 cm^{-1}) to 2.5 μm (4000 cm^{-1}) and contains
a great wealth of spectral data largely concerned with overtone or
combination vibrations. Taking into account the number of overtone
and combination frequencies possible from a large molecule it might
appear that this region would be extremely complex. In actual fact,
however, only the overtone or combination bands of vibrations involving hydrogen (such as C-H, O-H, N-H) are observed at appreciable
intensities. Because of their comparatively small absorptivities

TABLE 4-11 Structure-Spectra Correlation and Molar Absorptivity Data $(cm^2/mmol)$ for some Characteristic NIR Absorption Bands Compiled from Refs. 602 to 604

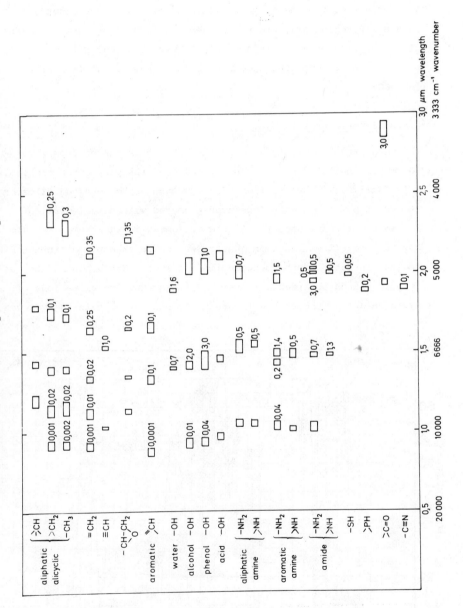

sample paths must be longer than those required for fundamental
bands. In Table 4-11 the assignment, wavelength position and molar
absorptivities of some characteristic absorption bands observable
in the NIR region are summarized. Keeping in mind the origin of the
NIR absorption bands it is possible to use this region for analytical purposes in much the same way as the conventional IR region
[601-603].

4.5.2 Experimental

The path length commonly used in the NIR region ranges from 1 to
10 mm for undiluted substances depending on the particular problem.
This facilitates the testing of solids such as hot-pressed films or
stacks of polymer sheets (preferably coated with some immersion
liquid). Cells are used for liquid or semisolid substances and are
commonly made from glass or quartz (which are transparent in the NIR
region) in path lengths between 1mm and 10 cm. Since most of the absorption bands observable in the NIR region involve vibrations of
hydrogen atoms, the best solvents are those not containing hydrogen,
for example, CCl_4 or CS_2. Other solvents that have been used are

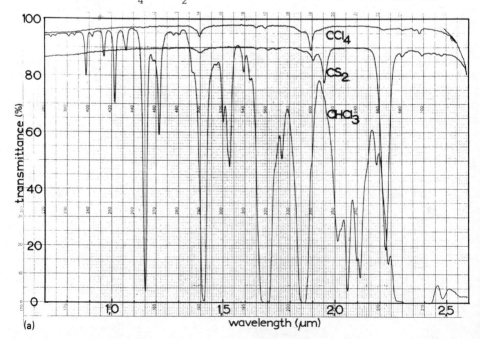

(a)

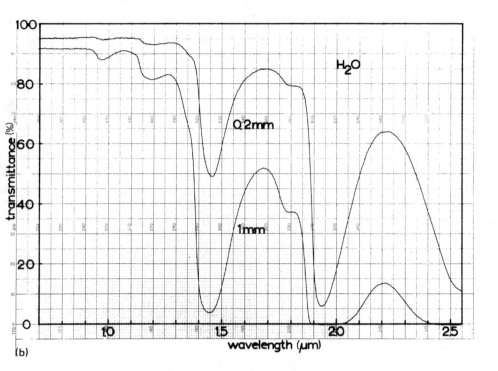

(b)

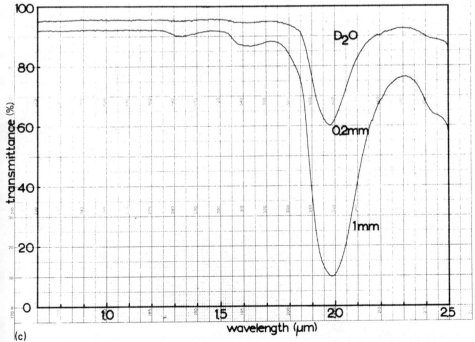

(c)

FIGURE 4-92 NIR spectra of some common solvents (path length 5 cm
if not otherwise stated).

chloroform, methylene chloride, and dioxane. Water-soluble organic
compounds may be studied complementarily in H_2O and D_2O solutions
The absorption spectra of some of these solvents are shown in Fig.
4-92.

4.5.3 Inharmonicity

Actual molecular vibrations are not strict harmonic oscillators and
their energy levels are not equally spaced. Consequently, transitions
between nonadjacent energy levels and interactions between vibrations
giving rise to overtone and combination bands, respectively, are possible.

The intensity of NIR overtone and combination bands depends on
the inharmonicity of the molecular vibrations responsible for the
absorption of the fundamental frequencies. The deviations from simple
harmonic motions of atoms or molecular groups increase with the amplitude of oscillation, which is a function of the masses and force
constants involved. Thus, owing to the comparatively large inharmonicity of vibrations involving the hydrogen atom, the corresponding
overtone and combination bands can be observed at appreciable intensities.

The degree of inharmonicity can be derived from the difference
in energy levels, which is expressed by

$$E = hc\bar{\nu}\left[\left(v + \frac{1}{2}\right) - x\left(v + \frac{1}{2}\right)^2\right] \qquad (4.73)$$

where $\bar{\nu}$ is the wavenumber of the fundamental vibration, v is the
vibrational quantum number which can assume integer values of 0, 1,
2, 3, ..., and x is the inharmonicity constant (for C-H vibrations
x has values of 0.01 to 0.05 [605]). Apart from mechanical inharmonicity, the intensity of an overtone also depends on electrical inharmonicity which is caused by nonlinear changes in dipole moment
during the vibration under consideration [602].

A detailed treatment of the accurate determination of inharmon-
icities has been given by Herzberg [606]. The transition probability

between two energy levels becomes progressively less for increasing
Δv, and overtone and combination band intensities are generally an
order of magnitude less than their lower-term preceding analogues.

4.5.4 Application of Near-Infrared Spectroscopy to the Investigation of Polymeric Structure

Absorption spectroscopy in the NIR offers a welcome alternative to
the conventional IR region because of the comparatively simple specimen preparation and the fact that absorption bands are often widely
spaced and a clear nonoverlapped band of known origin can be selected
for analytical purposes. NIR spectroscopy is increasingly applied in
the examination and analysis of polymeric materials and a discussion
of some relevant problems demonstrates the value of this method.

For the analysis of butadiene-styrene copolymers Miller and Willis
[607] have proposed a NIR method based on the evaluation of the aromatic and aliphatic C-H combination bands at 2.18 μm (4580 cm^{-1}) and
2.35 μm (4250 cm^{-1}), respectively. Styrene-acrylonitrile copolymers
of various composition have been studied by Takeuchi et al. [608]
and a linear relation has been established for the absorbance ratio
$A_{1.675\ \mu m}/A_{1.910\ \mu m}$ of the aromatic v(C-H) overtone (1.675 μm, or
5970 cm^{-1}) and the v(C$\equiv$N) + v_{as}(CH$_2$) combination vibration (1.910 μm,
or 5235 cm^{-1}) and the acrylonitrile content of the copolymer (see
Fig. 4-93).

Several accounts on NIR spectroscopic investigations of ethylene-
propene copolymers are available in the literature [609, 610]. Thus,
Bly et al. [611] report a NIR method for the determination of the
monomer concentration ratios in the 0 to 40% ethylene range regard-
less of the copolymer type. They obtained a linear relation for the
absorbance ratio $A_{2.312\ \mu m}/A_{2.275\ \mu m}$ and the ratio of ethylene/propene
content, where the 2.275 μm (4395 cm^{-1}) band has been assigned to a
combination of the v_{as}(CH$_3$) + δ_{as}(CH$_3$) vibrations and the intensity
of the 2.312 μm (4325 cm^{-1}) band to two overlapping combination bands
of the v_s(CH$_2$) + δ(CH$_2$) and v_s(CH$_3$) and δ_{as}(CH$_3$) vibrations. The NIR
method was found to be less sensitive to differences in crystallinity

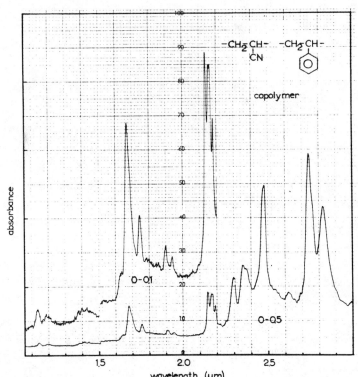

FIGURE 4-93   NIR spectrum of acrylonitrile-styrene copolymer film.

and randomness than methods based on fundamental IR vibrations [612-614].

Tosi [615] obtained a correlation between the frequency of the $\nu(CH_2)$ overtone at 1.725 µm (5800 cm^{-1}) and the copolymer composition and interpreted the results in terms of the methylenic sequence lengths of the ethylene-propene copolymers prepared by different catalysts.

NIR spectroscopy has also been proposed for the determination of the propylene oxide (PO)-ethylene oxide (EO) content in the corresponding mixed block copolymers. The absorbance ratio $A_{1.176\ µm}/A_{1.209\ µm}$ of the $\nu(CH_3)$ (1.176 µm, or 8503 cm^{-1}) and $\nu(CH_2)$ (1.209 µm, or 8271 cm^{-1}) second overtone bands yields satisfactory results but is primarily applicable to liquid samples which are measured without dilution [616]. When the first overtones of the $\nu(CH_3)$ and $\nu(CH_2)$

vibrations at 1.684 μm (5938 cm^{-1}) and 1.732 μm (5773 cm^{-1}), respectively, are used to characterize PO-EO content reasonable absorption intensities may be obtained for 10% copolymer solutions in CCl$_4$ at 2 cm path length [617].

An analytical technique to determine acrylonitrile-butadiene-styrene terpolymer modifiers in polyvinyl chloride is based on the evaluation of the aromatic ν(C-H) combination band in the 2.10 to 2.20 μm (4761-4545 cm^{-1}) region [618].

Miller and Willis have studied the progress in methylmethacrylate polymerization and the content of residual monomer in the final product by monitoring the overtone band of the ν(=CH$_2$) vibration at about 1.70 μm (5882 cm^{-1}) [607].

Durbetaki and Miles [619] obtained a linear relation between molar absorptivity and the average number molecular weight of free-radical-catalyzed polybutadiene using the first overtone of the ν(=C-H) vibration of the terminal -CH=CH$_2$ group at 1.636 μm or 6112 cm^{-1}.

The NIR spectroscopic determination of functional groups in epoxy resins in the liquid or solid state, in uncured or cured condition, as well as during the cure, has been thoroughly studied [620-622]. Epoxy values are determined with the 2.205 μm (4535 cm^{-1}) band, which has been assigned to a combination of the ν(C-H)-fundamental (3050 cm^{-1}) and the δ(CH$_2$)-fundamental (1460 cm^{-1}) vibrations. Difficulties were encountered in the determination of hydroxyl groups by the presence of hydroxyl in intra- and intermolecular hydrogen-bonded forms. The first overtones of the two forms absorb at 1.43 μm (6993 cm^{-1}) (intra) and 1.57 μm (6369 cm^{-1}) (inter), respectively, and transitions from one form to the other occur on heating or curing. However, taking advantage of the existence of an isosbestic point between the absorption peaks of the two hydrogen-bonded species an accurate method for the determination of hydroxyl values was developed. Furthermore the use of the sharp, intense band at 2.025 μm (4938 cm^{-1}) for the quantitative determination of primary NH$_2$ groups has been discussed briefly [620].

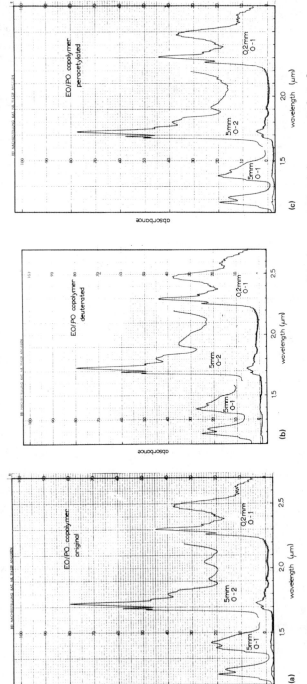

FIGURE 4-94 NIR spectra of EO-PO copolymer: (a) original, (b) deuterated, (c) peracetylated.

The NIR spectroscopic determination of the hydroxyl content in polymers (e.g., polyethers, polyesters) has been discussed by various authors [607, 617, 623-625]. The hydroxyl number in polyethers may be conveniently determined by the 2.05 μm (4880 cm^{-1}) combination band of the ν(OH) and δ(OH) vibrations. The assignment can be confirmed by chlorination [607], peracetylation, and deuteration whereupon this absorption band disappears (Fig. 4-94). The intensity of the hydroxyl combination band may be calibrated by mixtures of a polyether with a known hydroxyl number (determined by an independent method) and its peracetylated analogue (Fig. 4-95).

A number of papers have dealt with the determination of the water content in polymers [626-628] by calibration of the absorption intensity of the ν(OH) + δ(OH) combination vibration of water at 1.94 μm (5150 cm^{-1}). Another application of the NIR absorptions of water has been the study of water binding to globular proteins [629]. Thus, the

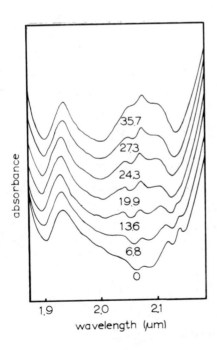

FIGURE 4-95 NIR spectra of polypropylene oxide mixtures with different hydroxyl numbers (see text).

differences in the bulk and protein bound forms of water have been
characterized by NIR difference spectroscopy and NIR spectroscopic
investigation of the various stages of protein dehydration.

The preferential orientation of the aliphatic chain and other
functional groups in several polymers (polyethylene, polyvinyl chloride, polyvinylidene chloride, polyvinyl alcohol) has been the subject
of dichroic measurements in the NIR region [630-632]. NIR investigations with polarized radiation and the differences observed for different conformations in natural and synthetic proteins and polypeptides have been discussed by Elliott [626], Fraser [309, 633] and
Hecht [627]. A deuteration method and its application to the study
of band assignment and band dichroism in wool and cellulose has been
reported by Bassett et al. [634]. The NIR spectra of polyamide-6 and
the N-deuterated analogue have been interpreted by Jokl [635].

Crandall et al. [636, 637] have studied the NIR spectra of several addition and condensation polymers and assigned certain absorption bands to functional groups of these polymers. The characteristic
carbonyl-, hydroxyl-, and NH- overtones observed in the spectra of
polyesters, polyamides, polyurethanes, polyamic acids, and urea and
phenol formaldehyde resins have been shown to be of great value for
identification purposes and in following polymerization processes.

Quantitative NIR spectroscopy has been recently applied to kinetic investigations of the oxygen-initiated high-pressure ethylene
polymerization at 508 and 513 K [638]. The kinetic parameters have
been derived from the intensity changes of the ethylene and polyethylene ν(CH) combination and overtone absorptions at 1.114 μm
(8977 cm^{-1}), 1.143 μm (8749 cm^{-1}), and 1.211 μm (8258 cm^{-1}), respectively (see also Fig. 4-96).

4.6   RESONANCE RAMAN SPECTROSCOPY

4.6.1   Introduction

The development of powerful laser light sources with increasingly
wide tuning ranges of the electromagnetic spectrum has opened new
possibilities for Raman spectroscopy via the resonance Raman effect

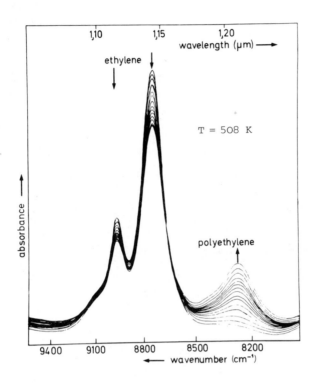

FIGURE 4-96   High-pressure $(3.5 \times 10^{8}$ Pa) ethylene polymerization
              progress characterized by second overtone NIR spectra.
              The spectra were recorded in time intervals of 10 min.
              [Reprinted with permission from F. W. Nees and M. Buback,
              Ber. Bunsenges. Phys. Chem. 80(10):1017 (1976).]

(RRE). It can be readily seen from Eq. (2.42) that the classical
bond polarizability theory only applies when the exciting laser frequency is well removed from that of the lowest electronic transition
of the molecule and the initial and final states of the molecule both
involve the ground electronic state. When the exciting frequency approaches that of an electronic absorption band of the scattering
molecule, certain Raman bands increase in intensity (pre-RRE), and
they are strongly enhanced when the frequency of the exciting radiation coincides with that of the electronic transition (rigorous RRE).
The RRE is also characterized by an apparent breakdown in the har-

monic oscillator selection rules such that overtones may appear with
intensities comparable with that of the fundamental. The selectively
enhanced Raman bands correspond to vibrational modes which involve
motions of that part of the molecule where the electronic transition
is localized. Vibrations which do not couple with the electronic state
will not be resonantly enhanced. Therefore, the RRE provides a means
whereby vibrations of a conjugated polymer backbone or of biological
chromophores, for example, can be distinguished from the vast majority of vibrational modes associated with the side groups or a complex
biological matrix, respectively. Consequently the resonant Raman spectrum is much simpler than the nonresonant spectrum and can be used as
a structure probe for chromophoric groupings. Moreover it is possible
to design resonance Raman labels, similar to spin labels, which can
be attached to selective sites of the molecule under examination (see
below).

A number of experimental difficulties are encountered in the quantitative measurement of Raman band intensities when the exciting frequency closely approaches the lowest allowed electronic transition
of the molecule:

1. Sample decomposition by local overheating
2. Allowance for the competition between scattering and absorption
3. Elimination of fluorescence

The elimination or suppression of these sources of errors by devices
which permit relative motion between the sample and the focused laser
beam (rotating sample, rotating cell and surface scanning techniques)
have been discussed in Sec. 4.1.2. The reduction of fluorescence background is based on time discrimination against fluorescence [639].
The technique involves excitation of the sample with a pulsed laser,
the pulse duration being of the order of nanoseconds or picoseconds
and time-adjusted gate electronics to permit the separation of the
resonance Raman from the fluorescence spectra [640].

Several important reviews on the theoretical and practical aspects
of the RRE have appeared [41, 641-643, 649] and the present section
briefly discusses some applications in the polymer field.

4.6.2   Application of the Resonance Raman Effect to Structural
        Studies of Polymers

Considerable RRE enhancements have been observed in the Raman spectra
of conjugated polyolefins, particularly for the vibrational modes at
about 1550 and 1200 cm^{-1}, which are assigned to in-phase stretching
modes of the carbon-carbon double and single bonds, respectively.
The carotenoid pigment β-carotene, for example, has been studied in
live carrot tissue and despite the complexity of the biological sample
the pigment clearly dominates the Raman spectrum, all other scattering
being attenuated by the intense absorption [643].

    The RRE has also proved of great value in structure investigations
of diacetylene polymers [388, 406, 644]. This unique class of crystalline polymers is synthesized by solid-state polymerization of
diacetylene compounds [645], according to the scheme outlined in

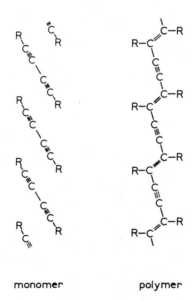

monomer   polymer

FIGURE 4-97   Arrangement of diacetylene monomer molecules and di-
              acetylene polymer molecules produced by solid-state
              polymerization (R = side groups). [Reproduced by per-
              mission from K. J. Ivin (ed.), Structural Studies of
              Macromolecules by Spectroscopic Methods, Wiley, New York
              1976.]

Fig. 4-97. X-ray analysis has shown [646] that the highly perfect
crystals contain a chain-extended polymer with a conjugated backbone.
The repeat unit consists of alternate double, single, triple and
single bonds (=RC-C≡C-CR=)$_n$ in a *trans* conformation. The polarization
properties of the reflection and absorption spectra of such polymers
indicate an intense optical absorption resulting from an electronic
excitation of the conjugated polymer backbone [647]. When the exciting laser frequency is close to this electronic excitation, resonant
Raman enhancement can be expected for vibrations which couple strongly
with the electronic states of the backbone. In Fig. 4-98 the Raman
spectra of single-crystal diacetylene polymers with different side
groups are shown. The assignment of the observed Raman lines has been
based on a simplified calculation of the vibrational frequencies of
a model diacetylene chain in which the side groups are represented

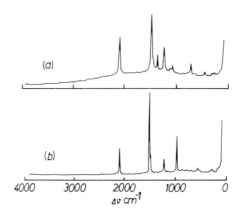

FIGURE 4-98   Raman spectra of diacetylene polymer single crystals for
               laser wavelength 647.1 nm (incident light polarized par-
               allel to the polymer chain, scattered light unpolarized).

               (a) Side group R: $-(CH_2)_3-O-CO-NH-\bigcirc$

               (b) Side group R: $-CH_2-O-SO_2-\bigcirc-CH_3$

               [Reproduced by permission from K. J. Ivin (ed.), Struc-
               tural Studies of Macromolecules by Spectroscopic Methods,
               Wiley, New York 1976.]

by single atoms with an appropriately chosen mass [644]. The Raman
bands at about 2100 and 1500 cm^{-1} which are predominantly $\nu(C\equiv C)$-
and $\nu(C=C)$-stretching vibrations occur in both spectra and correlate
well with the calculated frequencies. The remaining intense bands
which involve some bond-bending components and side-group motion are
slightly shifted to higher frequencies as compared with the calculated values [644]. Despite the presence of complex and different side
groups the spectra clearly reflect the identical bonding sequence of
the polymer backbones. The RRE also enables the structure and conformation of the backbone to be studied in amorphous or poorly crystalline polymers of this type where x-ray methods cannot easily
provide structural information [644].

Gerrard and Maddams [648] have studied in some detail the Raman
spectra of thermally degraded polyvinyl chloride samples. They could
show that the 1511 and 1124 cm^{-1} bands arise from a RRE involving the
conjugated polyene sequence formed during the thermal degradation.
This has been demonstrated by the band intensities, their dependence
on the wavelength of the exciting line in relation to the visible
absorption spectra of the degraded polymers, and the appearance of
strong harmonic and combination bands. Intensity measurements on the
two strongest resonance lines for a series of samples degraded to
known extents at various temperatures indicate that the average sequence length of the conjugated polyenes increases with increasing
time and temperature of degradation.

When ultraviolet (UV) Raman laser sources become generally available, the very important class of biological chromophores including
nucleic acid bases and aromatic side chains of proteins will become
accessible to resonance Raman studies [649]. The involved π-π^* and
n-π^* electronic transitions provide strong resonance enhancement of
in-plane stretching modes. The application of resonance Raman spectroscopy as a probe of peptide and protein conformation has been discussed by various authors [284, 288, 643].

The scope of resonance Raman spectroscopy can be extended by
using resonance Raman labels. The principle of the technique is to
attach small molecules with favorable resonance Raman characteristics

to particular sites of the molecule under investigation and report
back structural information via their Raman intensities [643]. For
further details of this technique as applied to the study of biological macromolecules, the reader is referred to the special literature
[643, 650].

4.7 KINETIC STUDIES

Qualitative and quantitative vibrational spectroscopy are widely applied in studies of polymer and polymerization reactions. Observation
of time- and temperature-dependent intensity changes of specific absorption bands may successfully contribute to the elucidation of reaction mechanisms and kinetics. Qualitative information can be obtained from the spectral assignment of intermediate and reaction products, while quantitative evaluation of the spectra recorded in suitable time intervals provides the basis for the calculation of reaction kinetics.

Although Raman spectroscopic investigations on conformation and
phase transition, degradation, and polymerization kinetics of certain
polymer systems have been reported [651-654], the majority of papers
is concerned with results derived from IR spectroscopic studies. A
wide variety of reaction types has been investigated so far [638,
655-664] and in what follows only a few examples will be discussed
in more detail.

Generally, vibrational spectroscopy may be conveniently applied
to study the progress of polymerization reactions by focusing on intensity changes of absorption bands which are either characteristic
of the monomeric starting material or the final polymeric product.
The time-dependent intensity decrease or increase, respectively,
measured at various temperatures will then yield the necessary data
to derive the order and activation energy of the investigated reaction.

A Raman spectroscopic investigation of the suspension polymerization of styrene has been discussed by Witke and Kimmer [654]. In
this study the ν(C=C) Raman line (1631 cm^{-1}) of monomeric styrene,

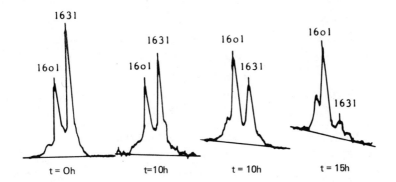

FIGURE 4-99   Raman spectra of a styrene suspension polymerization
system at various reaction times. [Reproduced with permission from K. Witke and W. Kimmer, Plaste u. Kautschuk,
23, 11:799 (1976)].

whose intensity is enhanced by conjugation with the aromatic ring
[665], has been found suitable to monitor the consumption of styrene
in the course of the polymerization (Fig. 4-99). Owing to the close
proximity and appropriate intensity, the aromatic ring vibration at
1601 cm^{-1} has been chosen as reference band for the quantitative
measurements. A general relationship between the intensity ratio
$Q = I_{1631 \text{ cm}^{-1}}/I_{1601 \text{ cm}^{-1}}$ and the styrene concentration, established
with the aid of calibration mixtures of styrene and polystyrene, was
then applied to the quantitative evaluation of the polymerization
progress.

The reaction kinetics of the thermal polymerization of aliphatic
bismaleimides have been studied by IR spectroscopy [666]. The monomers polymerize readily above their melting temperatures (about 393 K)
to highly cross-linked products without the need of a catalyst:

monomeric bismaleimide

$R = (-CH_2-)_n$ n = 6, 8, 10, 12

polybismaleimide   (4.74)

The polymerization process is accompanied by a gradual decrease
of the concentration of maleic double bonds and by the formation of
substituted succinic residues. However, the cured resin still contains
a certain percentage of residual double bonds. In the absence of aromatic or olefinic components [the absorption bands of their ν(=C-H)-
stretching vibrations would interfere] the intensity decrease of the
ν(=C-H) band of the maleimide ring at 3100 cm^{-1} is characteristic of
the polymerization progress. The doublet of this absorption band observed in the IR spectrum of the crystalline monomer coalesces into
a single band at the polymerization temperature above the melting
point (Fig. 4-100). The approximate invariance of the absorptivity
of the analytical absorption band was verified in solutions of the
monomer in CDCl$_3$ (0.04-0.8 mol/dm^3) and assumed to be transferable
to the polymerizing system at low conversion. The monomer was prepared as molten film between KBr plates with a 50 µm spacer, and the
analytical wavenumber range 3300 to 3000 cm^{-1} was recorded at the respective polymerization temperatures in regular time intervals. Er-

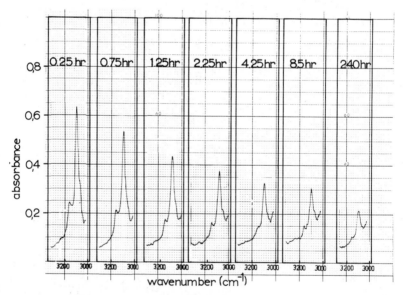

FIGURE 4-100   Polymerization of 1,6-bismaleimidohexamethylene at 473 K:
               intensity decrease of the analytical ν(=C-H) absorption
               band with reaction progress.

rors due to emission at these high temperatures were eliminated by
a double-chopper system with a phase difference of 90° (Sec. 3.1.1.1).
In the Figs. 4-101 and 4.102 log A (measured at the band maximum)
has been plotted versus time for the hexa- and dodecamethylene derivatives. Up to a conversion of 20 to 30%, the polymerization reaction is pseudo-first-order. With increasing conversion the viscos-
ity of the cross-linking system which is generally higher in the derivatives with short methylene sequences [667] increases. Thus, the
reaction rate is controlled by the small mobility of the reactive
sites and residual monomer molecules. This view is supported by the
bend in the log A versus time diagram for hexamethylenebismaleimide
at 473 K, which is far less pronounced at higher conversions in the
dodecamethylene derivative at 493 K. The activation energy may be
calculated from the specific reaction rates determined for at least

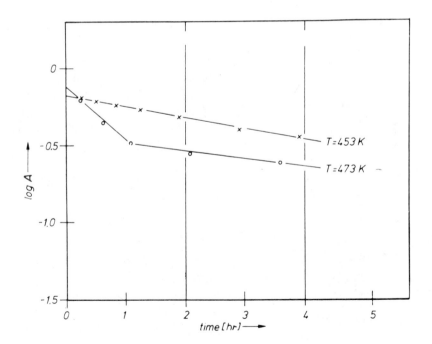

FIGURE 4-101 Polymerization progress of hexamethylenebismaleimide
          monitored by the absorbance of the ν(=C-H) absorption
          band at 453 and 473 K.

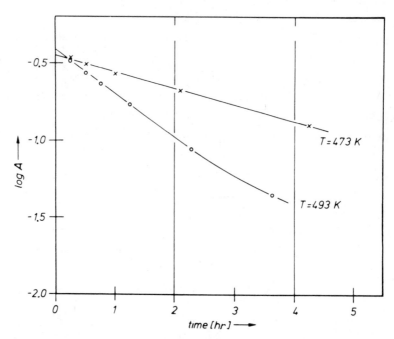

FIGURE 4-102   Polymerization progress of dodecamethylenebismaleimide
               monitored by the absorbance of the ν(=C-H) absorption
               band at 473 and 493 K.

two different temperatures with the aid of the Arrhenius equation:

$$E_a = \frac{R[\ln(k_2/k_1)]\,T_1 T_2}{T_1 - T_2} \tag{4.75}$$

The kinetic data for the polymerization of a homologous series of
aliphatic bismaleimides are summarized in Table 4-12.

Numerous ring-closure reactions have been studied and the degree
of cyclization determined by IR spectroscopy. In the synthesis of
poly-N,N'(4,4'-oxydiphenylene)pyromellitimide [668], pyromellitic
dianhydride and 4,4'-diaminodiphenylether are condensation polymerized
to yield a soluble open-chain prepolymer. In a subsequent thermal step
(up to 573 K) the prepolymer undergoes cyclization (imidization) to
the polyimide:

TABLE 4-12   Kinetic Data for the Polymerization of Aliphatic Bismaleimides with the General Formula

$$\text{H-C-C}\begin{smallmatrix}O\\\\O\end{smallmatrix}\text{N-(CH}_2)_n\text{-N}\begin{smallmatrix}O\\\\O\end{smallmatrix}\text{C-C-H}$$

n	453 K	463 K	473 K	493 K	E_a (kJ/mol)
6	4.8×10^{-5}	–	2.3×10^{-4}	–	140.2
8	2.2×10^{-5}	4.5×10^{-5}	–	–	121.4
10	1.7×10^{-5}	3.2×10^{-5}	–	–	114.3
12	–	–	7.0×10^{-5}	1.9×10^{-5}	94.2

(4.76)

In Fig. 4-103 the IR spectra of the intermediary poly(amide acid)
and the ring-closed final product are shown. In the spectrum of the
prepolymer the absorption bands at 3300, 1650, and 1540 cm^{-1} may be
assigned to the ν(NH), amide I, and amide II absorption bands of the
amide group. The shoulder at 1720 cm^{-1} belongs to the ν(C=O) vibra-

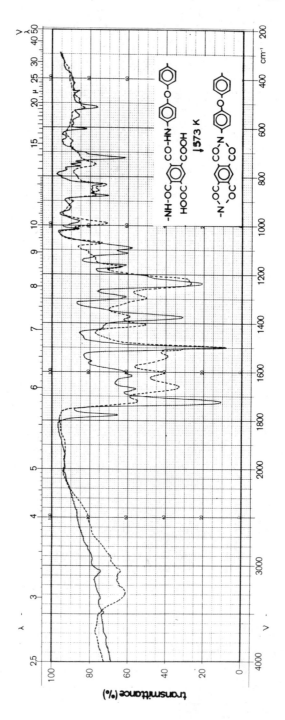

FIGURE 4-103   IR spectroscopic characterization of polyimide ring-closure reaction (see text): dashed line, prepolymer; solid line, ring-closed polymer.

tion of the carboxylic groups. Upon thermal treatment the absorptions at 3300, 1650, and 1540 cm^{-1} disappear almost completely and significant absorption bands occur newly or at increased intensity at 1780, 1720, 1380, and 720 cm^{-1}. These absorption bands are highly characteristic of the final product [669] and may be applied to assess the completeness of the ring-closure reaction.

Analogous investigations have been reported for other polymeric systems, e.g., the cyclization to polybenzothiazole [670], of polyhydroxamide [671] and the isomerization of o-cyanopolyamides [672], to mention only a few.

Among studies of polymer reactions, the kinetics of the photo- and thermal oxidation of polyolefins (e.g., polyethylene, polypropene, polybutene, polypentene, and polybutadiene) have received considerable attention [82, 673-681]. Figure 4-104 shows the spectra of polypentene before and after oxidation. The rates of formation of nonvolatile products under various experimental conditions (temperature, oxygen concentration) may be derived from the ν(C=O) absorption band of the carbonyl groups formed upon oxidation. In Fig. 4-105 the area under the ν(C=O) absorption band, normalized to uniform film thickness, is plotted versus time for various oxygen concentrations (measuring temperature 373 K). A kinetic scheme generally applicable to the thermal oxidation of polyolefins satisfactorily explains the experimental results. The activation energies for the major oxidation stages of the investigated polymers were estimated from various Arrhenius plots [677, 678].

Signal-to-noise ratio and resolution considerations limit the approach of rapidly scanning the entire spectrum on a dispersive instrument and prevent the application of conventional IR spectroscopy to the study of reactions with half-lives shorter than a few minutes. Hence, multiple experiments with reference to the monitored frequency are necessary if the complete picture of a rapid, complex reaction is desired. Since the introduction of rapid-scanning FTIR spectrometers however, the limitations by the time factor have been significantly reduced and chemical reactions with shorter half-lives have become amenable to spectroscopic investigation. Simultaneous measure-

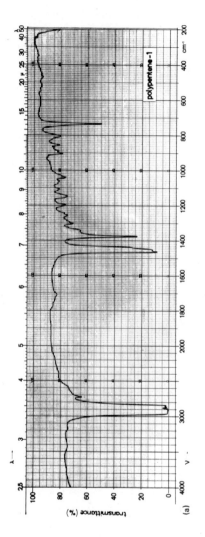

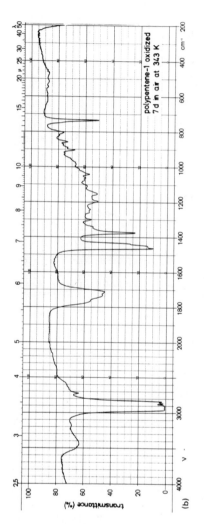

FIGURE 4-104   IR spectra of polypentene: (a) before oxidation, (b) after oxidation in air at 343 K for 7 days.

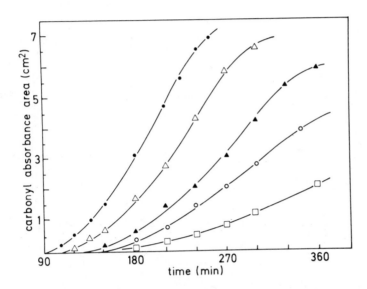

FIGURE 4-105   Plots of carbonyl absorbance area versus time for
poly(1-propylethylene) at 373 K under various oxygen
concentrations: rectangle, 10%; open circle, 25%;
solid triangle, 50%; open triangle, 75%; solid circle,
100%. [Reproduced from S. M. Gabbay, S. S. Stivala,
and L. Reich, Polymer, 16:749 (1975) by permission of
the publishers, IPC Business Press, Ltd.]

ment of reactant and product concentrations eliminates the necessity
of multiple experiments and the introduction of errors due to band
shifts during the reaction. The analysis time can be further cut down
by fully exploiting the data handling capabilities of the FTIR system.
Thus, if a reactant or product exhibits an absorption free of spectral interference a kinetic profile of this reaction species is readily represented by a plot of absorbance versus time. When no isolated
absorption bands of a particular component are available differential
spectra and differential profiles obtained by ratioing each spectrum
against a spectrum preceding it in time can provide both an improved
qualitative picture of the reaction and quantitative information
[682].

The complex nature of the reaction of isocyanates in the forma-

tion of plastic foams has been the subject of detailed FTIR investigations [683]. With the aid of a specially constructed ATR cell using
a germanium reflection element it was possible to monitor the IR spectra of the reaction mixture in short time intervals and to characterize the reaction progress (Fig. 4-106). The rate of consumption of
free isocyanate groups and the competitive formation of urethane,
isocyanurate, and urea linkages according to the scheme

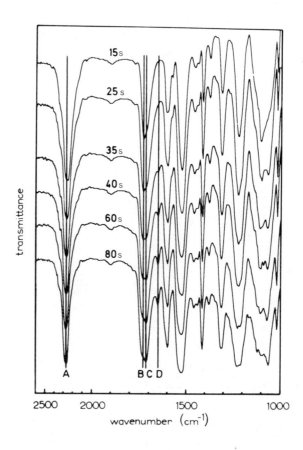

FIGURE 4-106   FTIR spectroscopic study of polyurethane foam formation:
               A, isocyanate ν(N=C=O) absorption (2275 cm^{-1});
               B, urethane ν(C=O) absorption (1725 cm^{-1});
               C, isocyanurate ν(C=O) absorption (1710 cm^{-1});
               D, urea ν(C=O) absorption (1640 cm^{-1}). (Courtesy of
               G. Bayer, Bayer AG, Leverkusen, West Germany.)

$$RNCO + R'OH \longrightarrow RNHCOOR'$$

$$3\ RNCO \longrightarrow \text{(isocyanurate ring)} \qquad (4.77)$$

$$2\ RNCO + H_2O \longrightarrow RNHCONHR + CO_2$$

can be deduced from the decrease of the $\nu(N{=}C{=}O)$ absorption band at $2275\ cm^{-1}$ (A), and the increase of the $\nu(C{=}O)$ absorption bands at $1725\ cm^{-1}$ (B), $1710\ cm^{-1}$ (C), and $1640\ cm^{-1}$ (D), respectively (Fig. 4-106).

4.8   COPOLYMERS

Basically, copolymers may be defined as polymers with imperfect chain structures. However, these defects are in most cases intentionally introduced to modify the chemical and physical properties of the products. By suitable choice of catalyst, composition of feed mixture, reaction conditions, and nature of monomeric starting materials, these properties can be varied in a wide range. The irregularities are due to either chemically different repeating units or different sterical configurations of equivalent repeating units. Taking into account the additional variants of length and distribution of homosequences for the simplest case of a binary system, the following basic types of copolymers can be distinguished:

1. Random $-M_1M_1M_2M_1M_2M_2M_2M_1M_1M_2M_2M_1M_2M_2M_2-$

2. Alternating $-M_1M_2M_1M_2M_1M_2M_1M_2M_1M_2M_1M_2M_1M_2M_1-$

3. Block $-M_1M_1M_1M_1M_2M_2M_2M_2M_2M_1M_1M_1M_1M_1M_1-$

4. Graft $-M_1M_1M_1M_1M_1M_1M_1M_1M_1M_1M_1M_1M_1M_1M_1-$

$$
\begin{array}{ccc}
M_2 & M_2 & M_2 \\
M_2 & M_2 & M_2 \\
M_2 & M_2 & M_2 \\
M_2 & M_2 & M_2 \\
M_2 & M_2 & \\
    & M_2 & \\
\end{array}
$$

For a uniform mathematical treatment of copolymerization a quantitative measure of the relative monomer reactivities is required. From
the four propagation reactions of a radical copolymerization in a binary system [684]:

$$\ldots M_1 \cdot + M_1 \xrightarrow{\ k_{11}\ } \ldots M_1 M_1 \cdot \qquad (4.78)$$

$$\ldots M_1 \cdot + M_2 \xrightarrow{\ k_{12}\ } \ldots M_1 M_2 \cdot \qquad (4.79)$$

$$\ldots M_2 \cdot + M_1 \xrightarrow{\ k_{21}\ } \ldots M_2 M_1 \cdot \qquad (4.80)$$

$$\ldots M_2 \cdot + M_2 \xrightarrow{\ k_{22}\ } \ldots M_2 M_2 \cdot \qquad (4.81)$$

the well-known copolymerization equation has been derived [685-688]:

$$\frac{dM_1}{dM_2} = \frac{M_1}{M_2} \frac{r_1 M_1 + M_2}{r_2 M_2 + M_1} = \frac{(M_1/M_2)r_1 + 1}{(M_2/M_1)r_2 + 1} \qquad (4.82)$$

with $r_1 = k_{11}/k_{12}$ and $r_2 = k_{22}/k_{21}$.

This equation relates dM_1/dM_2, the composition of the copolymer
being formed at any instant from a polymerizing mixture of two monomers to their concentration M_1 and M_2 by means of two measurable parameters r_1 and r_2. The copolymerization parameters characterize the
reactivity of the monomers at the given reaction conditions and make
it possbile to predict theoretically the composition of the copolymer.
Various methods for the determination of the copolymerization parameters with the aid of the differential as well as the integrated form
of Eq. (4.82) have been discussed in the literature [686, 688-700].
The monomer reactivity values can be assigned the following general
interpretation [701]:

1. If $(r_1 M_1 + M_2)/(r_2 M_2 + M_1) = 1$, the initial polymer possesses the
    same composition as the monomer mixture.
2. If $r_1 = r_2 = 1$, the initial polymer is of the same composition as
    the monomer feed for all concentrations of M_1 and M_2.
3. If $r_1 < 1$, $r_2 > 1$, i.e., monomer M_2 is more reactive than monomer
    M_1 with both types of growing reactive centers, the initial co-

polymer composition will contain an excess of M_2 over M_1.

4. If $r_1 > 1$, $r_2 < 1$, i.e., monomer M_2 is less reactive than monomer M_1 for both types of active centers, the initial copolymer composition will contain predominantly M_1 monomer.

5. If $r_1 = r_2 < 1$, an active site of the type M_1 prefers reaction with the "other" monomer M_2, and vice versa, resulting in alternating type of copolymer.

Generally speaking blocklike copolymers are characterized by high products of reactivity ratios ($r_1 r_2 > 1$), copolymers of the alternating type have low products ($r_1 r_2 < 1$), and for $r_1 r_2 = 1$ the copolymer is completely random. Two aspects of copolymer structure are of central interest:

1. The composition of the copolymer

2. The sequence distribution of the different monomers in the polymer chain

Commonly, copolymers with long homosequences (block and graft copolymers) can hardly be distinguished by their vibrational spectra from mixtures of the corresponding homopolymers. For statistical and alternating copolymers the superposition principle is no longer applicable, and the vibrations of the homosequences will be modified by the occurrence of structurally different monomer units. It is obvious that the vibrations of a monomer unit in the polymer chain depend on the nature of the neighboring groups, and as a result of inter- and intramolecular interactions, certain absorption bands exhibit characteristic frequency and intensity changes depending on the composition and sequence distribution of the copolymer under investigation. Thus, the study of the vibrational spectra should yield valuable information about these parameters [702]. Correlation of the FTIR spectra of a series of acrylonitrile-styrene copolymers, for example, with NMR sequence data revealed that certain infrared bands vary in frequency with the copolymer composition whereas others vary with sequence distribution [703]. However, the requirements for absorption bands used to determine copolymer composition and sequence distribution are practically opposite.

For the application of IR and Raman spectroscopy to composition analysis the following requirements should be satisfied:

1. Proportionality between the intensity of the analytical band and the content of a given monomer in the copolymer

2. Knowledge of the assignment of the chosen calibration band and insensitivity of this band to the sequence distribution

3. Availability of standards of similar constitution whose composition has been determined by independent methods (radiochemical calibration with ^{14}C-labeled copolymers [704, 705], NMR spectroscopy [706], calibration with model compounds [707, 708], or homopolymer mixtures [709-711])

A method to check the insensitivity of absorption bands toward sequence distribution has been described by Harwood [712]. In practice, the absorption bands characteristic of highly localized vibrations [for example, $\nu(C\equiv N)$, $\nu(C=O)$, $\nu(C=C)$, stretching or bending vibrations of CH_3-, C_6H_5-] will be adequate [713]. Numerous examples of compositional analysis of copolymers by IR and Raman spectroscopy are available in the literature [25, 53, 705, 713, 714-724] and two applications have been treated in Sec. 4.2.

The proper selection of sequence-sensitive absorption bands is closely connected with the study of spectral changes induced by conformational, configurational, and constitutional defects in the polymer chain (e.g., appearance of new or forbidden bands due to defect vibrations or weakened selection rules). Influence of sequence length distribution has been observed mainly for stereoregularity, conformational regularity, and crystallinity bands. Although the dependence between the length of regular blocks and IR absorption frequency and intensity has been treated theoretically [29, 713, 725-729], most results so far obtained are empirical and only minimum threshold values of block length for the appearance of certain absorption bands are reported [727, 730-734].

In favorable cases absorption bands can be assigned in the vibrational spectra which are representative for certain sequences in the polymer chain. Only a limited number of copolymers have been characterized completely by their sequence distribution data. Ethylene-propylene copolymers have received most attention, probably because of their technical importance. The best results have been obtained from combined sequence analysis of the ethylene and propylene units in the 700 to 1500 cm^{-1} region [705, 735-739].

Detailed IR sequence analyses have also been reported for methyl-methacrylate-acrylonitrile [733, 740], vinyl chloride-isobutene [741], vinylidene chloride-isobutene [742], tetrafluoroethylene-trifluoro-chloroethylene [743], and various styrene copolymers [744-747]. The microstructure of vinyl chloride (VC)-vinylidene chloride (VDC) co-polymers has been characterized by the IR [748] and Raman [721] spectra of polymers with differing comonomer ratio. Germar [748] investigated the $\delta(CH_2)$ absorption band (1400-1500 cm^{-1}), which is very sensitive to the nature of the neighboring repeating units :

$-CH_2-CH_2-CH_2-\ldots$	1470 cm^{-1}	polyethylene
$-CHCl-CH_2-CHCl-\ldots$	1434 cm^{-1}	VC-VC
$-CCl_2-CH_2-CHCl-\ldots$	1420 cm^{-1}	VDC-VC
$-CCl_2-CH_2-CCl_2-\ldots$	1405 cm^{-1}	VDC-VDC

From quantitative evaluation of the band complex the probability of the three possible polymerization steps and the sequence length distribution have been determined. In analogy Meeks and Koenig [721] have found good correlation between the relative intensities of certain Raman bands and the concentrations of different comonomer sequences:

-VC-VC-	1320 cm^{-1}
-VC-VC-VC-	1167 cm^{-1}
-VDC-VDC-	2979 cm^{-1}
-VDC-VDC-VDC-VDC-	887 cm^{-1}

Many aspects of composition and sequence analysis have also been reviewed in publications by Hummel [749], Schnell [750], Tosi [705], Harwood [746], Schmolke [735], and Kissin [713].

REFERENCES

1. H. Kriegsmann, R. Heess, R. Reich, and O. Nillius, Z. Chem. 7, 12:449 (1967).

2. R. P. Bauman, *Absorption Spectroscopy*, Wiley, New York 1963.

3. W. Brügel, *Einführung in die Ultrarotspektroskopie*, Steinkopff, Darmstadt 1962.

4. I. Kössler, *Methoden der Infrarot-Spektroskopie in der chemischen Analyse*, Akademische Verlagsgesellschaft, Leipzig 1961.

5. J. Derkosch, *Absorptionsspektralanalyse im ultravioletten, sichtbaren und infraroten Gebiet*, Akademische Verlagsgesellschaft, Frankfurt am Main 1967.

6. W. J. Potts, *Chemical Infrared Spectroscopy, vol. 1, Techniques*, Wiley, New York 1963.

7. K. Kiss-Eröss, in *Comprehensive Analytical Chemistry*, ed. G. Svehla, Elsevier, Amsterdam 1976.

8. R. N. Jones, R. Venkataraghavan, and J. W. Hopkins, Spectrochim. Acta 23A:935 (1967).

9. R. N. Jones, R. Venkataraghavan, and J. W. Hopkins, Spectrochim. Acta 23A:941 (1967).

10. M. Yasumi, Bull. Chem. Soc. Jap. 28:489 (1955).

11. J. P. Hawranek, P. Neelakantan, R. P. Young, and R. N. Jones, Spectrochim. Acta 32A:75 (1976).

12. T. Hirschfeld, Appl. Spectrosc. 29:523 (1975).

13. D. A. Ramsay, J. Amer. Chem. Soc. 74:72 (1952).

14. H. Krempl, Thesis, Technical University, Munich 1952.

15. N. Wright, Ind. Eng. 13:1 (1941).

16. J. J. Heigl, M. F. Bell, and J. V. White, Anal. Chem. 19:293 (1947).

17. G. Pirlot, Bull. Soc. Chim. Belg. 58:28 (1949).

18. F. Oswald, Z. Elektrochem. 58:345 (1954).

19. H. U. Pohl and D. O. Hummel, Makromol. Chem. 113:190 (1968).

20. R. D. B. Fraser and E. Suzuki, Anal. Chem. 38:1770 (1966).

21. R. N. Jones, D. A. Ramsay, D. S. Keir, and K. Dobriner, J. Amer. Chem. Soc. 74:80 (1952).

22. E. B. Wilson, Jr., and A. J. Wells, J. Chem. Phys. 14:578 (1946).

23. R. P. Bauman, Appl. Spectrosc. 13:156 (1959).

24. M. V. Sefton, and E. W. Merrill, J. Appl. Polym. Sci. 20:157 (1976).

25. A. Miyake, J. Chem. Soc. Jap. Ind. Chem. Sec. 62:1443 (1959).

26. R. N. Jones, J. Amer. Chem. Soc. 74:2681 (1952).

27. B. C. Stace, in *Laboratory Methods in Infrared Spectroscopy*, ed. R. G. Miller, Heyden, London 1965, p. 112.

28. A. E. Martin, Trans. Faraday Soc. 47:1182 (1951).

29. R. Zbinden, *Infrared Spectroscopy of High Polymers*, Academic Press, New York 1964.

30. C. J. Henniker, *Infrared Spectroscopy of Industrial Polymers*, Academic Press, London 1967.

31. J. Haslam and H. A. Willis, *Identification and Analysis of Plastics*, Iliffe Books Ltd., London 1965.

32. J. Dechant, *Ultrarotspektroskopische Untersuchungen an Polymeren*, Akademie Verlag, Berlin 1972.

33. R. G. Miller and B. C. Stace (eds.), *Laboratory Methods in Infrared Spectroscopy*, Heyden, London 1972.

34. J. Goubeau and L. Thaler, Angew. Chem. Beiheft 41 (1941).

35. W. Otting, *Der Ramaneffekt und seine analytische Anwendung*, Springer, Berlin 1952.

36. D. H. Rank, R. W. Scott, and M. R. Fenske, Anal. Chem. 14:816 (1942).

37. R. F. Stamm, Anal. Chem. 17:318 (1945).

38. W. E. L. Grossman, Anal. Chem. 48:261R (1976).

39. B. R. Loy, R. W. Chrisman, R. A. Nyquist, and C. L. Putzig, Appl. Spectrosc. 33, 2:174 (1979).

40. J. Brandmüller and H. Moser, *Einführung in die Ramanspektroskopie*, Steinkopff, Darmstadt 1962.

41. H. A. Szymanski (ed.), *Raman Spectroscopy: Theory and Practice*, Plenum Press, New York 1970.

42. W. Kiefer, in *Advances in Infrared and Raman Spectroscopy*, vol. 3, eds. R. J. H. Clark and R. E. Hester, Heyden, London 1976.

43. D. D. Tuncliff and A. C. Jones, Spectrochim. Acta 18:579 (1962).

44. W. Kiefer, W. J. Schmid, and J. A. Topp, Appl. Spectrosc. 29:434 (1975).

45. R. J. H. Clark, Spex Speaker 18:1 (1973).

46. J. A. Koningstein and B. F. Gächter, J. Opt. Soc. Amer. 63:892 (1973).

47. D. E. Irish and H. Chen, Appl. Spectrosc. 25:1 (1971).

48. A. K. Covington, M. L. Hassal, and D. E. Irish, J. Solution Chem. 3:629 (1974).

49. G. E. Walrafen, J. Chem. Phys. 52:4176 (1970).

50. D. J. Turner, J. Chem. Soc. Faraday Trans. 170:1346 (1974).

51. W. Kiefer, Appl. Spectrosc. 27:253 (1973).

52. A. K. Covington and J. M. Thain, Appl. Spectrosc. 29:386 (1975).

53. H. J. Sloane and R. Bramston-Cook, Appl. Spectrosc. 27:217 (1973).

54. P. L. Wanchek and L. E. Wolfram, Appl. Spectrosc. 30:542 (1976).

55. D. O. Hummel and F. Scholl, *Infrared Analysis of Polymers, Resins and Additives, vol. II*, Verlag Chemie, Weinheim 1973.

56. P. Arnold and H. A. Willis, in *Polymer Science*, ed. A. D. Jenkins, North Holland, Amsterdam 1972, p. 1561.

57. F. Scholl, in *Kunststoff-Handbuch, vol. I*, eds. R. Vieweg and D. Braun, Carl Hanser, München 1975, p. 796.

58. G. M. Kline (ed.), *Analytical Chemistry of Polymers, vol. II*, Interscience, New York 1962.

59. E. Schröder, J. Franz, and E. Hagen, *Ausgewählte Methoden der Plastanalytik*, Akademie Verlag, Berlin 1976.

60. J. Urbański, W. Czerwiński, K. Janicka, F. Majewska, and H. Zowall, *Handbook of Analysis of Synthetic Polymers and Plastics*, Ellis Horwood Ltd., Chichester 1977.

61. R. Schmolke, W. Kimmer, and W. Sauer, Acta Polymerica 30:432 (1979).

62. K. D. Ledwoch, in *Handbuch der Infrarotspektroskopie*, ed. H. Volkmann, Verlag Chemie, Weinheim 1972, p. 377.

63. D. O. Hummel and F. Scholl, *Atlas der Polymer- und Kunststoff-analyse*, vol. I, *Polymere: Struktur und Spektren*, Verlag Chemie, Weinheim, 1978.

64. W. Kimmer and R. Schmolke, Plaste und Kautschuk 19:260 (1972).

65. L. D. Moore, W. W. Meyer, and W. J. Frazer, Appl. Polym. Symp. 7:67 (1968).

66. B. D. Gesner, Appl. Polym. Symp. 7:53 (1968).

67. G. D. Grant and A. L. Smith, Anal. Chem. 30:1016 (1958).

68. J. D. Lady, G. M. Bower, R. E. Adams, and F. P. Byrne, Anal. Chem. 31:1100 (1959).

69. E. R. Shull, Anal. Chem. 32:1627 (1960).

70. M. V. Zeller, Perkin-Elmer Infrared Bulletin 18, 1973.

71. C. W. Young, P. C. Servais, C. C. Currie, and M. J. Hunter, J. Amer. Chem. Soc. 70:3758 (1948).

72. H. Kriegsmann, Z. Elektrochem. 65:336 (1961).

73. R. E. Richards and H. W. Thompson, J. Roy. Chem. Soc. (London) 1949:124.

74. J. L. Koenig, Appl. Spectrosc. 29:293 (1975).

75. J. L. Koenig and M. K. Antoon, Appl. Opt. 17:1374 (1978).

76.   T. Hirschfeld, Anal. Chem. 48:721 (1976).

77.   J. L. Koenig, L. D'Esposito, and M. K. Antoon, Appl. Spectrosc. 31:292, 518 (1977).

78.   D. L. Tabb and J. L. Koenig, Macromolecules 8:929 (1975).

79.   A. H. Willbourn, J. Polym. Sci. 34:569 (1959).

80.   S. C. Lin, B. J. Bulkin, and E. M. Pearce, J. Polym. Sci. (Polym. Chem. Ed.), 17:3121 (1979).

81.   M. M. Coleman and P. C. Painter, in *Applications of Polymer Spectroscopy*, ed. E. G. Brame, Jr., Academic Press, New York 1978, p. 135.

82.   M. M. Coleman and P. C. Painter, J. Macromol. Sci. Revs. Macromol. Chem. C16, 2:197 (1978).

83.   P. C. Painter and J. L. Koenig, Biopolymers 15:229 (1976).

84.   D. O. Hummel and F. Scholl, *Atlas der Kunststoffanalyse, vol. I*, Carl Hanser, München 1968.

85.   W. L. Truett, Appl. Spectrosc. 21:400 (1967).

86.   J. W. Cassels, Appl. Spectrosc. 22:477 (1968).

87.   T. Takeuchi, S. Tsuge, and K. Ito, Jap. Analyst 18:383 (1969).

88.   D. Z. Gross, Z. Anal. Chem. 248:40 (1969).

89.   W. L. Truett, Polym. Prep. Am. Chem. Soc. Div. Polym. Chem. 18:107 (1977).

90.   G. Leukroth, Gummi, Asbest und Kunststoffe 10:794 (1974).

91.   S. C. Pattacini, Perkin-Elmer Infrared Bulletin 52, 1975.

92.   R. L. Levy and M. Lederer, Chromatogr. Rev. 8:48 (1966).

93.   M. P. Stevens, *Characterization and Analysis of Polymers by Gas Chromatography*, Marcel Dekker, New York 1969.

94.   D. Noffz, W. Benz, and W. Pfab, Z. Anal. Chem. 235:121 (1968).

95.   W. H. McFadden, *Techniques of Combined Gas Chromatography-Mass Spectrometry: Application in Organic Analysis*, Interscience, New York 1973.

96.   D. Braun and G. Nixdorf, Kunststoffe 62:318 (1972).

97.   E. Stahl and V. Bruederle, Adv. Polym. Sci. 30:1 (1979).

98.   E. Stahl and L. S. Oey, Angew. Makromol. Chem. 44:107 (1975).

99.   E. Stahl and L. S. Oey, Z. Anal. Chem. 275:187 (1975).

100. G. M. Brauer, in *Techniques and Methods of Polymer Evaluation, vol. 2*, Marcel Dekker, New York 1970, p. 41.

101. J. Q. Walker, Chromatographia 5:547 (1972).

102. R. W. McKinney, in *Ancillary Techniques of Gas Chromatography*, eds. L. S. Ettre and W. H. McFadden, Interscience, New York 1969.

103. J. Voigt, Kunststoffe 54:2 (1964).

104. A. Barlow, R. S. Lehrle, and J. C. Robb, Polymer 2:27 (1961).

105. F. W. Willmott, J. Chromatogr. Sci. 7:101 (1969).

106. C. Oertli, C. Bühler, and W. Simon, Chromatographia 6:499 (1973).

107. D. L. Fanter, R. C. Levy, and C. J. Wolff, Anal. Chem. 44:43 (1972).

108. N. E. Vanderborgh and W. T. Ristau, Anal. Chem. 45:1529 (1973).

109. E. Kiran and J. K. Gillham, J. Appl. Polym. Sci. 20:931 (1976).

110. G. A. Junk, Int. J. Mass Spectrom. Ion Phys. 8:1 (1972).

111. R. A. Flath, in *Guide to Modern Methods of Instrumental Analysis*, ed. T. H. Gouw, Interscience, New York 1972, p. 323.

112. D. O. Hummel, H. D. Schüddemage, and K. Rübenacker, in *Polymer Spectroscopy*, ed. D. O. Hummel, Verlag Chemie, Weinheim 1974, p. 355.

113. T. Morimoto, K. Takeyama, and F. Konishi, J. Appl. Polym. Sci. 20:1967 (1976).

114. S. A. Liebman, D. H. Ahlstrom, and P. R. Griffith, Appl. Spectrosc. 30:355 (1976).

115. P. R. Griffith, in *Fourier Transform Infrared Spectroscopy*, *vol. 1*, eds. J. R. Ferraro and L. J. Basile, Academic Press, New York 1978, p. 143.

116. Nicolet Instrument Corporation, GC/FTIR Application Note, 1978.

117. D. W. Vidrine and D. R. Mattson, Appl. Spectrosc. 32:502 (1978).

118. S. Krimm, J. Chem. Phys. 22:567 (1954).

119. F. Boerio and J. L. Koenig, J. Chem. Phys. 52:3425 (1970).

120. K. Holland-Moritz, E. Sausen, and D. O. Hummel, Colloid Polym. Sci. 254:342 (1976).

121. V. Zamboni and G. Zerbi, J. Polym. Sci. C7:153 (1964).

122. I. M. Ward, Trans. Faraday Soc. 53:1406 (1957).

123. W. Heinen, J. Polym. Sci. 38:545 (1959).

124. J. C. Woodbrey and Q. A. Trementozzi, J. Polym. Sci. C8:113 (1965).

125. G. Heidemann and H. J. Nettelbeck, Faserforsch. Textiltechn. 18:183 (1967).

126. D. Kuehl and P. R. Griffith, J. Chromatographic Sci. 17:471 (1979).

127. T. Miyazawa, K. Fukushima, S. Sugano, and Y. Masuda, in *Conformation of Biopolymers, vol. II*, ed. G. N. Ramachandran, Academic Press, New York 1967, p. 557.

128. M. M. Sushchinskii, *Raman Spectra of Molecules and Crystals*, Keter, New York, 1972, p. 273.

129. T. Miyazawa, J. Chem. Phys. 32:1647 (1960).

130. T. Miyazawa, J. Chem. Phys. 35:693 (1961).

131. S. Krimm, J. Mol. Biol. 4:528 (1962).

132. J. Dechant and C. Ruscher, Faserforsch. Textiltechn. 16:180 (1965).

133. S. Krimm and Y. Abe, Proc. Nat. Acad. Sci. U.S.A. 69:2788 (1972).

134. V. N. Nikitin and E. J. Pokrovskij, Dokl. Akad. Nauk SSR 95:109 (1954).

135. T. Okada and L. Mandelkern, J. Polym. Sci. A2, 5:239 (1967).

136. H. G. Zachmann and H. A. Stuart, Makromol. Chem. 44/46:622 (1961).

137. G. Natta, P. Corradin , and M. Cesari, Atti. Accad. Naz. Lincei Cl. Sci. Fis. Mat. Nat. Rend. 22:11 (1957).

138. R. G. Quynn, J. L. Riley, D. A. Young, and H. D. Noether, J. Appl. Polym. Sci. 2:166 (1959).

139. H. Hendus and G. Schnell, Kunststoffe 51:60 (1961).

140. I. I. Novak, Vysokomol. Soedin 5:1645 (1963).

141. H. W. Starkweather and R. E. Moynihan, J. Polym. Sci. 22:363 (1956).

142. I. Sandeman and A. Keller, J. Polym. Sci. 19:401 (1956).

143. G. R. Strobl and W. Hagedorn, J. Polym. Sci. Polym. Phys. Ed. 16:1181 (1978).

144. C. Painter, M. Watzek, and J. L. Koenig, Polymer 18:1169 (1978).

145. M. M. Coleman, P. C. Painter, D. L. Tabb, and J. L. Koenig, J. Polym. Sci. Polym. Lett. Ed. 12:577 (1974).

146. L. D'Esposito and J. L. Koenig, J. Polym. Sci. Polym. Phys. Ed. 14:1731 (1976).

147. P. C. Painter, J. Havens, W. W. Hart, and J. L. Koenig, J. Polym. Sci. Polym. Phys. Ed. 15:1237 (1977).

148. P. C. Painter, J. Runt, M. M. Coleman, and I. R. Harrison, J. Polym. Sci. Polym. Phys. Ed. 16:1253 (1978).

149. J. Dechant and R. Danz, Plaste und Kautschuk 19:250 (1972).

150. S. K. Bahl, D. D. Cornell, F. J. Boerio, and G. E. McGraw, J. Polym. Sci. Polym. Lett. Ed. 12:13 (1974).

151. A. Miyake, J. Polym. Sci. 38:479, 497 (1959).

152. T. R. Manley and D. A. Williams, J. Polym. Sci. C22:1009 (1969).

153. R. Danz, J. Dechant, and C. Ruscher, Faserforsch. Textiltechn. 21:251 (1970).

154. D. Grime and I. M. Ward, Trans. Faraday Soc. 54:959 (1958).

155. F. J. Boerio, S. K. Bahl, and G. E. McGraw, J. Polym. Sci. Polym. Phys. Ed. 14:1029 (1976).

156. R. de P. Daubeny and C. W. Bunn, Proc. Roy. Soc. (London) A226:531 (1954).

157. I. M. Ward, Text. Res. J. 31:650 (1961).

158. G. Farrow and I. M. Ward, Polymer 1:330 (1960).

159. J. L. Koenig and M. J. Hannon, J. Macromol. Sci. Phys. B1, 1: 119 (1967).

160. I. H. Hall and M. G. Pass, Polymer 17:807 (1976), 18:825 (1977).

161. I. M. Ward and M. A. Wilding, Polymer 18:327 (1977).

162. P. Simak, Makromol. Chem. 178:2927 (1977).

163. B. Schneider, P. Schmidt, and O. Wichterle, Collect. Czech. Chem. Commun. 27:1749 (1962).

164. C. Ruscher and J. Dechant, Faserforsch. Textiltechn. 15:481 (1964):

165. P. Simak, Angew. Makromol. Chem. 28:75 (1973).

166. L. N. Ovander, Opt. Spektrosk. 12:711 (1962).

167. N. F. Brockmeier, J. Appl. Polym. Sci. 12:2129 (1968).

168. G. Natta and P. Corradini, Atti. Accad. Naz. Lincei Mem. Cl. Sci. Fis. Mat. Nat. 4:73 (1955).

169. G. Natta, Makromol. Chem. 35:94 (1960).

170. G. Natta, M. Peraldo, and G. Allegra, Makromol. Chem. 75:215 (1964).

171. F. Danusso and G. Gianotti, Makromol. Chem. 80:1 (1964).

172. J. P. Luongo and R. Salovey, J. Polym. Sci. A2, 4:997 (1966).

173. F. J. Boerio and J. L. Koenig, J. Macromol. Sci. Rev. Macromol. Chem. C7, 2:209 (1972).

174. G. Natta, P. Corradini, and I. W. Bassi, Makromol. Chem. 21:246 (1956).

175. V. F. Holland and R. L. Miller, J. Appl. Phys. 35:3241 (1964).

176. A. Cojazzi, A. Malta, A. Celotti, and A. Zanelli, Makromol. Chem. 177:915 (1976).

177. G. Goldbach and G. Peitscher, J. Polym. Sci. Polym. Lett. Ed. 6:783 (1968).

178. J. Y. Decroix, M. Moser, and M. Boudeulle, Eur. Polym. J. 11:357 357 (1975).

179. M. Moser and M. Boudeulle, J. Polym. Sci. Polym. Phys. Ed. 14: 1161 (1976).

180. I. D. Rubin, J. Polym. Sci. A2, 5:1323 (1967).

181. J. P. Luongo and R. Salovey, J. Polym. Sci. B, 3:513 (1965).

182. K. Holland-Moritz, E. Sausem, P. Djudovic, M. M. Coleman, and
      P. C. Painter, J. Polym. Sci. Polym. Phys. Ed. 17:25 (1979).

183. K. Holland-Moritz and E. Sausen, J. Polym. Sci. Polym. Phys.
      Ed. 17:1 (1979).

184. K. Holland-Moritz, J. Appl. Polym. Sci. Appl. Polym. Symp.
      34:49 (1978).

185. A. Turner-Jones, Makromol. Chem. 71:1 (1964).

186. G. Trafara, R. Koch, K. Blum, and D. O. Hummel, Makromol. Chem.
      177:1089 (1976).

187. G. Trafara, Makromol. Chem. 181 (1980), in press.

188. K. Holland-Moritz, P. Djudoviv, and D. O. Hummel, Progr. Col-
      loid Polym. Sci. 57:206 (1975).

189. M. J. Gall, P. J. Hendra, P. J. Peacock, C. J. Cudby, and
      H. A. Willis, Spectrochim. Acta 28A:1485 (1972).

190. K. Holland-Moritz, I. Modrić, K. U. Heinen, und D. O. Hummel,
      Kolloid-Z. Z. Polym. 251:913 (1973).

191. K. Holland-Moritz, Colloid Polym. Sci. 253:922 (1975).

192. S. Krimm, Adv. Polym. Sci. 2:51 (1960).

193. K. Holland-Moritz, E. Sausen, and D. O. Hummel, Colloid Polym.
      Sci. 254:976 (1976).

194. M. J. Gall, P. J. Hendra, D. S. Watson, and C. J. Peacock,
      Appl. Spectrosc. 25:423 (1971).

195. F. J. Boerio and J. L. Koenig, J. Chem. Phys. 52:3425 (1970).

196. W. T. Astbury, Kolloid-Z. 69:346 (1934).

197. J. Mann and H. J. Marrinan, Trans. Faraday Soc. 52:481 (1956).

198. D. S. Trifan and J. F. Terenzi, J. Polym. Sci. Polym. Lett. Ed.
      28:443 (1958).

199. C. Y. Liang and R. H. Marchessault, J. Polym. Sci. 35:529 (1959).

200. Y. Kinoshita, Makromol. Chem. 33:1 (1959).

201. A. Miyake, J. Polym. Sci. 44:223 (1960).

202. L. B. Weisfeld, J. R. Little, and W. E. Wolstenholme, J. Polym.
      Sci. 56:455 (1962).

203. R. H. Marchessault, Pure Appl. Chem. 5:107 (1962).

204. G. Wlodarsky, Polimery 9:40 (1964).

205. B. Ranby, Das Papier 18:593 (1964).

206. Yu. M. Bojarcuk, L. Ya. Rappoport, V. N. Nikitin, and
      N. P. Apukhtina, Polym. Sci. USSR 7:859 (1965).

207. J. Rubin, J. Polym. Sci. B, 5:1135 (1967).

208. T. Tanaka, T. Yokoyama, and Y. Yamaguchi, J. Polym. Sci. A1, 6:2137 (1968).

209. E. Bessler and G. Bier, Makromol. Chem. 122:30 (1969).

210. R. W. Seymour, G. M. Estes, and S. L. Cooper, Macromolecules 3:579 (1970).

211. K. Ogura and H. Sobue, Polym. J. 3:153 (1972).

212. R. W. Seymour and S. L. Cooper, Macromolecules 6:48 (1973).

213. K. Frigge and J. Dechant, Faserforsch. Textiltechn. 26:547 (1975).

214. C. S. P. Sung and N. S. Schneider, Macromolecules 8:68 (1975).

215. F. J. Kolpak and J. Blackwell, Macromolecules 9:273 (1976).

216. W. Rutenkolk and R. Bonart, Colloid Polym. Sci. 254:190 (1976).

217. V. W. Srichatrapimuk and S. L. Cooper, J. Macromol. Sci. Phys. B15, 2:267 (1978).

218. L. R. Schroeder and S. L. Cooper, J. Appl. Phys. 47:4310 (1976).

219. H. W. Siesler, Colloid Polym. Sci. 17:93 (1977).

220. A. H. Nissan, Macromolecules 9:840 (1976).

221. G. C. Pimentel and A. L. McClellan, *The Hydrogen Bond*, W. H. Freeman, San Francisco 1960.

222. A. S. N. Murthy and C. N. R. Rao, Appl. Spectrosc. Rev. 2:69 (1968).

223. C. N. R. Rao, *Chemical Application of Infrared Spectroscopy*, Academic Press, New York, 1963.

224. J. E. Del Bene, J. Chem. Phys. 58:926, 3139 (1973).

225. J. E. Del Bene, J. Amer. Chem. Soc. 95:5460 (1973).

226. G. M. Badger, Rev. Pure Appl. Chem. 7:55 (1957).

227. C. A. Coulsen, in *Hydrogen Bonding*, eds. D. Hadzi and W. H. Thompson, Pergamon Press, London 1959, p. 339.

228. N. D. Sokolov, Anal. Chem. 10:497 (1965).

229. J. C. West and S. L. Cooper, J. Polym. Sci. Polym. Symp. 60: 127 (1977).

230. H. Zimmermann, Angew. Chem. 76:1 (1964).

231. E. R. Lippincott, J. N. Finck, and R. Schroeder, in *Hydrogen Bonding*, eds. D. Hadzi and W. H. Thompson, Pergamon Press, London 1959, p. 361.

232. E. Heilbronner, H. H. Günthard, and R. Gerdil, Helv. Chim. Acta 39:1171 (1956).

233. C. G. Cannon, Spectrochim. Acta 10:341 (1958).

234. F. Gerson, Helv. Chim. Acta 44:471 (1961).

235. D. Hadzi and W. H. Thompson (eds.), *Hydrogen Bonding*, Pergamon Press, London 1959.

236. H. E. Hallam, in *Infrared Spectroscopy and Molecular Structure*, ed. M. Davies, Elsevier, New York 1963, p. 405.

237. P. Schuster, G. Zundel, and C. Sandorfy (eds.), *The Hydrogen Bond*, North Holland, New York 1976.

238. W. C. Hamilton and I. Ibers, *Hydrogen Bonding in Solids*, W. A. Benjamin, New York 1968.

239. S. N. Vinogradov and R. H. Linnell, *Hydrogen Bonding*, Van Nostrand-Reinhold, New York 1971.

240. M. D. Joesten and L. J. Schaad, *Hydrogen Bonding*, Marcel Dekker, New York 1974.

241. P. A. Kollamn and L. C. Allen, Chem. Rev. 72:283 (1972).

242. L. C. Allen, J. Amer. Chem. Soc. 97:6921 (1975).

243. W. Luck, Naturwissenschaften 54:601 (1967).

244. S. Bratoz, in *Advances in Quantum Chemistry, vol. 3*, ed. P. O. Lowdin, Academic Press, New York 1967.

245. E. F. Gross, in *Hydrogen Bonding*, eds. D. Hadzi and W. H. Thompson, Pergamon Press, London (1959).

246. W. J. Hurley, J. D. Kuntz, and G. E. Leroi, J. Amer. Chem. Soc. 88:3199 (1966).

247. R. J. Jakobsen, J. W. Brasch, and Y. Mikawa, Appl. Spectrosc. 22:641 (1968).

248. R. F. Lake and H. W. Thompson, Proc. Roy. Soc. (London) A291: 469 (1966).

249. G. Statz and E. Lippert, Ber. Bunsenges. Phys. Chem. 71:673 (1967).

250. K. Itoh and T. Shimanouchi, Biopolymers 5:921 (1967).

251. R. M. Badger, J. Chem. Phys. 8:288 (1940).

252. M. D. Joesten and R. S. Drago, J. Amer. Chem. Soc. 84:3817 (1962).

253. S. Singh, A. S. N. Murthy, and C. N. R. Rao, Trans. Faraday Soc. 62:1056 (1966).

254. G. C. Pimentel and C. H. Sederholm, J. Chem. Phys. 24:639 (1956).

255. H. Fritzsche, Ber. Bunsenges. Phys. Chem. 68:459 (1964).

256. S. M. Huggins and G. C. Pimentel, J. Chem. Phys. 60:1615 (1964).

257. R. C. Lord and R. Merrifield, J. Chem. Phys. 21:166 (1953).

258. E. D. Becker, Spectrochim. Acta 17:436 (1961).

259. K. Nakamoto, M. Margoshes, and R. E. Rundle, J. Amer. Chem. Soc. 77:6480 (1955).

260. T. Granstad, Spectrochim. Acta 19:497 (1963).

261. B. A. Zadoroznnyi and I. K. Ishchenko, Opt. Spektrosk. 19:551 (1965).

262. Yu. M. Bojarcuk and V. N. Nikitin, Dokl. Akad. Nauk. SSR 19: 397 (1964).

263. M. Tichy, Advan. Org. Chem. 5:115 (1965).

264. R. D. B. Fraser and T. P. McRae, *Conformation in Fibrous Proteins and Related Synthetic Polypeptides*, Academic Press, New York 1973.

265. A. G. Walton and J. Bláckwell, *Biopolymers*, Academic Press, New York 1973.

266. A. V. R. Warrier and S. Krimm, J. Chem. Phys. 52:4316 (1970).

267. A. V. R. Warrier and S. Krimm, Macromolecules 3:709 (1970).

268. A. Allerhand and P. von Ragué Schleyer, J. Amer. Chem. Soc. 85:1715 (1963).

269. W. J. Macknight and M. Yang, J. Polym. Sci. Polym. Symp. 42: 817 (1973).

270. R. Blinc and D. Hadzi, Spectrochim. Acta 16:852 (1960).

271. G. Zundel and H. Metzger, Spectrochim. Acta 23A:759 (1967).

272. E. G. Weidemann and G. Zundel, Z. Phys. 198:288 (1967).

273. N. Joop and H. Zimmermann, Z. Elektrochem. Ber. Bunsenges. Phys. Chem. 66:541 (1962).

274. C. Berthomien and C. Sandorfy, J. Mol. Spectrosc. 15:22 (1965).

275. A. Foldes and C. Sandorfy, J. Mol. Spectrosc. 20:262 (1966).

276. Y. Maréchal and A. Witkowski, Theor. Chim. Acta (Berlin) 9:116 (1967).

277. C. A. Dementeva, A. B. Jogansen, and G. A. Kurskci, Opt. Spektrosk. 29:868 (1970).

278. A. E. Mirsky and L. Pauling, Proc. Nat. Acad. Sci. U.S.A. 22: 439 (1936).

279. W. Kauzmann, *Denaturation of Proteins and Enzymes*, Johns Hopkins Press, Baltimore 1954.

280. G. E. Schulz, Angew. Chem. 89:24 (1977).

281. J. L. Koenig, J. Polym. Sci. D:59 (1972).

282. J. L. Koenig, in *Advances in Raman Spectroscopy*, ed. J. P. Mathieu, Heyden, London 1973, p. 265.

283. V. Fawcett and D. A. Long, J. Mol. Spectrosc. 1:352 (1973).

284. E. W. Small and W. C. Peticolas, Biopolymers 10:1377 (1971).

285. B. B. Doyle, E. G. Bendit, and E. R. Blout, Biopolymers 14:937 (1975).

286. H. T. Miles and J. Frazier, Biochem. Biophys. Res. Commun. 14:21 (1964).

287. A. M. Bellocq, R. C. Lord, and R. Mendelsohn, Biochim. Biophys. Acta 257:280 (1972).

288. N. T. Yu, B. H. Jo, and D. C. O'Shea, Arch. Biochem. Biophys. 156:71 (1973).

289. M. N. Siamwiza, R. C. Lord, M. C. Chen, T. Takamatsu, I. Harada, H. Matsuura, and T. Shimanouchi, Biochemistry 14:4870 (1975).

290. M. C. Chen and R. C. Lord, J. Amer. Chem. Soc. 98:990 (1976).

291. I. Harada and R. C. Lord, Spectrochim. Acta 26A:2305 (1970).

292. Y. Kyogoku, R. C. Lord, and A. Rich, J. Amer. Chem. Soc. 89: 496 (1967).

293. C. P. Beetz, Jr., and G. Ascarelli, Biopolymers 15:2299 (1976).

294. R. D. B. Fraser, J. Chem. Phys. 21:1511 (1953).

295. W. C. Price and R. D. B. Fraser, Proc. Roy. Soc. (London) B141:66 (1953).

296. H. W. Siesler, Polymer 15:146 (1974).

297. M. Tsuboi, J. Polym. Sci. 59:139 (1962).

298. E. M. Bradbury, A. Elliott, and R. D. B. Fraser, Trans. Faraday Soc. 56:1117 (1960).

299. S. Krimm, in *Infrared Spectroscopy and Molecular Structure*, ed. M. Davies, Elsevier, Amsterdam 1963, p. 270.

300. R. G. Snyder, J. Chem. Phys. 47:1316 (1967).

301. B. E. Read and R. S. Stein, Macromolecules 1:116 (1968).

302. P. G. Schmidt, J. Polym. Sci. A, 1:1271 (1963).

303. J. L. Koenig, S. W. Cornell, and D. E. Witenhafer, J. Polym. Sci. A2, 5:301 (1967).

304. J. L. Koenig, L. Wolfram, and J. Grasselli, Div. Polym. Sci. Amer. Chem. Soc. N. Y. Polym. Prep. 10, 2:959 (1969).

305. H. Tadokoro, S. Seki, I. Nitta, and R. Yamadera, J. Polym. Sci. 28:244 (1958).

306. L. E. Wolfram, J. G. Grasselli, and J. L. Koenig, Appl. Spectrosc. 24:263 (1970).

307. P. A. Flournoy and W. J. Schaffers, Spectrochim. Acta 22:5 (1966).

308. J. Dechant, Faserforsch. Textiltechn. 25:24 (1974).

309. R. D. B. Fraser, J. Chem. Phys. 24:89 (1956).

310. M. Beer, Proc. Roy. Soc. (London) A236:136 (1956).

311. R. D. B. Fraser, J. Chem. Phys. 29:1428 (1958).

312.   P. H. Hermans and P. Platzek, Kolloid-Z. 88:68 (1939).

313.   P. H. Hermans, *Contributions to the Physics of Cellulosic Fibers*, Elsevier, Amsterdam 1946, p. 138.

314.   R. S. Stein, J. Polym. Sci. 31:327 (1958).

315.   R. J. Samuels, *Structured Polymer Properties*, Interscience, New York 1974.

316.   Z. W. Wilchinsky, J. Appl. Phys. 30:792 (1959).

317.   L. E. Alexander, *X-ray Diffraction Methods in Polymer Science*, Interscience, New York 1969.

318.   R. J. Samuels, J. Polym. Sci. A, 3:1741 (1965).

319.   A. Elliott, *Infrared Spectra and Structure of Organic Long-Chain Polymers*, Arnold, London 1969.

320.   H. G. Ingersoll, J. Appl. Phys. 17:924 (1946).

321.   M. Kakudo and N. Kasai, *X-ray Diffraction by Polymers*, Kodansha Scientific Books, Tokyo 1972, p. 254.

322.   R. S. Stein, J. Polym. Sci. 50:339 (1961).

323.   R. D. B. Fraser, J. Chem. Phys. 28:1113 (1958).

324.   S. Nomura, H, Kawai, I. Kimura, and M. Kagiyama, J. Polym. Sci. A2, 5:479 (1967).

325.   I. Sandeman and A. Keller, J. Polym. Sci. 19:401 (1956).

326.   B. E. Read, in *Structure and Properties of Oriented Polymers*, ed. I. M. Ward, Applied Science, London 1975, p. 150.

327.   A. Cunningham, G. R. Davies, and I. M. Ward, Polymer 15:743 (1974).

328.   P. Hendra, in *Polymer Spectroscopy*, ed. D. O. Hummel, Verlag Chemie, Weinheim 1974, p. 151.

329.   M. C. Tobin, *Laser Raman Spectroscopy*, Wiley, New York 1971.

330.   L. A. Woodward and D. A. Long, Trans. Faraday Soc. 45:1131 (1949).

331.   S. P. S. Porto, J. Opt. Soc. Amer. 56:1585 (1966).

332.   F. J. Boerio and J. L. Koenig, J. Macromol. Sci. Rev. Macromol. Chem. 7:209 (1972).

333.   P. J. Hendra and H. A. Willis, Chem. Commun. 1968:225.

334.   R. G. Snyder, J. Mol. Spectrosc. 37:353 (1971).

335.   B. Fanconi, B. Tomlinson, L. A. Nafie, W. Small, and W. L. Peticolas, J. Chem. Phys. 51:9 (1969).

336.   F. J. Boerio and R. A. Bailey, J. Polym. Sci. Polym. Lett. Ed. 12:433 (1974).

337.   V. B. Carter, J. Mol. Spectrosc. 34:356 (1970).

338. J. L. Derouault, P. J. Hendra, M. E. A. Cudby, and H. A. Willis, Chem. Commun. 1972:1187.

339. S. W. Cornell and J. L. Koenig, J. Appl. Phys. 39:4883 (1968).

340. D. I. Bower, J. Polym. Sci. Polym. Phys. Ed. 10:2135 (1972).

341. C. G. Cannon, Spectrochim. Acta 16:302 (1960).

342. P. J. Hendra, D. S. Watson, M. E. A. Cudby, H. A. Willis and P. Holliday, Chem. Commun. 1970:1048.

343. T. Miyazawa, K. Fukushima, and Y. Ideguchi, J. Chem. Phys. 37: 2764 (1962).

344. J. Maxfield and I. W. Shepherd, Polymer 16:505 (1975).

345. J. L. Koenig and A. C. Angood, J. Polym. Sci. A2, 8:1787 (1970).

346. H. Tadokoro, in *Polymer Spectroscopy*, ed. D. O. Hummel, Verlag Chemie, Weinheim 1974, p. 1.

347. K. van Werden, Thesis, University of Cologne, Cologne, 1976.

348. H. W. Siesler, Makromol. Chem. 176:2451 (1975).

349. C. R. Bohn, J. R. Schaefgen, and W. O. Statton, J. Polym. Sci. 55:531 (1961).

350. C. Ruscher and R. Schmolke, Faserforsch. Textiltechn. 14:459 (1963).

351. G. Hinrichsen, J. Polym. Sci. C, 38:303 (1972).

352. M. Takahashi and Y. Nukushina, Sen-i-Gakkaishi, 16:622 (1960).

353. H. W. Siesler, Colloid Polym. Sci. 255:321 (1977).

354. W. Schauler and U. Kashani, Faserforsch. Textiltechn. 26:270 (1975).

355. R. Schmolke, Faserforsch. Textiltechn. 16:514 (1965).

356. I. B. Klimenko and L. V. Smirnov, Vysokomol. Soedin. 5, 10:1520 (1963).

357. C. Gentilhomme, A. Piguet, J. Rosset, and C. Eyraud, Bull. Soc. Chim. Fr. 5:901 (1960).

358. C. Y. Liang and S. Krimm, J. Chem. Phys. 27:327 (1957).

359. C. J. Heffelfinger and R. L. Burton, J. Polym. Sci. 47:289 (1960).

360. C. J. Heffelfinger and P. G. Schmidt, J. Appl. Polym. Sci. 9: 2661 (1965).

361. J. L. Koenig and S. W. Cornell, J. Polym. Sci. C, 22:1019 (1969).

362. H. Tadokoro, K. Tatsuka, and S. Murahashi, J. Polym. Sci. 59: 413 (1962).

363. G. W. Urbanczyk, Faserforsch. Textiltechn. 27:183 (1976).

364. A. Cunningham, I. M. Ward, H. A. Willis, and V. Zichy, Polymer 15:749 (1974).

365. C. Y. Liang and S. Krimm, J. Mol. Spectrosc. 3:554 (1959).

366. J. Purvis, D. I. Bower, and I. M. Ward, Polymer 14:400 (1973).

367. D. I. Bower, in *Structure and Properties of Oriented Polymers*, ed. I. M. Ward, Applied Science, London 1975, p. 187.

368. J. Purvis, D. I. Bower, and I. M. Ward, Polymer 14:398 (1973).

369. M. E. R. Robinson, D. I. Bower, and W. F. Maddams, J. Polym. Sci. Polym. Phys. Ed. 16:2115 (1978).

370. C. H. Bamford, A. Elliott, and W. E. Hanby, *Synthetic Polypeptides*, Academic Press, New York 1956.

371. E. G. Bendit, Biopolymers 4:561 (1966).

372. R. G. Quynn and R. Steele, Nature 173:1240 (1954).

373. S. L. Aggarwal, G. P. Tilley, and O. J. Sweeting, J. Appl. Polym. Sci. 1:91 (1959).

374. M. Tasumi and T. Shimanouchi, Spectrochim. Acta 17:731 (1961).

375. B. Z. Volchek and V. N. Nikitin, Sov. Phys. Tech. Phys. 2:1705 (1957).

376. C. Ruscher and R. Schmolke, Faserforsch. Textiltechn. 14:340 (1963)

377. I. W. Shepherd, in *Advances in Infrared and Raman Spectroscopy, vol. 3*, eds. R. J. H. Clark and R. E. Hester, Heyden, London 1976, p. 127.

378. A. Peterlin (ed.), *Plastic Deformation of Polymers*, Marcel Dekker, New York 1971.

379. C. Y. Liang, in *Newer Methods of Polymer Characterization*, ed. Bacon Ke, Interscience, New York 1964, p. 33.

380. H. Tadokoro, *Structure of Crystalline Polymers*, Wiley, New York 1979.

381. G. R. Bird and E. R. Blout, J. Amer. Chem. Soc. 81:2499 (1959).

382. G. Spach, Compt. Rend. 249:667 (1959).

383. E. Iizuka, Polym. J. 7:650 (1975).

384. V. A. Platonov, G. D. Litovcenko, V. G. Kulicihin, M. V. Sablygin, T. A. Belousova, N. S. Pozalkin, V. D. Kalmykova, and S. P. Papkov, Chim. Volokna 4:36 (1975).

385. V. I. Vettegren and I. I. Novak, J. Polym. Sci. Polym. Phys. Ed. 11:2135 (1973).

386. L. Penn and F. Milanovich, Polymer 20:31 (1979).

387. Y. Uemura and R. S. Stein, J. Polym. Sci. A2, 10:1691 (1972).

388. V. K. Mitra, W. M. Risen, Jr., and R. H. Baughman, J. Chem. Phys. 66:2731 (1977).

389. R. A. Evans and H. E. Hallam, Polymer 17:839 (1976).

390. R. P. Wool and W. O. Statton, J. Polym. Sci. Polym. Phys. Ed. 12:1575 (1974).

391. R. P. Wool, J. Polym. Sci. Polym. Phys. Ed. 14:1921 (1976).

392. S. Sikka, Ph. D. Thesis, University of Utah, Salt Lake City 1977.

393. R. P. Wool and W. O. Statton, in *Applications of Polymer Spectroscopy*, ed. E. G. Brame, Jr., Marcel Dekker, New York 1978, p. 185.

394. D. K. Roylance and K. L. DeVries, J. Polym. Sci. B, 9:443 (1971).

395. H. W. Siesler, in *Proceedings of the 5th European Symposium on Polymer Spectroscopy*, ed. D. O. Hummel, Verlag Chemie, Weinheim 1979, p. 137.

396. K. K. R. Mocherla and W. O. Statton, J. Appl. Polym. Sci. Appl. Polym. Symp. 31:183 (1977).

397. T. M. Stoeckel, J. Blasius, and B. Crist, J. Polym. Sci. Polym. Phys. Ed. 16:485 (1978).

398. R. Jakeways, T. Smith, I. M. Ward, and M. A. Wilding, J. Polym. Sci. Polym. Lett. Ed. 14:41 (1976).

399. S. Onogi and T. Asada, J. Polym. Sci. C, 16:1445 (1967).

400. H. W. Siesler, J. Polym. Sci. Polym. Lett. Ed. 17:453 (1979).

401. S. N. Zhurkov, V. I. Vettegren, V. E. Korsukov, and I. I. Novak, *Fracture* (Proceedings of the Second Int. Conference on Fracture, Brighton, England 1969), ed. P. L. Pratt, Chapman and Hall, London 1969, p. 545.

402. A. I. Gubanov and V. A. Kosobukin, Mech. Polym. 4:586 (1968).

403. V. M. Voroboyev, I. V. Razumovska, and V. I. Vettegren, Polymer 19:1267 (1978).

404. I. Sakurada, T. Ito, and K. Nakamae, J. Polym. Sci. C, 15:75 (1966).

405. S. N. Zhurkov, V. A. Zakrevskyi, V. E. Korsukov, and V. S. Kuksenko, J. Polym. Sci. A2, 10:1509 (1972).

406. D. N. Batchelder and D. Bloor, J. Polym. Sci. Polym. Phys. Ed. 17:569 (1979).

407. M. G. Brereton, G. R. Davies, R. Jakeways, T. Smith, and I. M. Ward, Polymer 19:17 (1978).

408. M. Yokouchi, Y. Sakakibara, Y. Chatani, H. Tadokoro, T. Tanaka, and K. Yoda, Macromolecules 9:266 (1976).

409. B. Stambaugh, J. L. Koenig, and J. B. Lando, J. Polym. Sci. Polym. Phys. Ed. 17:1053, 1063 (1979).

410. S. N. Zhurkov and V. E. Korsukov, J. Polym. Sci. Polym. Phys. Ed. 12:385 (1974).

411. H. H. Kausch, *Polymer Fracture*, Springer Verlag, Heidelberg 1978.

412. K. L. DeVries, D. K. Roylance, and M. L. Williams, J. Polym. Sci. A1, 8:237 (1970).

413. R. Bonart and F. Schultze-Gebhardt, Angew. Makromol. Chem. 22:41 (1972).

414. D. J. Meier (ed.), *Molecular Basis of Transitions and Relaxations*, Midland Macromolecular Institute Monographs, No. 4, 1976.

415. Z. Mencik, J. Polym. Sci. Polym. Phys. Ed. 13:2173 (1975).

416. H. Fischer and D. O. Hummel, in *Polymer Spectroscopy*, ed. D. O. Hummel, Verlag Chemie, Weinheim 1974, p. 289.

417. U. Alter and R. Bonart, Colloid. Polym. Sci. 254:348 (1976).

418. G. Zerbi, M. Gussoni, and F. Ciampelli, Spectrochim. Acta 23A:301 (1967).

419. I. M. Ward, *Mechanical Properties of Solid Polymers*, Wiley, New York 1971.

420. W. Holzmüller, Adv. Polym. Sci. 26:1 (1978).

421. E. H. Andrews and P. E. Reed, Adv. Polym. Sci. 27:1 (1978).

422. K. J. Heritage, J. Mann, and L. Roldan-Gonzales, J. Polym. Sci. A, 1:671 (1963).

423. S. Suzuki, Y. Iwashita, T. Shimanouchi, and M. Tsuboi, Biopolymers 4:337 (1966).

424. S. K. Dirlikov and J. L. Koenig, Appl. Spectrosc. 33:551 (1979).

425. J. C. Bevington, Adv. Polym. Sci. 2:1 (1960).

426. D. Kato, J. Appl. Phys. 47:1072 (1976).

427. M. Takeda, R. E. S. Iavazzo, D. Garfunkel, I. H. Scheinberg, and J. T. Edsall, J. Amer. Chem. Soc. 80:3818 (1958).

428. S. Krimm, C. Y. Liang, and G. B. B. M. Sutherland, J. Chem. Phys. 25:549, 778 (1956).

429. S. Krimm, J. Chem. Phys. 32:1780 (1960).

430. O. Redlich, Z. Phys. Chem. B28:371 (1935).

431. R. Yamadera, H. Tadokoro, and S. Murahashi, J. Chem. Phys. 41:1233 (1964).

432. T. Onishi and S. Krimm, J. Appl. Phys. 32:2320 (1961).

433. M. Kobayashi, Bull. Chem. Soc. Jap. 34:560 (1961).

434. H. Tadokoro, N. Nishiyama, S. Nozakura, and S. Murahashi, J. Polym. Sci. 36:553 (1959).

435. W. W. Daniels and R. E. Kitson, J. Polym. Sci. 33:161 (1958).

436. R. Jeffries, Polymer 8:1 (1967).

437. P. Schmidt and B. Schneider, Collect. Czech. Chem. Commun. 28:2685 (1963).

438. H. Siesler, H, Krässig, F. Grass, K. Kratzl, and J. Derkosch, Angew. Makromol. Chem. 42:139 (1975).

439. D. K. Ramsden and J. C. Moore, Polymer 18:185 (1977).

440. A. Koshimo, J. Appl. Polym. Sci. 9:81 (1965).

441. H. T. Lokhande, E. H. Daruwalla, and M. R. Padhye, J. Appl. Polym. Sci. 21:2943 (1978).

442. O. Sepall and S. G. Mason, Can. J. Chem. 39:1934 (1961).

443. J. Dechant, Faserforsch. Textiltechn. 18:239 (1967).

444. B. Ranby, Papier 18:593 (1964).

445. J. Mann and H. J. Marrinan, J. Polym. Sci. 32:357 (1958).

446. Y. Sumi, R. D. Hale, and B. G. Ranby, Tappi 46:126 (1963).

447. V. J. Frilette, J. Hanle, and H. Mark, J. Amer. Chem. Soc. 70:1107 (1948).

448. J. K. Smith, W. J. Kitchen, and D. B. Mutton, J. Polym. Sci. C, 2:499 (1963).

449. J. Mann and H. J. Marrinan, Trans. Faraday Soc. 52:492 (1956).

450. R. J. E. Cumberbirch and R. Jeffries, J. Appl. Polym. Sci. 11:2083 (1967).

451. J. Dechant, Faserforsch. Textiltechn. 19:365 (1968).

452. H. W. Siesler, (unpublished results).

453. H. Tadokoro, S. Seki, and I. Nitta, J. Polym. Sci. 22:563 (1956).

454. S. Krimm, C. Y. Liang, and G. B. B. M. Sutherland, J. Polym. Sci. 22:227 (1956).

455. L. Glatt, D. S. Weber, C. Seaman, and J. W. Ellis, J. Chem. Phys. 18:413 (1950).

456. C. W. Bunn, Nature 161:929 (1948).

457. E. G. Bendit, Nature 193:236 (1962).

458. Yu. N. Cirgadze, The Hydrogen Bond, Academy of Sciences USSR, Institute of Physical Chemistry, 310 (1964).

459. G. Gurato, A. Fichera, F. Z. Grandi, R. Zannetti, and P. Canal, Makromol. Chem. 175:953 (1974).

460. F. Druschke, H. W. Siesler, G. Spilgies, and H. Tengler, Polym. Eng. Sci. 17:93 (1977).

461. H. W. Siesler, Structural Aspects of IR-spectroscopic Investigations of Deuteration in Polymeric Materials, Paper presented at the 4th European Symposium on Polymer Spectroscopy, Strasbourg, France, March 1976.

462. D. R. Paul, Ber. Bunsenges. Phys. Chem. 83:294 (1979).

463. A. Peterlin, J. Macromol. Sci. (Phys.) B11, 1:57 (1975).

464. J. Crank and G. S. Park (eds.), *Diffusion in Polymers*, Academic Press, London 1968.

465. P. Meares, in *Proceedings of the International Symposium on Macromolecules Rio de Janeiro, 1974*, ed. E. B. Mano, Elsevier, New York 1975, p. 131.

466. J. S. Vrentas and J. L. Duda, Macromolecules 9:785 (1976).

467. H. Horacek, Ber. Bunsenges. Phys. Chem. 83:352 (1979).

468. H. W. Siesler, unpublished results.

469. M. Tasumi and G. Zerbi, J. Polym. Sci. B5:985 (1967).

470. M. Tasumi, T. Shimanouchi, H. Kenjo, and S. Ikeda, J. Polym. Sci. A1, 4:1011 (1966).

471. Y. Kikuchi and S. Krimm, J. Macromol. Sci. (Phys.) B4:461 (1970).

472. J. H. C. Ching and S. Krimm, Macromolecules 8:894 (1975).

473. G. Natta, SPE J. 15:373 (1959).

474. G. Natta, M. Farina, and M. Peraldo, Atti. Accad.Naz. Lincei Cl. Sci. Fis. Mat. Nat. Rend. 25:424 (1958).

475. M. Peraldo, Gazz. Chim. Ital. 89:798 (1959).

476. T. Miyazawa and Y. Ideguchi, Bull. Chem. Soc. Jap. 36:1125 (1963).

477. C. Y. Liang, M. R. Lytton, and C. J. Boone, J. Polym. Sci. 54:523 (1961).

478. M. P. McDonald and I. M. Ward, Polymer 2:341 (1961).

479. C. Y. Liang and M. R. Lytton, J. Polym. Sci. 61:45 (1962).

480. H. Tadokoro, T. Kitazawa, S. Nozakura, and S. Murahashi, Bull. Chem. Soc. Jap. 34:1209 (1961).

481. M. Peraldo and M. Farina, Chem. Ind. 42:1349 (1960).

482. G. Natta, M. Farina, and M. Peraldo, Makromol. Chem. 38:13 (1960).

483. H. Tadokoro, M. Kobayashi, M. Ukita, K. Yasufuku, S. Murahashi, and T. Torii, J. Chem. Phys. 42:1432 (1965).

484. J. G. Murray, J. Zymonis, E. R. Santee, and H. Harwood, Polym. Prepr. Am. Chem. Soc. Div. Polym. Chem. 14, 2:1157 (1973).

485. G. G. Wanless and J. P. Kennedy, Polymer 6:111 (1965).

486. H. Tadokoro, S. Nozakura, T. Kitazawa, A. Yasuhara, and S. Murahashi, Bull. Chem. Soc. Jap. 32:313 (1959).

487. H. Tadokoro, Y. Nishiyama, S. Nozakura, and S. Murahashi, Bull. Chem. Soc. Jap. 34:381 (1961).

488. A. B. Willenberg, Makromol. Chem. 177:3625 (1976).

489. M. A. Golub and J. J. Shipman, Spectrochim. Acta 16:1165 (1960).

490. M. A. Golub and J. J. Shipman, Spectrochim. Acta 20:701 (1964).

491. L. Porri and M. Aglietto, Makromol. Chem. 177:1465 (1976).

492. M. G. Fatica and H. J. Harwood, J. Macromol. Sci. (Chem). A12, 8:1099 (1978).

493. C. Y. Liang, F. G. Pearson, and R. H. Marchessault, Spectrochim. Acta 17:568 (1961).

494. H. Tadokoro, S. Murahashi, R. Yamadera, and T. Kamei, J. Polym. Sci. A, 1:3029 (1963).

495. S. Enomoto, M. Asahina, and S. Satoh, J. Polym. Sci. A1, 4:1373 (1966).

496. S. Krimm, V. L. Volt, J. J. Shipman, and A. R. Berens, J. Polym. Sci. A, 1:2621 (1963).

497. S. Narita, S. Ichinohe, and S. Enomoto, J. Polym. Sci. 37:281 (1959).

498. T. Shimanouchi and M. Tasumi, Bull. Chem. Soc. Jap. 34:359 (1961).

499. S. Narita, S. Ichinohe, and S. Enomoto, J. Polym. Sci. 37:263 (1959).

500. H. Nagai, J. Appl. Polym. Sci. 7:1697 (1963).

501. M. Mihailov, S. Dirlikov, N. Peeva, and Z. Georgieva, Makromol. Chem. 176:789 (1975).

502. P. Schmidt, B. Schneider, S. Dirlikov, and M. Mihailov, Eur. Polym. J. 11:229 (1975).

503. S. K. Dirlikov and J. L. Koenig, Appl. Spectrosc. 33:555 (1979).

504. G. Zerbi and P. J. Hendra, J. Mol. Spectrosc. 27:17 (1968).

505. H. Tadokoro, M. Kobayashi, Y. Kawaguchi, A. Kobayashi, S. Murahashi, J. Chem. Phys. 38:703 (1963).

506. A. Novak and E. Whalley, Trans. Faraday Soc. 55:1484 (1959).

507. A. Novak and E. Whalley, Can. J. Chem. 37:1710 (1959).

508. H. Tadokoro, Y. Chatani, T. Yoshihara, S. Tahara, and S. Murahashi, Makromol. Chem. 73:109 (1964).

509. T. Yoshihara, H. Tadokoro, and S. Murahashi, J. Chem.Phys. 41:2902 (1964).

510. M. Yokoyama, H. Ochi, A. M. Ueda, and H. Tadokoro, J. Macromol. Sci. Phys. B7, 3:465 (1973).

511. H. J. Nettelbeck, Thesis, Technical University, Aachen, 1966.

512. W. Stach, K. Holland-Moritz, and H. W. Siesler, Progr. Colloid Polym. Sci. (in press).

513. G. Heidemann and H. Zahn, Makromol. Chem. 62:123 (1963).

514. S. Suzuki, Y. Iwashita, and T. Shimanouchi, Biopolymers 4:337 (1966).

515. E. Correns and J. Dechant, Faserforsch. Textiltechn. 19:393 (1968).

516. J. Dechant. Faserforsch. Textiltechn. 19:491 (1968).

517. G. Vergoten, G. Fleury, and Y. Moschetto, in *Advances in IR and Raman Spectroscopy, vol. 4*, eds. R. J. H. Clark and R. E. Hester, Heyden, London 1978, p. 195.

518. J. K. Haken and R. L. Werner, Spectrochim. Acta 27A:343 (1971).

519. J. E. Stewart and F. J. Linning, J. Res. Nat. Bur. Stand. Sect. A, 71:19 (1967).

520. J. P. Luongo, J. Polym. Sci. A2, 10:1119 (1972).

521. I. Matsubara and J. H. Magill, J. Polym. Sci. Polym. Phys. Ed. 11:1173 (1973).

522. R. G. Jones, E. A. Nicol, J. R. Birch, G. W. Chantry, J. W. Fleming, H. A. Willis, and M. A. Cudby, Polymer 17:153 (1976).

523. I. Modric, K. Holland-Moritz, and D. O. Hummel, Colloid Polym. Sci. 254:342 (1976).

524. M. Goldstein, M. E. Seeley, H. A. Willis, and V. J. I. Zichy, Polymer 14:530 (1973).

525. M. S. Mathur and N. A. Weir, J. Mol. Struct. 15:459 (1973).

526. T. R. Manley and C. G. Martin, Polymer 12:524 (1971).

527. T. Miyazawa, K. Fukushima, and Y. Ideguchi, J. Polym. Sci. Polym. Lett. Ed. 1:385 (1967).

528. S. J. Spells and I. W. Shepherd, J. Chem. Phys. 66:1427 (1977).

529. H. Tadokoro, Y. Chatani, H. Kusanagi, and M. Yokoyama, Macromolecules 3:441 (1970).

530. S. Krimm and J. Jakes, Macromolecules 4:605 (1971).

531. W. Stach, Thesis, University of Cologne, 1977.

532. S. Krimm and C. Y. Liang, J. Polym. Sci. 22:95 (1956).

533. D. O. Hummel, *Infrared Spectra of Polymers in the Medium and Long Wavelength Regions*, Interscience, New York 1966.

534. C. Y. Liang and S. Krimm, J. Polym. Sci. 31:513 (1958).

535. C. Y. Liang and S. Krimm, J. Chem. Phys. 25:563 (1956).

536. G. Zerbi and G. Masetti, J. Mol. Spectrosc. 22:284 (1967).

537. T. R. Manley and D. A. Williams, J. Polym. Sci. C; 22:1009
      (1969).

538. W. Frank and D. Knaupp, Ber. Bunsenges. Phys. Chem. 79:1041
      (1975).

539. E. Amrhein and F. H. Müller, Kolloid-Z. Z. Polym. 226:97 (1968).

540. R. T. Bailey, A. J. Hyde, and J. J. Kim, Adv. Raman Spectrosc.
      1:296 (1972).

541. G. Zerbi and M. Sacchi, Macromolecules 6:692 (1973).

542. G. Masetti, F. Cabassi, G. Morelli, and G. Zerbi, Macromol-
      ecules 6:700 (1973).

543. H. Frischkorn and E. Amrhein, Ber. Bunsenges. Phys. Chem. 74:
      880 (1970).

544. H. Tadokoro, M. Kobayashi, M. Yoshidome, K. Tai, and D. Makino,
      J. Chem. Phys. 49:3359 (1968).

545. H. A. Willis, M. E. Cudby, G. W. Chantry, E. A. Nicol, and
      J. W. Fleming, Polymer 16:74 (1975).

546. H. Tadokoro and Y. Chatani, Macromolecules 3:441 (1970).

547. K. W. Johnson and J. F. Rabolt, J. Chem. Phys. 58:4536 (1973).

548. P. L. Gordon, C. Huang, R. C. Lord, and I. V. Yannas, Macro-
      molecules 7:954 (1974).

549. J. Haigh, A. Ali, and G. J. Davis, Polymer 16:714 (1975).

550. H. Tadokoro, Y. Takahashi, Y. Chatani, and H. Kakida, Makro-
      mol. Chem. 109:96 (1967).

551. H. Kakida, D. Makino, Y. Chatani, M. Kobayashi, and H. Tadokoro,
      Macromolecules 3:569 (1970).

552. I. Matsubara, Y. Itoh, and M. Shinomiya, J. Polym. Sci. B, 4:47
      (1966).

553. J. L. Koenig and D. Druesdow, J. Polym. Sci. A2, 7:1489 (1969).

554. F. J. Boerio and J. L. Koenig, J. Chem. Phys. 52:4826 (1970).

555. L. Piseri, B. M. Powell, and G. Dolling, J. Chem. Phys. 58:
      158 (1973).

556. T. R. Manley, in *Introduction to the Spectroscopy of Biological
      Polymers*, Academic Press, New York 1976, p. 119.

557. G. W. Chantry, H. M. Evans, J. W. Fleming, and H. A. Gebbie,
      Infrared Phys. 9:31 (1969).

558. D. Makino, M. Kobayashi, and H. Tadokoro, Spectrochim. Acta
      31A:1481 (1975).

559. S. J. Spells, I. W. Shepherd, and C. J. Wright, Polymer 18:905
      (1977).

374 Applied Spectroscopy

560. J. L. Koenig, in *Introduction to the Spectroscopy of Biological Polymers*, ed. D. W. Jones, Academic Press, New York 1976, p. 81.

561. R. F. Schaufele, Macromol. Rev. 4:67 (1970).

562. R. F. Schaufele and T. Shimanouchi, J. Chem. Phys. 47:3605 (1967).

563. S. I. Mizushima and T. Shimanouchi, J. Amer. Chem. Soc. 71:1320 (1949).

564. F. Khoury, B. Fanconi, J. D. Barnes, and L. H. Bolz, J. Chem. Phys. 59:5849 (1973).

565. H. Olf and B. Fanconi, J. Chem. Phys. 59:534 (1973).

566. G. R. Strobl and R. Eckel, J. Polym. Sci. Polym. Phys. Ed. 14:913 (1976).

567. W. L. Peticolas, G. W. Hilber, J. L. Lippert, A. Peterlin, and H. Olf, Appl. Phys. Lett. 18:87 (1971).

568. H. Olf, A. Peterlin, and W. Peticolas, J. Polym. Sci. Polym. Phys. Ed. 12:359 (1974).

569. S. L. Hsu, S. Krimm, S. Krause, and G. S. Yeh, J. Polym. Sci. Polym. Lett. Ed. 14:195 (1976).

570. P. J. Hendra and H. A. Majid, J. Mat. Sci. 10:1871 (1975).

571. J. L. Koenig and D. L. Tabb, J. Macromol. Sci. Phys. B9, 1:141 (1974).

572. P. J. Hendra and E. P. Marsden, J. Polym. Sci. Polym. Lett. Ed. 15:259 (1977).

573. M. Folkes, A. Keller, J. Stejny, P. Goggin, G. V. Fraser, and P. J. Hendra, Colloid Polym. Sci. 253:354 (1975).

574. B. Fanconi and J. Crissmann, J. Polym. Sci. Polym. Lett. Ed. 13:421 (1975).

575. P. J. Hendra, H. P. Jobic, and K. Holland-Moritz, J. Polym. Sci. Polym. Lett. Ed. 13:365 (1975).

576. H. D. Noether, J. Mat. Sci. 11:1971 (1976).

577. S. L. Hsu and S. Krimm, J. Appl. Phys. 47:4265 (1976), 48:4013 (1977).

578. D. H. Reneker and B. Fanconi, J. Appl. Phys. 46:4144 (1975).

579. S. Krimm and M. Bank, J. Chem. Phys. 42:4059 (1965).

580. M. Bank and S. Krimm, J. Appl. Phys. 39:4951 (1968).

581. G. Ellinghorst, Thesis, University of Cologne, Cologne 1968.

582. G. D. Dean and D. H. Martin, Chem. Phys. Lett. 1:415 (1967).

583. J. E. Bertie and E. Whalley, J. Chem. Phys. 41:575 (1964).

584. W. Frank, H. Schmidt, and W. Wulf, J. Polym. Sci. Polym. Symp. 61:317 (1977).

585. A. O. Frenzel and J. P. Butler, J. Opt. Soc. Amer. 54:1059 (1964).

586. E. M. Amrhein and F. H. Müller, Kolloid-Z. Z. Polym. 234:1078 (1969).

587. M. Tasumi and T. Shimanouchi, J. Chem. Phys. 43:1245 (1965).

588. M. Tasumi and S. Krimm, J. Chem. Phys. 46:755 (1967).

589. H. P. Großmann and W. Frank, Polymer 18:341 (1977).

590. P. C. Hägele and W. Pechold, Kolloid-Z. Z. Polym. 241:977 (1970).

591. L. Beck and P. C. Hägele, Colloid Polym. Sci. 254:28 (1976).

592. R. G. Snyder, S. J. Krause, and J. R. Scherer, J. Polym. Sci. Polym. Phys. Ed. 16:1593 (1978).

593. G. Zerbi, Pure Appl. Chem. 26:499 (1971).

594. G. Zerbi, in Phonons, ed. M. A. Nusimovici, Flammarion Sciences, Paris 1971, p. 248.

595. C. Schmid, Colloid Polym. Sci. 257:561 (1979).

596. D. H. Martin, Adv. Phys. 14:39 (1965).

597. B. Szigeti, J. Phys. Chem. Solids 24:225 (1963).

598. L. Genzel, in Optical Properties of Solids, eds. S. Nudelman and S. S. Mitra, Plenum Press, New York 1969.

599. J. T. Houghton and S. D. Smith, Infrared Physics, Clarendon Press, Oxford 1966.

600. G. Zerbi, L. Piseri, and F. Cabassi, Mol. Phys. 22:241 (1971).

601. K. B. Whetsel, Appl. Spectrosc. Rev. 2, 1:1 (1968).

602. W. Kaye, Spectrochim. Acta 6:257 (1954).

603. R. T. M. Fraser, Analysis in the Near-Infrared, Explosives Research and Development Establishment Waltham, England, Technical Note No. 27 (1971).

604. R. F. Goddu and D. A. Delker, Anal. Chem. 32:140 (1960).

605. G. Gauthier, J. Phys. Radium 14:19 (1953).

606. G. Herzberg, Molecular Spectra and Molecular Structure, vol. II, Van Nostrand, Princeton, 1945.

607. R. G. J. Miller and H. A. Willis, J. Appl Chem. 6:385 (1956).

608. T. Takeuchi, S. Tsuge, and Y. Sugimura, J. Polym. Sci. A1, 6: 3415 (1968).

609. G. Bucci and F. Simonazzi, Chim. Ind. (Milan) 44:262 (1962).

610. H. V. Drushel and F. A. Iddings, Anal. Chem. 35:28 (1963).

611. R. M. Bly, P. E. Kiener, and B. A. Fries, Anal. Chem. 38:217 (1966).

612. J. A. Brown, M. Tyron, and J. Mandel, Anal. Chem. 35:2172 (1963).

613. F. Ciampelli, G. Bucci, T. Simonazzi, and A. Santambrogio, Chim. Ind. (Milan) 44:489 (1962).

614. J. van Schooten, E. W. Duck, and R. Berkenbosch, Polymer 2:357 (1961).

615. C. Tosi, Makromol. Chem. 112:303 (1968).

616. G. Weis, Fette, Seifen, Anstrichmittel 70, 5:355 (1968).

617. W. Kremer and G. Mischer, (unpublished results).

618. W. H. Greive and D. D. Doepken, Polym. Eng. Sci. 1:19 (1968).

619. A. J. Durbetaki and C. M. Miles, Anal. Chem. 37:1231 (1965).

620. H. Dannenberg, SPE Trans. 78 (Jan. 1963).

621. G. S. Kituchina and V. V. Zarkov, Plast. Massy. 7:74 (1976).

622. R. F. Goddu and D. A. Delker, Anal. Chem. 30:2013 (1953).

623. C. L. Hilton, Anal. Chem. 31:1610 (1959).

624. E. A. Burns and A. Muraca, Anal. Chem. 31:397 (1959).

625. H. A. Willis and R. G. J. Miller, Spectrochim. Acta 14:119 (1959).

626. A. Elliott, W. E. Hanby, and B. R. Malcolm, Brit. J. Appl. Phys. 5:377 (1954).

627. K. T. Hecht and D. L. Wood, Proc. Roy. Soc. (London) 235:174 (1956).

628. E. R. S. Jones, J. Sci. Instrum. 30:132 (1953).

629. N. Ressler, C. Ziauddin, C. Vygantas, W. Jansen, and K. Karachorlu, Appl. Spectrosc. 30:295 (1976).

630. V. A. Suckov and I. I. Novak, Vysokomol. Soedin. 11:2753 (1969).

631. L. Glatt and J. W. Ellis, J. Chem. Phys. 16:551 (1948).

632. L. Glatt and J. W. Ellis, J. Chem. Phys. 19:449 (1951).

633. R. D. B. Fraser, J. Opt. Soc. Amer. 48:1017 (1958).

634. K. H. Bassett, C. Y. Liang, and R. H. Marchessault, J. Polym. Sci. A, 1:1687 (1963).

635. J. Jokl, Collect. Czech. Chem. Commun. 28:3305 (1963).

636. E. W. Crandall, E. L. Johnson, and C. H. Smith, J. Appl. Polym. Sci. 19:897 (1975).

637. E. W. Crandall and A. N. Jagtap, J. Appl. Polym. Sci. 21:449 (1977).

638. F. W. Nees and M. Buback, Ber. Bunsenges. Phys. Chem. 80:1017 (1976).

639. T. Hirschfeld, J. Opt. Soc. Amer. 63:1309 (1973).

640. W. Kiefer, Chem. Instrum. 3:21 (1971).

641. P. P. Shorygin, Sov. Phys. Usp. 16:99 (1973).

642. R. J. H. Clark and R. E. Hester (eds.), *Advances in Infrared and Raman Spectroscopy*, vol. 1, Heyden, London 1975, p. 98, 143.

643. T. G. Spiro, in *Chemical and Biochemical Applications of Lasers*, ed. C. B. Moore, Academic Press, New York 1974.

644. D. Bloor, F. H. Preston, D. J. Ando, and D. N. Batchelder, in *Structural Studies of Macromolecules by Spectroscopic Methods*, ed. K. J. Ivin, Wiley-Interscience, London 1976, p. 91.

645. G. Wegner, Makromol. Chem. 154:35 (1972).

646. E. Hadicke, E. C. Mez, C. H. Krauch, G. Wegner, and J. Kaiser, Angew. Chem. Int. Ed. 10:266 (1971).

647. D. Bloor, J. Ando, F. H. Preston, and G. C. Stevens, Chem. Phys. Lett. 24:407 (1974).

648. D. L. Gerrard and W. F. Maddams, Macromolecules 8:54 (1975).

649. Y. Nishimura, A. Y. Hirikawa, and M. Tsuboi, in *Advances in Infrared and Raman Spectroscopy*, vol. 5, eds. R. J. H. Clark and R. E. Hester, Heyden, London 1978, p. 217.

650. P. R. Carey, H. Schneider, and H. J. Bernstein, Biochem. Biophys. Res. Commun. 47:588 (1972).

651. A. L. Bortnichuk, A. D. Stepukhovich, I. S. Rabinovich, and A. D. Veselova, Z. Prikl. Spektrosk. 20, 2:255 (1974).

652. A. J. Hartley and I. W. Shepherd, J. Polym. Sci. Polym. Phys. Ed. 14:643 (1976).

653. S. L. Hsu and S. Krimm, J. Polym Sci. Polym. Phys. Ed. 14:521 (1976).

654. K. Witke and W. Kimmer, Plaste u. Kautschuk 23, 11:799 (1976).

655. C. E. Faezel and E. A. Verchot, J. Polym. Sci. 25:351 (1957).

656. Z. Jedlinski and R. Hippe, Polimery 6:238 (1965).

657. A. Ueno and C. Schuerch, J. Polym. Sci. B, 3:53 (1965).

658. R. Fountain and T. W. Haas, J. Appl. Polym. Sci. 19:1767 (1975).

659. V. I. Bukhgalter, L. N. Pirozhnaya, B. I. Sazhin, and N. I. Sergeyeva, Polym. Sci. USSR 6:139 (1964).

660. E. J. Slowinsky and G. C. Claver, J. Polym. Sci. 17:269 (1955).

661. R. A. Spurr, B. M. Hanking, and J. W. Rowen, J. Polym. Sci. 37:431 (1959).

662. M. M. Koton, V. V. Kudryavtsev, N. A. Adrova, K. K. Kalninsh, A. M. Dubnova, and V. M. Svetlichnyi, Vysokomol. Soedin. A16, 9:2081 (1974).

663. M. V. Shishkina, M. M. Teplyakov, V. P. Chebotaryev, and V. V. Korshak, Makromol. Chem. 175:3475 (1974).

664. M. Rinaudo and A. Domard, Biopolymers 15:2185 (1976).

665. G. Schreier and G. Peitscher, Raman Newsletter 34:6 (1971).

666. D. O. Hummel, K. U. Heinen, H. Stenzenberger, and H. Siesler, J. Appl. Polym. Sci. 18:2015 (1974).

667. H. Stenzenberger, Appl. Polym. Symp. 22:77 (1973).

668. C. E. Sroog, Encyclopedia Polym. Sci. Tech. 11:247 (1969).

669. D. O. Hummel, Farbe und Lack 74, 1:11 (1968).

670. P. M. Hergenrother and H. H. Levine, J. Polym. Sci. A1, 4:2341 (1966).

671. A. Ja. Chernihov, A. N. Shabadash, M. P. Noskova, V. A. Isaeva, and M. N. Jakovlev, XXIII Int. Symp. on Macromolecules, Madrid, Sept. 15-20, Preprints, vol. 1, 445 (1974).

672. H. Deibig, M. Plachky, and M. Sander, Angew. Makromol. Chem. 32:131 (1973).

673. Yu. A. Pentin, B. N. Tarasevich, and B. S. Eltefson, Russian J. Phys. Chem. 46, 8:1207 (1972).

674. S. S. Stivala, L. Reich, and P. G. Kelleher, Makromol. Chem. 59:28 (1963).

675. B. R. Jadrnicek, S. S. Stivala, and L. Reich, Macromolecules 5:20 (1972).

676. S. S. Stivala, G. Yo, and L. Reich, J. Appl. Polym. Sci. 13: 1289 (1969).

677. S. M. Gabbay, S. S. Stivala, and L. Reich, Polymer 16:749 (1975).

678. S. M. Gabbay and S. S. Stivala, Polymer 17:137 (1976).

679. G. Menzel, Angew. Makromol. Chem. 40/41:405 (1974).

680. P. Blais, D. J. Carlsson, and D. M. Wiles, J. Polym. Sci. A1, 10:1077 (1972).

681. H. C. Beachell and J. C. Spitsbergen, J. Polym. Sci. 62:73 (1962).

682. J. O. Lephardt and G. Vilcins, Appl. Spectrosc, 29:221 (1975).

683. G. Bayer and T. Werner, paper presented at the Symposium on Rechnerunterstützte Spektroskopie, Bodenseewerk Perkin Elmer, Überlingen, April 6-9, 1979.

684. H. Dostal, Monatsh. Chem. 69:424 (1934).

685. T. Alfrey and G. Goldfinger, J. Chem. Phys. 12:205 (1944).

686. F. T. Wall, J. Amer. Chem. Soc. 63:1862 (1941).

687. F. T. Wall, J. Amer. Chem. Soc. 66:2050 (1944).

688. F. R. Mayo and F. M. Lewis, J. Amer. Chem. Soc. 66:1595 (1944).

689. M. Fineman and S. D. Ross, J. Polym. Sci. 5:259 (1950).

690. G. E. Ham, J. Polym. Sci. 45:169 (1960).

691. I. Skeist, J. Amer. Chem. Soc. 68:1781 (1946).

692. D. Braun and H. G. Keppler, Makromol. Chem. 78:100 (1964).

693. V. Jaacks, Makromol. Chem. 161:161 (1972).

694. S. P. Chang, T. K. Miwa, and W. H. Tallent, J. Polym. Sci. A1, 7:471 (1969).

695. D. Braun, W. Brendlein, and G. Mot, Eur. Polym. J. 9:1007 (1973).

696. V. E. Meyer and G. G. Lowry, J. Polym. Sci. A1, 3:2843 (1965).

697. V. E. Meyer and G. G. Lowry, J. Polym. Sci. A1, 4:2819 (1966). A1, 3:2819 (1966).

698. G. Heublein, R. Wondraczek, H. Toparkus, and H. Berndt, Faserforsch. Textiltechn. 26:537 (1975).

699. T. Kelen and F. Tüdös, J. Macromol. Sci. Chem. A9, 1:1 (1975).

700. G. Heublein, *Zum Ablauf ionischer Polymerisationsreaktionen*, Akademie Verlag, Berlin 1975.

701. H. V. Boenig, *Unsaturated Polyesters, Structure and Properties*, Elsevier, New York 1964.

702. C. Tosi and G. Zerbi, Chim. Ind. (Milan) 55:334 (1973).

703. J. G. Grasselli and L. E. Wolfram, Appl. Opt. 17, 9:1386 (1978).

704. N. J. Wegemer, J. Appl. Polym. Sci. 14:573 (1970).

705. C. Tosi and F. Ciampelli, Adv. Polym. Sci. 12:87 (1973).

706. R. C. Ferguson, Macromolecules 4:324 (1971).

707. G. Natta, D. Dall'Asta, G. Mazzanti, and F. Ciampelli, Kolloid-Z. 182:50 (1962).

708. V. L. Khodjaeva and V. A. Mamedova, Vysokomol. Soedin. A9:447 (1967).

709. Yu. V. Kissin, Yu. Ya. Goldfarb, B. A. Krentsel, and Ho. Uylien, Eur. Polym. J. 8:487 (1972).

710. Ho. Uylien, Yu. V. Kissin, Yu. Ya. Goldfarb, and B. A. Krentsel, Vysokomol. Soedin. A14:2229 (1972).

711. K. Gehrke, A. Bledzki, and J. Ulbricht, Plaste und Kautschuk, 17:251 (1970).

712. H. J. Harwood, T. C. Ang, R. G. Bauer, N. W. Johnston, and K. Shimizu, Polym. Prepr. Am. Chem. Soc. Div. Polym. Chem. 7:980 (1966).

713. Yu. V. Kissin, Advan. Polym. Sci. 15:91 (1974).

714. W. Kimmer and R. Schmolke, Plaste und Kautschuk, 19:260 (1972).

715. S. Karayenev, G. Kostow, R. Milina, and M. Mikhailov, Vysokomol. Soedin. A16:2162 (1974).

716.   M. M. Sharabash and R. L. Guile, J. Macromol. Sci. Chem. A10: 1021 (1976).

717.   T. Takeuchi and S. Mori, Anal. Chem. 37:589 (1965).

718.   A. R. French, J. V. Benham, and T. J. Pullukat, Appl. Spectrosc. 28:477 (1974).

719.   F. S. Dainton and W. D. Sisley, Trans. Faraday Soc. 59:1385 (1963).

720.   H. U. Pohl and D. O. Hummel, Makromol. Chem. 115:141 (1968).

721.   M. Meeks and J. L. Koenig, J. Polym. Sci. A2, 9:717 (1971).

722.   M. J. Gall and P. J. Hendra, The Spex Speaker 16 (1971).

723.   F. Kobayashi and K. Matsuya, J. Polym. Sci. A, 1:111 (1963).

724.   E. G. Brame, Jr., J. E. Barry, and F. J. Toy, Jr., Anal. Chem. 44:2022 (1972).

725.   Yu. V. Kissin and E. I. Vizen, Vysokomol. Soedin. A16:1385 (1974).

726.   T. Uno and K. Machida, Spectrochim. Acta A24:1741 (1968).

727.   I. V. Kumpanenko and K. S. Kazanski, Vysokomol. Soedin. A15: 594 (1973).

728.   J. Jakes, J. Mol. Spectrosc. 30:167 (1969).

729.   Yu. V. Kissin and L. A. Rishina, Eur. Polym. J. 12:757 (1976).

730.   V. L. Folt, J. J. Shipman and S. Krimm, J. Polym. Sci. 61:17 (1962).

731.   M. Kobayashi, K. Tsumura, and H. Tadokoro, J. Polym. Sci. A2, 6:1493 (1968).

732.   M. Kobayashi, K. Akita, and H. Tadokoro, Makromol. Chem. 118: 324 (1968).

733.   R. Schmolke, H. Herma, and V. Gröbe, Faserforsch. Textiltechn. 16:589 (1965).

734.   P. Simak and E. Ropte, Makromol. Chem. Suppl. 1:507 (1975).

735.   R. Schmolke, in *Ultrarotspektroskopische Untersuchungen an Polymeren*, ed. J. Dechant, Akademie Verlag, Berlin 1972, p. 195.

736.   G. V. Fraser, P. J. Hendra, J. H. Walker, M. E. A. Cudby, and H. A. Willis, Makromol. Chem. 173:205 (1973).

737.   C. Tosi, P. Corradini, A. Valvassori, and F. Ciampelli, J. Polym. Sci. C, 22:1085 (1969).

738.   G. Bucci and T. Simonazzi, J. Polym. Sci. C, 7:203 (1964).

739.   P. J. Corish and M. E. Tunnicliffe, J. Polym. Sci. C, 7:187 (1964).

740.   D. O. Hummel and C. Schneider, Kunststoffe 50:427 (1960).

741.   D. Brück and D. O. Hummel, Makromol. Chem. 163:271 (1973).

742.   U. Johnsen and W. Lesch, Kolloid-Z. Z. Polym. 233:863 (1969).

743.   M. Iwasaki, M. Aoki, and K. Okuhara, J. Polym. Sci. 26:116
       (1957).

744.   E. Ropte, Ber. Bunsenges. Phys. Chem. 70:317 (1966).

745.   H. Hendus, K. H. Illers, and E. Ropte, Kolloid-Z. Z. Polym.
       216:110 (1967).

746.   H. J. Harwood, Angew. Chem. 77:393, 1124 (1965).

747.   K. Yanagisawa, N. Ashikari, T. Kanemitsu, and A. Nishioka,
       Chem. High Polymers (Tokyo) 21:312, 319 (1964).

748.   H. Germar, Makromol. Chem. 84:36 (1965).

749.   D. O. Hummel, Pure Appl. Chem. 11:497 (1965).

750.   G. Schnell, Ber. Bunsenges. Phys. Chem. 70:297 (1966).